XV Conference on Theoretical Nuclear Physics in Italy (TNPI 2016)

Journal of Physics: Conference Series Volume 981

Pisa, Italy
20 – 22 April 2016

ISBN: 978-1-5108-5955-5
ISSN: 1742-6588

Printed from e-media with permission by:

Curran Associates, Inc.
57 Morehouse Lane
Red Hook, NY 12571

proceedings.com

Some format issues inherent in the e-media version may also appear in this print version.

Printed by Curran Associates, Inc. (2018)

For permission requests, please contact the Institute of Physics
at the address below.

Institute of Physics
Dirac House, Temple Back
Bristol BS1 6BE UK

Phone: 44 1 17 929 7481
Fax: 44 1 17 920 0979

techtracking@iop.org

Additional copies of this publication are available from:

Curran Associates, Inc.
57 Morehouse Lane
Red Hook, NY 12571 USA
Phone: 845-758-0400
Fax: 845-758-2633
Email: curran@proceedings.com
Web: www.proceedings.com

Table of contents

Volume 981

TNPI2016 - XV Conference on Theoretical Nuclear Physics in Italy

20–22 April 2016, Pisa, Italy

Accepted papers received: 21 February 2018
Published online: 12 March 2018

Preface

TNPI2016 - XV Conference on Theoretical Nuclear Physics in Italy

TNPI2016 - XV Conference on Theoretical Nuclear Physics in Italy

Peer review statement

Papers

Applications of nuclear physics to a wider context: from molecules to stars passing through hypernuclei

Lorenzo Fortunato.....1

Chiral Nucleon-Nucleus Potentials at N^3LO

Paolo Finelli, Matteo Vorabbi and Carlotta Giusti.....11

A self-consistent equation of motion multiphonon method for odd mass nuclei

G De Gregorio, F Knapp, N Lo Iudice and P Veselý.....19

Direct reactions of weakly-bound nuclei within a one dimensional model

L Moschini, A Vitturi and AM Moro.....25

Ab initio optical potentials and nucleon scattering on medium mass nuclei

A. Idini, C. Barbieri and P. Navrátil.....33

$^{7}Li(d,p)^{8}Li$ transfer reaction in the NCSM/RGM approach

F Raimondi, G Hupin, P Navrátil and S Quaglioni.....40

Effects of the coupling of quasiparticle and collective vibrations on the properties of ^{120}Sn

Enrico Vigezzi.....50

Bridging the gap: effective interactions from microscopic dynamics

Omar Benhar.....58

Nuclear matter calculations with chiral interactions

Domenico Logoteta, Ignazio Bombaci and Alejandro Kievsky.....70

Meson properties in asymmetric matter

Andrea Mammarella and Massimo Mannarelli.....78

Spatially modulated chiral condensates

Stefano Carignano.....86

A Unified Equation of State on a Microscopic Basis : Implications for Neutron Stars Structure and Cooling

G. F. Burgio.....93

Three–Dimensional parton structure of light nuclei

Sergio Scopetta, Alessio Del Dotto, Leonid Kaptari, Emanuele Pace, Matteo Rinaldi and Giovanni Salmè.....105

Nucleon 3D structure from double parton scattering: a Light-Front quark model analysis

Matteo Rinaldi.....115

Spin degeneracy of Hadronic molecules in the heavy quark region

Yasuhiro Yamaguchi.....123

Heavy flavour in high-energy nuclear collisions: a theoretical overview

Andrea Beraudo.....129

Initial State fluctuations from midperipheral to ultracentral collisions in a transport approach

Salvatore Plumari, Francesco Scardina, Vincenzo Minissale, Lucia Oliva and Vincenzo Greco.....141

Ab initio calculations for non-strange and strange few-baryon systems

Winfried Leidemann.....149

Recent progress in *ab-initio* studies of nuclear reactions of astrophysical interest with $A \leq 3$

Laura E Marcucci.....161

Numerical studies of the Bethe–Salpeter equation for a two-fermion bound state

W. de Paula, T. Frederico, G. Salmè and M. Viviani.....171

1/2 spin-isospin fermions close to the unitary limit

A. Kievsky and M. Gattobigio.....179

IOP Conf. Series: Journal of Physics: Conf. Series **981** (2018) 011001 doi:10.1088/1742-6596/981/1/011001

Preface

The present volume of the *Journal of Physics: Conference Series* contains the invited and contributed papers presented at the XV Conference on Theoretical Nuclear Physics in Italy (TNPI2016) held during April $20^{\rm th}$ – $22^{\rm nd}$ 2016 at the Physics Department "Enrico Fermi" of the University of Pisa (Italy).

The aim of this conference is to bring together Italian theorists working in various fields of nuclear physics and related research areas to discuss their latest results and compare their points of view in a lively and informal way. This offers the opportunity to promote collaborations between different groups. There were about 40 participants at the conference, coming from various Italian Universities, from various "Sezioni" of the Istituto Nazionale di Fisica Nucleare (INFN) and from some non-Italian Institutions.

The TNPI2016 conference has covered recent developments in the following main topics:

- Nuclear Structure
- Nuclear Matter and Nuclear Dynamics
- Structure of Hadrons and Hadronic Matter
- Relativistic Heavy Ion Collisions and QGP
- Few-Nucleon Systems
- Nuclear Astrophysics

We deeply thank the authors for their high-level contributions and we are convinced that this volume of the *Journal of Physics: Conference Series* will be of great interest for the international community of people working in theoretical nuclear physics.

Each submitted contribution has been peer reviewed and the Editors are very grateful to the referees for their careful support.

We gratefully acknowledge the financial support of the University of Pisa, and the logistic support of the INFN Sezione di Pisa, who helped to make the conference possible and successful

The Editors of the TNPI2016 Proceedings:
Ignazio Bombaci (Università di Pisa)
Angela Bonaccorso (INFN, Sezione di Pisa)
Alejandro Kievsky (INFN, Sezione di Pisa)
Laura Elisa Marcucci (Università di Pisa)
Michele Viviani (INFN, Sezione di Pisa)

Pisa, $23^{\rm rd}$ of December 2017

IOP Conf. Series: Journal of Physics: Conf. Series **981** (2018) 011001 doi:10.1088/1742-6596/981/1/011001

XV Conference on Theoretical Nuclear Physics in Italy

Pisa, April 20 – 22, 2016

Main Topics

- Nuclear Structure
- Nuclear Matter and Nuclear Dynamics
- Structure of Hadrons and Hadronic Matter
- Relativistic Heavy Ion Collisions and QGP
- Few-Nucleon Systems
- Nuclear Astrophysics

Organizing Committee

Ignazio Bombaci (Univ. Pisa)
Angela Bonaccorso (INFN Pisa)
Alejandro Kievsky (INFN Pisa)
Laura Elisa Marcucci (Univ. Pisa)
Michele Viviani (INFN Pisa)

Lucia Lilli and Claudia Tofani, INFN Pisa (Secretaries)
TNPI2016@pi.infn.it
http://agenda.infn.it/event/TNPI2016

IN SUPREMÆ DIGNITATIS 1343
UNIVERSITÀ DI PISA

INFN Istituto Nazionale di Fisica Nucleare

IOP Conf. Series: Journal of Physics: Conf. Series **981** (2018) 011002 doi:10.1088/1742-6596/981/1/011002

List of participants

ALBERICO, Wanda Maria (Università di Torino)
BECATTINI, Francesco (Università di Firenze)
BENHAR, Omar (INFN Sezione di Roma)
BERAUDO, Andrea (INFN Sezione di Torino)
BOMBACI, Ignazio (Università di Pisa)
BONACCORSO, Angela (INFN Sezione di Pisa)
BURGIO, Fiorella (INFN Sezione di Catania)
CARIGNANO, Stefano (LNGS INFN Assergi)
CASTORINA, Paolo (Università di Catania)
COLASANTE, Damaso (INFN Sezione di Roma II)
COLÒ, Gianluca (Università di Milano)
DE GREGORIO, Giovanni (INFN Sezione di Napoli)
DRAGO, Alessandro (Università di Ferrara)
FERRETTI, Jacopo (INFN Sezione di Roma)
FINELLI, Paolo (Università di Bologna)
FORTUNATO, Lorenzo (Università di Padova)
GARGANO, Angela (INFN Sezione di Napoli)
GNECH, Alex (Università di Pisa)
GROSSI, Eduardo (Università di Firenze)
IDINI, Andrea (University of Surrey)
ITACO, Nunzio (Seconda Università di Napoli)
KARPENKO, Iurii (INFN Sezione di Firenze)
KIEVSKY, Alejandro (INFN Sezione di Pisa)
LEIDEMANN, Winfried (Università di Trento)
LOGOTETA, Domenico (INFN Sezione di Pisa)
MAMMARELLA, Andrea (LNGS INFN Assergi)
MARCUCCI, Laura Elisa (Università di Pisa)
MATERA, Francesco (Università di Firenze)
MOSCHINI, Laura (Università di Padova)
PLUMARI, Salvatore (LNS INFN Catania)
RAIMONDI, Francesco (University of Surrey)
RINALDI, Matteo (Universitat de Valencia)
ROCA MAZA, Javier (INFN Sezione di Milano)
ROSATI, Sergio (Università di Pisa)
SANTOPINTO, Elena (INFN Sezione di Genova)
SCOPETTA, Sergio (Università di Perugia)
SINGH, Jagjit (Università di Padova)
VIGEZZI, Enrico (INFN Sezione di Milano)
VITTURI, Andrea (Università di Padova)
VIVIANI, Michele (INFN Sezione di Pisa)
YAMAGUCHI, Yasuhiro (INFN Sezione di Genova)

IOP Conf. Series: Journal of Physics: Conf. Series **981** (2018) 011003 doi:10.1088/1742-6596/981/1/011003

Peer review statement

All papers published in this volume of *Journal of Physics: Conference Series* have been peer reviewed through processes administered by the proceedings Editors. Reviews were conducted by expert referees to the professional and scientific standards expected of a proceedings journal published by IOP Publishing.

IOP Conf. Series: Journal of Physics: Conf. Series **981** (2018) 012001 doi:10.1088/1742-6596/981/1/012001

Applications of nuclear physics to a wider context: from molecules to stars passing through hypernuclei

Lorenzo Fortunato

Dipartimento di Fisica e Astronomia "G.Galilei", via Marzolo 8, I-35131 Padova, Italy
I.N.F.N., Sezione di Padova, via Marzolo 8, I-35131 Padova, Italy

E-mail: fortunat@pd.infn.it

Abstract. In this contribution I will review some of the researches that are currently being pursued in Padova (mainly within the In:Theory and Strength projects), focusing on the interdisciplinary applications of nuclear theory to several other branches of physics, with the aim of contributing to show the centrality of nuclear theory in the Italian scientific scenario and the prominence of this fertile field in fostering new physics.

In particular, I will talk about: i) the recent solution of the long-standing "electron screening puzzle" that settles a fundamental controversy in nuclear astrophysics between the outcome of lab experiments on earth and nuclear reactions happening in stars; the application of algebraic methods to very diverse systems such as: ii) the supramolecular complex H2@C60, i.e. a diatomic hydrogen molecule caged in a fullerene and iii) to the spectrum of hypernuclei, i.e. systems made of a Lambda particles trapped in (heavy) nuclei.

1. Introduction

The aim of the present manuscript is to summarize part of the research lines that have been recently pursued in Padova. While most of these could be classified as nuclear theory (Strength project of the INFN section in Padova), interdisciplinary research has been pursued in a broader field that goes from nuclear astrophysics to molecular and to subnuclear physics. We will give three examples, one related to the solution of the famous electron screening puzzle in nuclear astrophysics[1, 2], one related to application of algebraic theories to diatomic molecules caged in a fullerene [3] and one to an algebraic theory for Λ particles in nuclei [4].

2. Electron screening puzzle

Nuclear reactions in stars happen at energies around the Gamow peak $E_G << E_{C.B.}$, much lower than the energy of the Coulomb barrier, at the presence of an electron plasma that creates a screening effect that enhances the reaction cross-sections at these ultra-low energies. At these energies tunneling of the potential barrier becomes dominant and the typical exponential behaviour of the cross-section with energy can be removed by defining the S-factor as:

$$S(E) = E\sigma(E)e^{2\pi\eta(E)} \tag{1}$$

where $\eta = Z_1Z_2e^2/\hbar v$ is the Sommerfeld parameter that depends on the charges of the two reactants and their relative velocity. Target materials in laboratory also have electrons bound in atoms, molecules or crystals' bands. Therefore the laboratory cross-sections, σ_s, and that

IOP Conf. Series: Journal of Physics: Conf. Series **981** (2018) 012001 doi:10.1088/1742-6596/981/1/012001

Table 1. Theoretical adiabatic limits, U_e^{adlim}, and experimental values of the electron screening potentials, U_e^{exp}, for a number of reactions. See Ref. [1] for more details and references.

	Reaction	U_e^{adlim} (eV)	U_e^{exp} (eV)
(1)	^{2}H$(d,t)^1$H	14	13.2±1.8
(2)	^{2}H$(d,^3$He$)$n	14	11.7 ±1.6
(3)	^{3}He$(d,p)^4$He	65	109±9
(4)	^{3}He$(d,p)^4$He	120	219±7
(5)	^{3}He$(^3$He,2p$)^4$He	240	294±47
(6)	^{3}He$(^3$He,2p$)^4$He	240	305±90
(7)	^{6}Li$(d,\alpha)^4$He	175	330±120
(8)	^{6}Li$(d,\alpha)^4$He	175	380±250
(9)	^{6}Li$(d,\alpha)^4$He	175	320±50
(10)	^{6}Li$(p,\alpha)^3$He	175	440±150
(11)	^{6}Li$(p,\alpha)^3$He	175	470±150
(12)	^{6}Li$(p,\alpha)^3$He	175	355±100
(13)	^{7}Li$(p,\alpha)^4$He	175	300±160
(14)	^{7}Li$(p,\alpha)^4$He	175	300±280
(15)	^{7}Li$(p,\alpha)^4$He	175	425±60
(16)	^{9}Be$(p,\alpha_0)^6$Li	240	900±50
(17)	^{9}Be$(p,\alpha_0)^6$Li	240	676±86
(18)	^{10}B$(p,\alpha_0)^7$Be	340	430±80
(19)	^{10}B$(p,\alpha_0)^7$Be	340	240±200
(20)	^{11}B$(p,\alpha_0)^8$Be	340	430±80
(21)	^{11}B$(p,\alpha_0)^8$Be	340	472±120

in the stellar environment usually differ with each other and they are enhanced with respect to the cross-sections obtained from bare nuclei σ_b (these could, in principle, be measured in beam-on-beam experiments). It is mandatory that the bare values should be extracted from beam-on-target experiments, in order to understand the effects of screening and to be able to relate the laboratory and the stellar cross-sections. An enhancement factor $f_{lab}(E)$ is usually defined [5]:

$$f_{lab}(E) = \frac{\sigma_s(E)}{\sigma_b(E)} = \frac{S_s(E)}{S_b(E)} \sim \exp\left[\pi\eta \frac{U_e{}^{(lab)}}{E}\right], \tag{2}$$

that depends on the electron screening potential in laboratory experiments. The idea is that one measures the laboratory value, σ_s, obtains the bare cross-section and from that guesses the stellar value by applying a plasma enhancement factor, $\sigma_{pl}(E) = \sigma_b(E) f_{pl}$, according to the Debye-Hückel theory, that depends on temperature, density and other variables of the stellar environment. Often it is impossible to account for the large values measured in direct experiments with available atomic physics models and this discrepancy has become known as the 'electron screening puzzle'. The resolution of this conundrum is very important for applications of nuclear reactions to astrophysics and deserves special attention. In Ref. [1], we have shown how this puzzle can be explained in terms of nuclear structure effects, namely clusterization of one or both reaction partners. The idea that nuclei exhibiting a more or less pronounced cluster

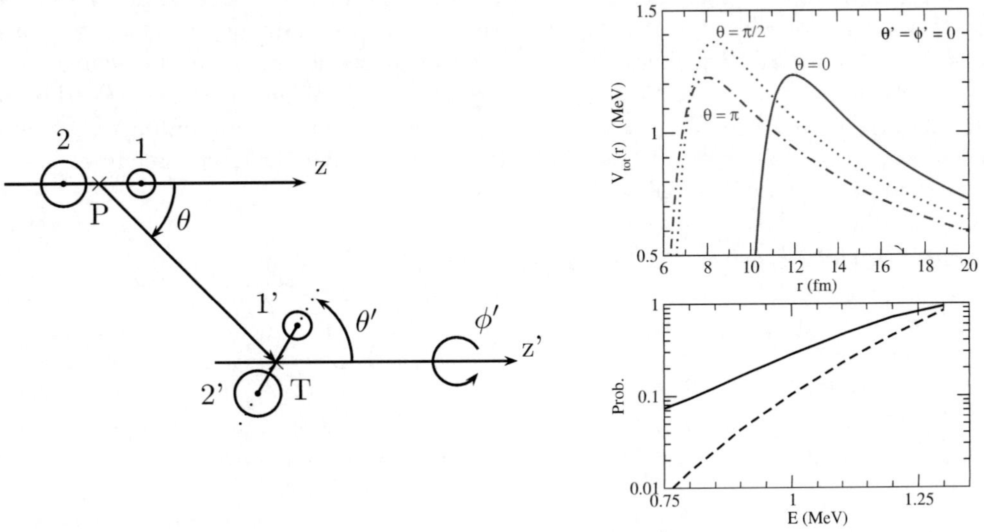

Figure 1. Coordinate system; ion-ion potentials with three different orientations as a function of c.m. distance to show the change in barrier height and position; penetration probability of the clusterized systems (black) compared with sphere-on-sphere (dashed).

structure are related to the electron screening comes from the observation that the list of nuclei with large discrepancies between the experimental and the theoretical (atomic physics models) values of the screening potentials contains exactly those nuclei for which a cluster structure has been proposed or identified (See Table I). If the effect is absent (or almost absent) in the first group of reactions involving Z=1 species, it becomes important for reaction involving ^{6,7}Li and Be that are known to possess cluster structures in their ground states. This phenomenon occurs in those nuclei where a sort of molecular bound state between two clusters is energetically favored with respect to a spherical shell-model configuration. Not all nuclei have cluster states, nor cluster configurations, whenever they occur, necessarily take up to 100% probability, but when they are present, they significantly alter the dynamics of fusion reactions (and others such as breakup, radiative capture processes, etc. [6, 7]).

The position and shape of the potential barrier thus depend on the geometry of the cluster structure (left panel of Fig. 1 for an example of parameterization of two dicluster nuclei impinging on each other), i.e. on the sum of Coulomb and nuclear potentials between each pair of fragments as

$$V_{tot}(r,\theta,\theta',\phi') = \sum_{i,j=1}^{2} \left(\frac{Z_i Z_j e^2}{r_{ij}} + V_N(r_{ij}) \right), \tag{3}$$

where the function depends on the relative inter-cluster distance d, that we keep constant (and equal to ~ 3.85 fm obtained from the cluster model formula (A.4) of Ref. [7]). The tunneling probability can be calculated, at energies below the barrier, in the WKB approximation as $P = e^{-2G}$, where the Gamow tunneling factor is given by

$$G(E,\theta,\theta',\phi') = \frac{\sqrt{2m}}{\hbar} \int_a^b \sqrt{V_{tot}(r,\theta,\theta',\phi') - E}\; dr\ . \tag{4}$$

The angle-averaged penetration probability as a function of bombarding energy is displayed in the last panel of Fig. 1. It is clearly larger than the analogous calculations for sphere-on-sphere and shows that the effects of clusterization on nuclear fusion at stellar energies *cannot be disregarded.* These effects become less pronounced at higher energies. Further studies and experiments are needed before one can say that we have a full understanding of these phenomena, but a first significant advance toward the resolution of the electron screening puzzle has been achieved by recognizing the role of cluster effects.

3. Endohedrally confined diatomic molecules

As an example of application to the realm of molecular physics, we will discuss a new algebraic theory of endohedrally confined diatomic molecules. The buckminsterfullerene molecule, or simply fullerene C_{60}, is composed of 60 carbon atoms placed at the vertices of a icosahedral solid. The interior is almost hollow, except for the tails of the electron wavefunctions, and large enough to host other small chemical species, such as atoms and diatomic molecules. These are said to be endohedrally confined and the whole system is called a supramolecular complex, indicated with the @ symbol. For example the notation $H_2@C_{60}$ refers to a diatomic hydrogen molecule trapped inside a fullerene as in Fig. 2. These systems are not bound by electric charge as in ionic molecules, nor by the mutual exchange of electrons as in covalent molecule, nor by some effective dipole-dipole (or higher order) potential as in Van der Waals molecules. They are simply trapped by spatial confinement and the dimension of the spherical cage (the icosahedral symmetry can be disregarded to all practical effects, unless one is looking for tiny splittings due to the highly symmetrical environment) is small enough that the motion of the trapped molecule inside it exhibits quantization and this quantization quite significantly alters the spectral properties. These systems have been produced thanks to an ingenious series of chemical reactions, known as 'molecular surgery' [9], that open a hole in the fullerene, force the gas in by applying pressure and then seal the molecule back to its original shape. Thus endohedrally confined species can be synthesized in the range of milligrams, well enough for experiments.

Infrared absorption, nuclear magnetic resonance and inelastic neutron scattering experiments have been performed and the status of both spectroscopic knowledge and traditional theoretical models is quite advanced (See [10, 11, 3] and references therein). We have seen the opportunity to simplify and generalize the theoretical approaches to this system with an algebraic theory, inspired to common treatments in the nuclear physics domain. The theoretical treatment starts from the description of the free diatomic molecule as in the Vibron model of Iachello and Levine[8]. The rotations and vibrations of such a molecule can very well be described with a

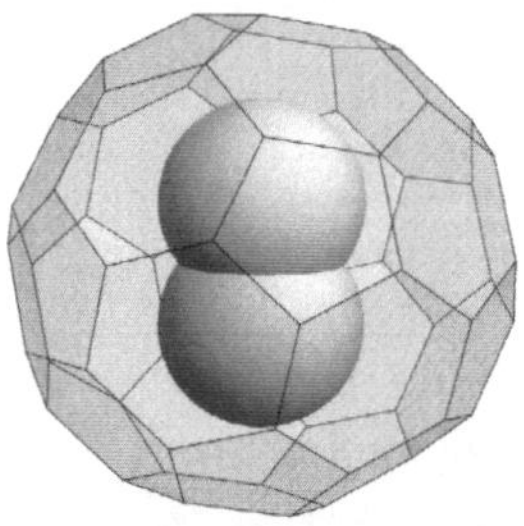

Figure 2. Pictorial representation of a H_2 molecule trapped inside a fullerene (electron in carbon atoms not shown)

IOP Conf. Series: Journal of Physics: Conf. Series **981** (2018) 012001 doi:10.1088/1742-6596/981/1/012001

$u(4)$ Lie algebra arising from a scalar (s, $L = 0$) and a vector (p, $L = 1$) bosons. The dynamical symmetries originating in the unitary algebra $u(4)$ and ending in the orthogonal algebra of the angular momentum $so(3)$, describe a large class of possible hamiltonians ranging from rigid structures with prevalent vibrational motion to easily deformable floppy molecules. The free hydrogen molecule is well described in the so-called $so(4)$ limit of the vibron model [12]. We have coupled this well-known analytic description to another Lie algebra, $u(3)$ that is used to model the effect of spatial confinement inside the cage. The three components of the vector boson q (with $L = 1$) are related to the displacement of the molecule as a whole from the equilibrium point at the center of the cage.

The dynamical symmetry that we consider thus amounts to

$$\begin{array}{ccccccc} u_p(4) \oplus u_q(3) & \supset & so_p(4) \oplus u_q(3) \supset & so_p(3) \oplus so_q(3) & \supset so_{pq}(3) & \supset so_{pq}(2) \\ N_p \quad N_q & & \omega & J \qquad L & \Lambda & M_\Lambda \end{array} , \tag{5}$$

where the quantum numbers associated with the Casimir operators of each algebra are also given. As usual, ω is replaced by the vibrational quantum number v through $v = \frac{1}{2}(N_p - \omega)$. The set of quantum numbers $(vJN_qL\Lambda)$ can be used to label the states of the system.

The total Hamiltonian is

$$\hat{H}_{endo} = \hat{H}_{u_p(4)} + \hat{H}_{u_q(3)} + \hat{H}_{Coupl} , \tag{6}$$

where the first term is the vibron model Hamiltonian for rotations and vibrations of a diatomic molecule, the second is the quantized motion of the molecular center-of-mass inside the cage, and the last term includes molecule-cage couplings.

The $u(4)$ vibron model Hamiltonian can be modeled as follows

$$\hat{H}_{u_p(4)} = \hat{H}_{so(4)} + \hat{H}_{Dun} , \tag{7}$$

in terms of a $so(4)$ dynamical symmetry plus higher-order Dunham-like corrections:

$$\hat{H}_{so(4)} = E_0 + \beta\,\hat{C}_2[so_p(4)] + \gamma\,\hat{C}_2[so_p(3)] , \tag{8}$$

$$\hat{H}_{Dun} = \gamma_2\hat{C}_2[so_p(3)]^2 + \kappa\,\hat{C}_2[so_p(4)]\hat{C}_2[so_p(3)] . \tag{9}$$

giving the energy formula

$$E_{u_p(4)} = E_0 + \beta\,\omega(\omega+2) + \gamma\,J(J+1) + \gamma_2\Big[J(J+1)\Big]^2 + \kappa\,\Big[\omega(\omega+2)J(J+1)\Big] , \tag{10}$$

where $\omega = N_p, N_p - 2, \ldots, 1$ *or* 0 and $J = 0, 1, \ldots, \omega$.

The $u_q(3)$ dynamical symmetry gives instead

$$\hat{H}_{u_q(3)} = a\,\hat{C}_1[u_q(3)] + b\,\hat{C}_2[u_q(3)] + c\,\hat{C}_2[so_q(3)] , \tag{11}$$

that can be immediately translated into eigenenergies, using the eigenvalues of Casimir operators:

$$E_{u_q(3)} = a\,N_q + b\,N_q^2 + c\,L(L+1) , \tag{12}$$

where N_q is the number of quanta and L is the orbital angular momentum.

Finally the cage-molecule couplings could be of several types, but we have selected those that might be relevant and proposed the following interaction:

$$\hat{H}_{Coupl} = \vartheta_{pq}[\hat{Q}_p^{(2)} \times \hat{Q}_q^{(2)}]^{(0)} + \vartheta_{pqw}\Big[\hat{C}_2[so_p(4)][\hat{Q}_p^{(2)} \times \hat{Q}_q^{(2)}]^{(0)} + [\hat{Q}_p^{(2)} \times \hat{Q}_q^{(2)}]^{(0)}\hat{C}_2[so_p(4)]\Big] +$$

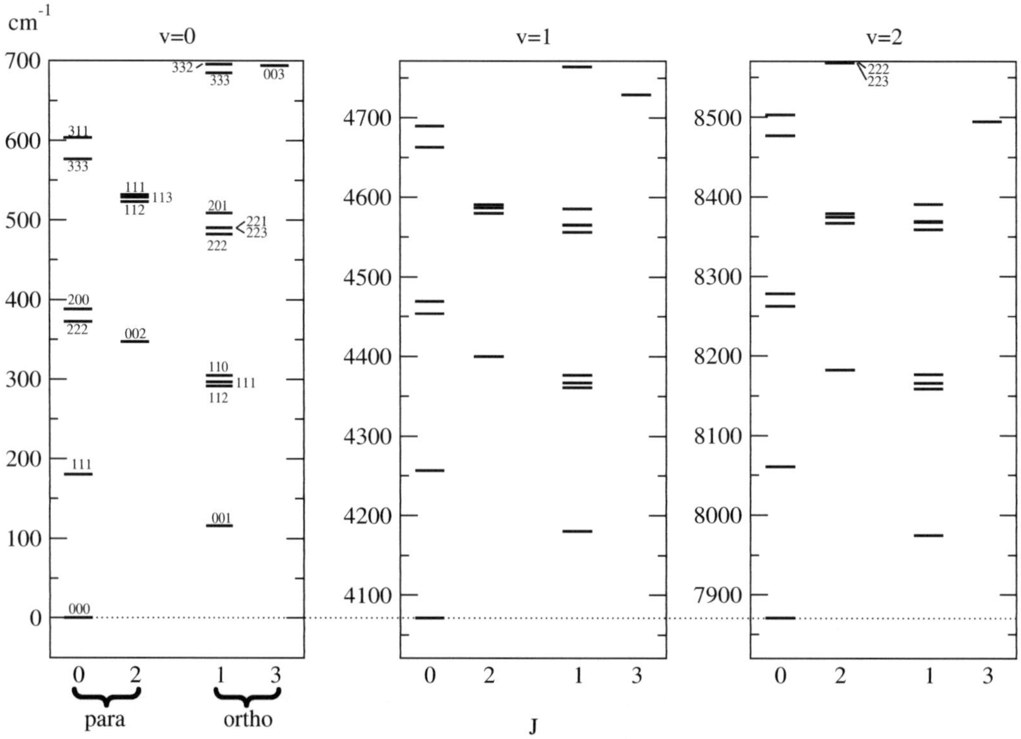

Figure 3. Theoretical rotational-vibrational spectrum of $H_2@C_{60}$. The three panels show the lowest states built on top of the vibrational excitations with $v = 0, 1$, and 2. States are divided into *para* and *ortho* states and labeled by J on the horizontal axis and $N_q L \Lambda$ on each state.

$$+v_{pq}\hat{C}_1[u_q(3)]\hat{C}_2[so_p(4)] \; , \tag{13}$$

where the parameters can be adjusted to model the effects of the coupling. The $[\hat{Q}_p^{(2)} \times \hat{Q}_q^{(2)}]^{(0)}$ quadrupole-quadrupole interaction lifts the degeneracy of $\Lambda \neq 0$ multiplets, giving the correct and somewhat unusual ordering seen in experiments. The matrix elements can be found in Ref. [3]. Formulas (10) and (12) plus the diagonalization of the coupling terms (13) give a practical and accurate way of modeling the caged H_2 molecule.

Extensive numerical investigations and minimization of theory-to-experiments least square differences, using a body of about 70 transition lines, has lead us to a set of parameters that improves the overall theoretical description with respect to competing theories (for example we have only 10 parameters for all possible vibrational quantum number v bands, while others have been fitting these bands separately with a larger number of parameters) and allows to suggest that a few published level assignments were probably incorrect. But the greatest advantage is that it gives a unique theoretical framework that contains all symmetry-allowed terms, even without the precise knowledge of each atom-atom interaction. The lowest portion of the spectrum, that differs in several respects from that of the free molecules (energy shifts and splitting of energy levels) is given in Fig. 3, where one can see several effects, for example the lowering of certain groups of states with increasing value of v or the increasing splitting with increasing values of L, etc.

Our theory is at the same time easier and more universal than previous fits based on direct solution of the Schroedinger equation and allows to make predictions for yet-to-be-measured states. It might seem that this type of theoretical models based on Lie algebras do not contain directly an expression for the interactions between the constituents, i.e. some $V_{ij}(r_{ij})$, and this is disturbing to many, but the advantage is that this information is not needed at all: if the hamiltonian complies with a certain symmetry (dynamical symmetries from higher/bigger algebra down to angular momentum SU(3) or SU(2)) the spectrum is automatically analytically known and the phenomenology boils down to the coefficients of the energy formulas. This is a very desirable tool.

4. Algebraic model for hypernuclei

We turn our attention now to applications in the realm of particle physics by describing an algebraic theory that applies to hyperons bound in a nuclear environment.

A series of experiments at KEK-PS SuperKaonSpectrometer measured the yield of Λ particles inside medium-mass or heavy nuclei. These particles have been produced in different angular momentum states inside the nucleus, but a complete theoretical understanding is still incomplete [13]. In particular, single-particle Λ states with orbital angular momentum ranging from $l = 0$ to $l = 3$ have been clearly identified in medium-heavy ${}^{89}_{\Lambda}$Y hypernucleus using the (π^+, K^+) reaction spectroscopy [14, 13]. These single-particle levels have been theoretically analyzed with the distorted-wave impulse approximation (DWIA) based on shell model calculations [15].

There have been nice theoretical works based on extensions of the shell-model that include that Λ degree of freedom and we can propose a model based on Lie algebras that is, in general, equivalent to an anharmonic oscillator. In this approach, single-particles levels are classified according to the underlying symmetries. We use a simple $su(3)$ bosonic symmetry for the anharmonic oscillator and an even simpler $su(2)$ fermionic symmetry for the inclusion of the hyperonic degree of freedom. The energy of each level is then given in terms of expectation values of Casimir operators associated with the following dynamical symmetry

$$u(3/2) \supset u_B(3) \times u_F(2) \supset so_B(3) \times su_F(2) \tag{14}$$

that is the only possible dynamical symmetry arising in this simple scheme.

We consider the following Hamiltonian:

$$\hat{H}_{\text{Hyp}} = \hat{H}_{\text{u(3)}} + \hat{H}_{\text{u(2)}} + \hat{V}_{\text{int}}, \tag{15}$$

where

$$\hat{H}_{\text{u(3)}} = \alpha\, \hat{C}_1(u(3)) + \beta\, \hat{C}_2(u(3)) + \gamma\, \hat{C}_2(so_L(3)) \tag{16}$$

represents the anharmonic oscillator with α, β, and γ as free parameters. The Casimir operators are given by $\hat{C}_1(u(3)) = \hat{N}$, $\hat{C}_2(u(3)) = \hat{N}^2$ and $C_2(so_L(3)) = \hat{\vec{L}}^2$. The spectrum is:

$$E_{\text{u(3)}} = \alpha N + \beta N(N+2) + \gamma L(L+1) \tag{17}$$

where N is the number of quanta and L is the orbital angular momentum of the Λ particle.

The fermionic part of the Hamiltonian is:

$$\hat{H}_{\text{u(2)}} = A\hat{C}_1(u(2)) + B\hat{C}_2(u(2)), \tag{18}$$

and the energy spectrum is

$$E_{\text{u(2)}} = A\langle C_1\rangle + B\langle C_2\rangle. \tag{19}$$

The representations of $u(2)$ are given by a pair of numbers $[\lambda_1, \lambda_2]$. Since there are only three possible fermionic states, the formula can take the values:

$$E_0 = 0, \qquad E_\Lambda = A + 2B, \qquad E_{\Lambda\Lambda} = 2A + 2B. \tag{20}$$

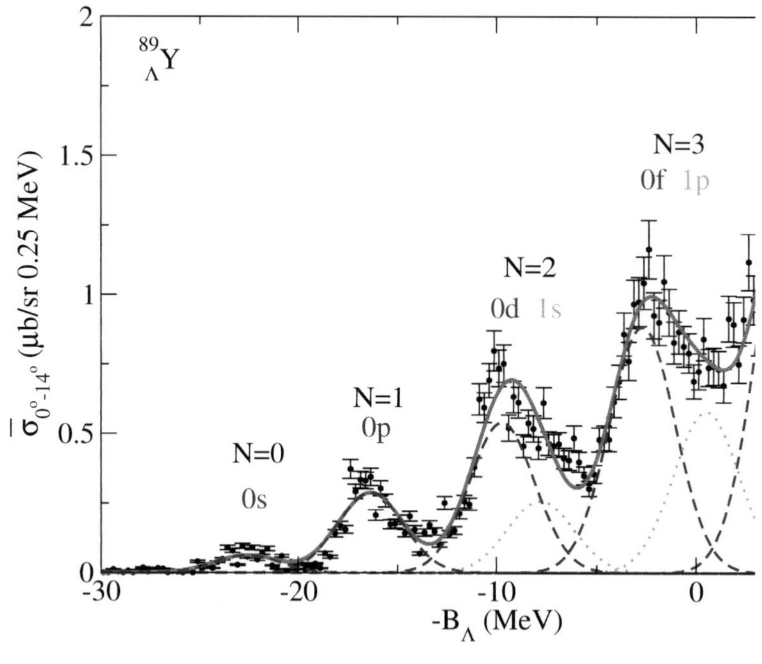

Figure 4. Experimental hypernuclear mass spectra of ${}^{89}_{\Lambda}$Y and fit based on the algebraic model.

a_{00}	1.03458	a_{11}	4.774	a_{20}	4.26848	a_{22}	8.93652
a_{31}	9.57975	a_{33}	14.6199	a_{42}	21.4563	a_{44}	22.7715
α	5.39547	β	0.506972	γ	-0.321663	E_Λ	-22.6373

Table 2. Parameters for the best fit of the empirical mass spectra of ${}^{89}_{\Lambda}$Y. The parameters α, β, γ and E_Λ are in MeV, while a_{NL} are in μb MeV.

The final energies for hypernuclei are thus

$$E_{\text{Hyp}} = \alpha N + \beta N(N+2) + \gamma L(L+1) + E_{n\Lambda}, \tag{21}$$

where the last term is given by Eq. (20), depending on the number of Λ particles in the system.

An application of our formula is made on the ${}^{89}_{\Lambda}$Y and ${}^{51}_{\Lambda}$V hypernuclei. The measured cross-sections (integrated in the 2°-14° range) for the (π^+, K^+) reaction leading to the formation of the hypernuclei show several peaks as a function of energy [13], that are interpreted as corresponding to different angular momentum states of the Λ particle.

Our fit has the same main ingredients of the work of Hotchi, namely gaussian functions with parametric width, but the advantage is that the centroids are related to our energy formula (details in [4]). In general we have fits of the same statistical value, but with a smaller number of parameters and with a theoretical framework for the peaks. Our results are summarized in Fig. 4 and 5 and Tables 2 and 3, where one can read off the best parameters to fit the data within the algebraic model.

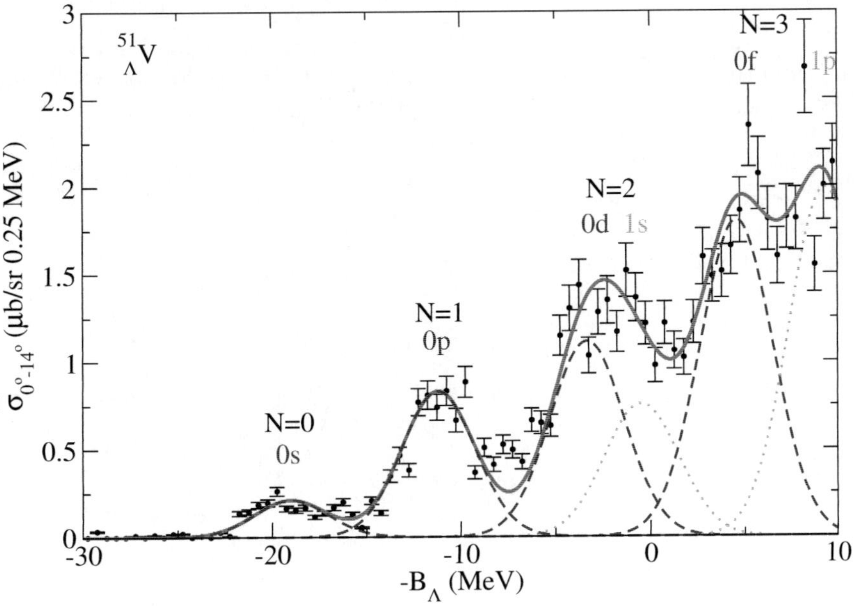

Figure 5. Same as Fig. 4, but for the $^{51}_{\Lambda}$V hypernucleus.

a_{00}	4.11173	a_{11}	16.2407	a_{20}	14.8101	a_{22}	21.8096
a_{31}	38.9592	a_{33}	35.3016				
α	7.2807	β	0.495241	γ	-0.476428	E_Λ	-19.003

Table 3. Same as Table. 2, but for the $^{51}_{\Lambda}$V hypernucleus.

One can see that our algebraic model nicely reproduces the data and gives an interpretation of peaks according to (an)harmonic oscillator quantum numbers, describing the major shells and the angular momentum energy shifts. One would ideally include higher order effects (spin-orbit coupling for example) with the aim of getting even better agreement with data, but the present level of accuracy seems adequate to the energy resolution that the data points display. Of course this model is very simple and it might become more and more incorrect as long as one goes close to the Λ separation threshold (inasmuch as the harmonic oscillator shell model has been abandoned in favour of the Wood-Saxon potential well), but it gives a nice way of classifying states according to a simple theoretical scheme and it can easily be extended to more species or to particles with higher spin.

5. Conclusions

We have shown applications of nuclear physics theories or models inspired to typical nuclear physics model in very different branches of physics like molecular physics, nuclear astrophysics and hypernuclear physics, with energies ranging from meV to keV to GeV. Our project, aimed at interdisciplinary applications of nuclear theories to other fields, has generated a large amount of new (and hopefully interesting) physics and we are determined to continue on this track. Our future endeavours will be the proposal of a new method to solve the non-relativistic Schroedinger equation, with the aim of applying it to atomic. molecular and nuclear physics cases.

Acknowledgments

All the collaborators and coauthors of the researches summarized in this paper are thankfully acknowledged. *In:Theory*, PRAT Project n. CPDA154713, Univ. Padova (Italy).

References

[1] Spitaleri C, Bertulani C A, Fortunato L and Vitturi A 2016 *Phys. Lett.* B **755** 275-278
[2] "Electron screening puzzle" solved (2016, February 26)
`http://phys.org/news/2016-02-electron-screening-puzzle.html`
[3] Fortunato L and Pérez-Bernal F 2016 *Phys. Rev.* A **94** 032508
[4] Fortunato L and Hagino K 2016 *arXiv:1604.07673v1 [nucl-th]*
[5] Rolfs C, Rodney W 1988 *Cauldrons in the Cosmos*, The University of Chicago Press (USA)
[6] Fortunato L and Vitturi A 2005 *Eur. Phys. J.* A **26**, 33-40
[7] Mason A, Chatterjee R, Fortunato L and Vitturi A 2009 *Eur. Phys. J.* A **39**, 107116
[8] Iachello F and Levine R D 1994 *Algebraic Theory of Molecules*, Topics in physical chemistry series. Oxford University Press, USA
[9] Komatsu K, Murata M and Murata Y 2005 *Science* **307** 238-240
[10] Mamone S, Min Ge, Hüvonen D, Nagel U, et al. 2009 *The Journal of Chemical Physics* **130**(8) 081103
[11] Horsewill A J, Rols S, Johnson M R, Murata Y, et al. 2010 *Phys. Rev.* B **82** 081410
[12] Frank A and Van Isacker P 2005 *Symmetry methods in molecules and nuclei*, S y G Editores (Mexico)
[13] Hotchi , et al. 2001 *Phys. Rev.* C **64** 044302
[14] Hashimoto O and Tamura H 2006 *Prog. Part. Nucl. Phys.* **57**, 564653
[15] Motoba T, Lanskoy D E, Millener D J and Yamamoto Y 2008 *Nucl. Phys.* **A804** 99

IOP Conf. Series: Journal of Physics: Conf. Series **981** (2018) 012002 doi:10.1088/1742-6596/981/1/012002

Chiral Nucleon-Nucleus Potentials at N^3LO

Paolo Finelli[1], Matteo Vorabbi[2], Carlotta Giusti[3]

[1]Dipartimento di Fisica e Astronomia, Università degli Studi di Bologna and INFN, Sezione di Bologna, Via Irnerio 46, I-40126 Bologna, Italy

[2]TRIUMF, 4004 Wesbrook Mall, Vancouver, British Columbia, V6T 2A3, Canada

[3]Dipartimento di Fisica, Università degli Studi di Pavia and INFN, Sezione di Pavia, Via A. Bassi 6, I-27100 Pavia, Italy

E-mail: finelli@bo.infn.it

Abstract. Elastic scattering is probably one of the most relevant tools to study nuclear interactions. In this contribution we study the domain of applicability of microscopic two-body chiral potentials in the construction of an optical potential. A microscopic complex optical potential is derived and tested performing calculations on ^{16}O at different energies. Good agreement with empirical data is obtained if a Lippmann-Schwinger cutoff at relatively high energies (above 500 MeV) is employed.

1. Introduction

Elastic proton scattering has been extensively studied over many decades and there now exist extensive measurements of cross sections and polarization observables for the elastic scattering of protons on a wide variety of stable nuclei. A successful framework to describe the nucleon-nucleus (NA) interaction in the elastic scattering is provided by the nuclear optical potential. With this method we can compute the scattering observables (the differential cross section, the analyzing power, and the spin rotation) for the elastic NA scattering across wide regions of the nuclear landscape.

The optical potential can be obtained microscopically. The calculation requires, in principle, the solution of the full nuclear many-body problem. In practice, with suitable approximations, microscopic optical potentials are usually derived from two basic quantities: the nucleon-nucleon (NN) t matrix and the matter distribution of the nucleus.

The theoretical framework employed in our work started almost 60 years ago with Ref. [1] where the authors developed the Watson multiple scattering approach in which the NA optical potential is expressed by a series expansion in terms of the free NN scattering amplitudes.

The NN potential is an essential ingredient in the NA scattering theory and its off-shell properties play an important role. To obtain a good description of these properties, the optical potential models have always employed "realistic" potentials, in which the experimental NN phase shifts are reproduced with a χ^2 per data $\simeq 1$. Recent available chiral potentials are developed at fourth order (N^3LO) in the chiral expansion parameter (i.e. external momentum over the chiral symmetry breaking scale, $\Lambda_\chi \approx 1$ GeV) and are used here and in Ref. [2] as a basic ingredient to compute the NN t matrix. In particular, in all the calculations presented in this paper we adopt the two different versions of chiral potentials developed by Entem and

IOP Conf. Series: Journal of Physics: Conf. Series **981** (2018) 012002 doi:10.1088/1742-6596/981/1/012002

Machleidt (EM, Ref. [3]), and Epelbaum, Glöckle, and Meißner (EGM, Ref. [4]). A future extension with the inclusion of the most recent potentials at N^4LO is in progress.

The second important ingredient of the NA scattering theory is the microscopic structure of the nuclear target, given by neutron and proton densities. These quantities are computed within the Relativistic Mean-Field (RMF) description [5] of spherical nuclei using a Density-Dependent Meson-Exchange (DDME) model, where the couplings between mesonic and baryonic fields are assumed as functions of the density itself [6].

2. Theoretical Framework

The general problem of the elastic scattering of a proton from a target nucleus of A nucleons can be stated in momentum space by the full $(A+1)$-body Lippmann-Schwinger (LS) equation

$$T = V + VG_0(E)T\,. \tag{1}$$

A reliable method to treat Eq. (1) is given by the spectator expansion [7] in which the multiple scattering theory is expanded in a finite series of terms where the target nucleons interact directly with the incident proton. In the standard approach to elastic scattering, Eq. (1) is separated into two equations. The first one is an integral equation for T

$$T = U + UG_0(E)PT\,, \tag{2}$$

where U is the optical potential operator, and the second one is an integral equation for U

$$U = V + VG_0(E)QU\,. \tag{3}$$

The operator V represents the external interaction and the total Hamiltonian for the $(A+1)$-nucleon system is given by

$$H_{A+1} = H_0 + V\,. \tag{4}$$

Of course, if we assume the presence of only two-body forces, the operator V is expressed as $V = \sum_{i=1}^{A} v_{0i}$, where the two-body potential v_{0i} describes the interaction between the incident proton and the ith target nucleon. $G_0(E)$ is the free propagator for the $(A+1)$-nucleon system. The operators P projects onto the elastic channel and, as a consequence, P and Q satisfy the following relation

$$P + Q = \mathbb{1}\,, \tag{5}$$

and P fulfills the condition

$$[G_0, P] = 0\,. \tag{6}$$

With these definitions, the elastic transition operator may be defined as $T_{\rm el} = PTP$. At the first order, the elastic scattering amplitude is given by

$$T_{\rm el}(\boldsymbol{k}', \boldsymbol{k}; E) = \frac{A}{A-1}\hat{T}(\boldsymbol{k}', \boldsymbol{k}; E)\,, \tag{7}$$

where the auxiliary elastic amplitude $\hat{T}$ is determined by the solution of the integral equation

$$\hat{T}(\boldsymbol{k}', \boldsymbol{k}; E) = \hat{U}(\boldsymbol{k}', \boldsymbol{k}; \omega) + \int d^3p \frac{\hat{U}(\boldsymbol{k}', \boldsymbol{p}; \omega)\hat{T}(\boldsymbol{p}, \boldsymbol{k}; E)}{E(k_0) - E(p) + i\epsilon}\,, \tag{8}$$

and the auxiliary first-order optical potential is defined by

$$\hat{U}(\boldsymbol{k}', \boldsymbol{k}; \omega) = (A-1)\,\langle \boldsymbol{k}', \Phi_A | t(\omega) | \boldsymbol{k}, \Phi_A \rangle\,. \tag{9}$$

IOP Conf. Series: Journal of Physics: Conf. Series **981** (2018) 012002 doi:10.1088/1742-6596/981/1/012002

t is any one of the free NN t_{0i} matrices and $|\Phi_A\rangle$ is the ground-state of the target. Defining the new variables

$$\boldsymbol{q} \equiv \boldsymbol{k}' - \boldsymbol{k}\,, \qquad \boldsymbol{K} \equiv \frac{1}{2}(\boldsymbol{k}' + \boldsymbol{k})\,, \tag{10}$$

some manipulations exploit the spin-structure of the t matrix

$$t_{pN}(\boldsymbol{\kappa}', \boldsymbol{\kappa}; \omega) = t^c_{pN}(\boldsymbol{\kappa}', \boldsymbol{\kappa}; \omega) + \frac{i}{2}\boldsymbol{\sigma} \cdot \boldsymbol{\kappa}' \times \boldsymbol{\kappa}\, t^{ls}_{pN}(\boldsymbol{\kappa}', \boldsymbol{\kappa}; \omega)\,. \tag{11}$$

and of the optical potential

$$\hat{U}(\boldsymbol{q}, \boldsymbol{K}; \omega) = \hat{U}^c(\boldsymbol{q}, \boldsymbol{K}; \omega) + \frac{i}{2}\boldsymbol{\sigma} \cdot \boldsymbol{q} \times \boldsymbol{K}\, \hat{U}^{ls}(\boldsymbol{q}, \boldsymbol{K}; \omega)\,, \tag{12}$$

where $\boldsymbol{\kappa}'$ and $\boldsymbol{\kappa}$ are relative momenta.

The first terms of Eqs. (11) and (12) correspond to the central spin-independent contributions and the second terms correspond to the spin-orbit contributions. Within this approximation, the optical potential exhibits nonlocality and off-shell effects through the dependence of t_{pN} upon $\boldsymbol{K}$. The energy ω at which the matrices t^c_{pN} and t^{ls}_{pN} are evaluated is fixed as

$$\omega = \frac{T_{lab}}{2} = \frac{1}{2}\frac{k^2_{lab}}{2m}\,, \tag{13}$$

where k_{lab} is the on-shell momentum of the projectile in the laboratory system.

From the conservation of the total angular momentum and parity, this spin operator can be expanded as

$$\hat{U}(\boldsymbol{k}', \boldsymbol{k}; \omega) = \frac{2}{\pi} \sum_{JLM} \mathcal{Y}^{L\frac{1}{2}}_{JM}(\hat{\boldsymbol{k}}')\, \hat{U}_{LJ}(k', k; \omega)\, \mathcal{Y}^{L\frac{1}{2}\dagger}_{JM}(\hat{\boldsymbol{k}})\,, \tag{14}$$

where $J = L \pm 1/2$ and $\mathcal{Y}^{L\frac{1}{2}}_{JM}$ is the standard spin-angular function. Inserting the expansion of Eq. (14) into Eq. (8) we obtain the same decomposition for the T matrix

$$\hat{T}(\boldsymbol{k}', \boldsymbol{k}; E) = \frac{2}{\pi} \sum_{JLM} \mathcal{Y}^{L\frac{1}{2}}_{JM}(\hat{\boldsymbol{k}}')\, \hat{T}_{LJ}(k', k; E)\, \mathcal{Y}^{L\frac{1}{2}\dagger}_{JM}(\hat{\boldsymbol{k}})\,, \tag{15}$$

where the partial-wave components of the transition operator for the elastic scattering are given by

$$\begin{aligned} \hat{T}_{LJ}(k', k; E) &= \hat{U}_{LJ}(k', k; \omega) \\ &+ \frac{2}{\pi} \int_0^\infty dp\, p^2 \frac{\hat{U}_{LJ}(k', p; \omega)\, \hat{T}_{LJ}(p, k; E)}{E(k_0) - E(p) + i\epsilon}\,, \end{aligned} \tag{16}$$

where

$$E(k_0) = \sqrt{k_0^2 + m^2_{proj}} + \sqrt{k_0^2 + m^2_{targ}}\,, \tag{17}$$

$$E(p) = \sqrt{p^2 + m^2_{proj}} + \sqrt{p^2 + m^2_{targ}}\,, \tag{18}$$

and m_{proj} and m_{targ} are the masses of the projectile and of the target, respectively. In terms of the partial wave components of the quantities $\hat{U}^c(\boldsymbol{k}', \boldsymbol{k}; \omega)$ and $\hat{U}^{ls}(\boldsymbol{k}', \boldsymbol{k}; \omega)$, we have

$$\hat{U}_{LJ}(k', k; \omega) = \hat{U}^c_L(k', k; \omega) + C_{LJ}\, \hat{V}^{ls}_L(k', k; \omega)\,, \tag{19}$$

where

$$C_{LJ} = \frac{1}{2}\left[J(J+1) - L(L+1) - \frac{3}{4}\right],$$
$$\hat{V}_L^{ls}(k', k; \omega) = \frac{k'k}{2L+1}\left[\hat{U}_{L+1}^{ls}(k', k; \omega) - \hat{U}_{L-1}^{ls}(k', k; \omega)\right]. \quad (20)$$

To obtain these results, the quantities $\hat{U}^c(\boldsymbol{k}', \boldsymbol{k}; \omega)$ and $\hat{U}^{ls}(\boldsymbol{k}', \boldsymbol{k}; \omega)$ are expanded in a manner similar to Eq. (14), with the difference that the partial wave components, $\hat{U}_L^c$ and $\hat{U}_L^{ls}$, are independent of J.

2.1. Observables

Under the assumptions of parity conservation and rotational invariance, the most general form of the full amplitude for the elastic proton scattering from a spin 0 nucleus is given by

$$M(k_0, \theta) = A(k_0, \theta) + \boldsymbol{\sigma} \cdot \hat{\boldsymbol{N}}\, C(k_0, \theta), \quad (21)$$

where $\hat{\boldsymbol{N}} = (\boldsymbol{k}' \times \boldsymbol{k})/(|\boldsymbol{k}'||\boldsymbol{k}|)$ and the amplitudes $A(k_0, \theta)$ and $C(k_0, \theta)$ are obtained from the partial wave solutions of Eq. (16) as

$$A(\theta) = \frac{1}{2\pi^2}\sum_{L=0}^{\infty}\left[(L+1)F_L^+(k_0) + LF_L^-(k_0)\right] P_L(\cos\theta), \quad (22)$$

$$C(\theta) = \frac{i}{2\pi^2}\sum_{L=1}^{\infty}\left[F_L^+(k_0) - F_L^-(k_0)\right] P_L^1(\cos\theta). \quad (23)$$

In Eqs. (22) and (23) an implicit dependence on k_0 is assumed. The functions $F_L^{\pm}$ denote F_{LJ} for $J = L \pm 1/2$, respectively, and are given in terms of the t matrix

$$F_{LJ}(k_0) = -\frac{A}{A-1}4\pi^2\mu(k_0)\hat{T}_{LJ}(k_0, k_0; E), \quad (24)$$

where the relativistic reduced mass is

$$\mu(k_0) = \frac{E_{proj}(k_0)\, E_{targ}(k_0)}{E_{proj}(k_0) + E_{targ}(k_0)}. \quad (25)$$

Three independent scattering observable will be considered: the unpolarized differential cross section, the analyzing power A_y, and the spin rotation Q. Their expressions as functions of the the amplitudes A and C are:

$$\frac{d\sigma}{d\Omega}(\theta) = |A(\theta)|^2 + |C(\theta)|^2, \quad (26)$$

$$A_y(\theta) = \frac{2\mathrm{Re}[A^*(\theta)\, C(\theta)]}{|A(\theta)|^2 + |C(\theta)|^2}, \quad (27)$$

$$Q(\theta) = \frac{2\mathrm{Im}[A(\theta)\, C^*(\theta)]}{|A(\theta)|^2 + |C(\theta)|^2}. \quad (28)$$

The Coulomb interaction between the incoming proton, with charge e, and the spin 0 target, with charge Ze, has been included following the algorithm outlined in Refs. [11, 12].

IOP Conf. Series: Journal of Physics: Conf. Series **981** (2018) 012002 doi:10.1088/1742-6596/981/1/012002

3. Results

3.1. NN amplitudes

In this section we present and discuss the theoretical results for the pp and pn components which are used to compute the central and the spin-orbit contributions of the NN t matrix. The NN elastic scattering amplitude for the scattering from a relative momentum $\boldsymbol{\kappa}$ to $\boldsymbol{\kappa}'$ is denoted by $M(\boldsymbol{\kappa}', \boldsymbol{\kappa}, \omega)$ and defined as follows

$$M(\boldsymbol{\kappa}', \boldsymbol{\kappa}, \omega) = \langle \boldsymbol{\kappa}' | M(\omega) | \boldsymbol{\kappa} \rangle = -4\pi^2 \mu \, \langle \boldsymbol{\kappa}' | t(\omega) | \boldsymbol{\kappa} \rangle \; , \tag{29}$$

where μ is the NN reduced mass. The most general form of this amplitude, consistent with invariance under rotation, time reversal, and parity is

$$\begin{aligned} M &= a + c(\boldsymbol{\sigma}_1 + \boldsymbol{\sigma}_2) \cdot \hat{\boldsymbol{n}} + m(\boldsymbol{\sigma}_1 \cdot \hat{\boldsymbol{n}})(\boldsymbol{\sigma}_2 \cdot \hat{\boldsymbol{n}}) \\ &+ (g+h)(\boldsymbol{\sigma}_1 \cdot \hat{\boldsymbol{l}})(\boldsymbol{\sigma}_2 \cdot \hat{\boldsymbol{l}}) + (g-h)(\boldsymbol{\sigma}_1 \cdot \hat{\boldsymbol{m}})(\boldsymbol{\sigma}_2 \cdot \hat{\boldsymbol{m}}) \, , \end{aligned} \tag{30}$$

where

$$\hat{\boldsymbol{l}} = \frac{\boldsymbol{\kappa}' + \boldsymbol{\kappa}}{|\boldsymbol{\kappa}' + \boldsymbol{\kappa}|} \, , \qquad \hat{\boldsymbol{m}} = \frac{\boldsymbol{\kappa}' - \boldsymbol{\kappa}}{|\boldsymbol{\kappa}' - \boldsymbol{\kappa}|} \, , \qquad \hat{\boldsymbol{n}} = \frac{\boldsymbol{\kappa} \times \boldsymbol{\kappa}'}{|\boldsymbol{\kappa} \times \boldsymbol{\kappa}'|} \, , \tag{31}$$

are the unit vectors defined by the NN scattering plane. The amplitudes a, c, m, g, and h can be expressed as complex functions of ω, $\boldsymbol{\kappa}$, and $\boldsymbol{\kappa}'$. From these equations one can obtain the explicit expressions for the pp and pn central and spin-orbit parts of the NN t matrix.

Calculations are performed using two different versions of the chiral potential at fourth order (N^3LO) based on the works of Entem and Machleidt [3] and Epelbaum et *al.* [4]. Entem and Machleidt (EM), who first presented a chiral potential at the fourth order, treat divergent terms in the two-pion exchange (2PE) contributions with dimensional regularization (DR), while Epelbaum, Glöckle, and Meißner (EGM) employ a spectral function regularization (SFR). In both cases the goal is to cut out the short-range part of the two pion exchange (2PE) contribution that has unphysically strong attraction. As a usual procedure, the nucleon-nucleon potential entering the LS equation is multiplied by a regulator function f^Λ

$$V(\boldsymbol{k}, \boldsymbol{k}') \to V(\boldsymbol{k}, \boldsymbol{k}') f^\Lambda(k, k') \tag{32}$$

where

$$f^\Lambda = \exp\left(-(k'/\Lambda)^{2n} - (k/\Lambda)^{2n}\right) \quad \text{with} \quad n = 2, 3 \, . \tag{33}$$

While Entem and Machleidt present results for three choices of the cutoff necessary to regulate the high-momentum components in the LS equation ($\Lambda = 450$, 500, and 600 MeV), Epelbaum *et al.* [4] allow also to study variations of the cutoff $\tilde{\Lambda}$ that regulates the 2PE contribution. In fact, in the latter approach one can choose between the following cutoff combinations:

$$\{\Lambda, \tilde{\Lambda}\} \quad = \quad \{450, 500\}, \{450, 700\}, \{550, 600\}, \{600, 600\}, \{600, 700\} \, . \tag{34}$$

In the following figures all the results are labelled by an acronym (to distinguish the authors) followed by the value of the LS cutoff (Λ). In the EGM case, for $\Lambda = 450$ and 600 MeV we plot bands that show how calculations can change respect to variations of the SFR cutoff $\tilde{\Lambda}$. In Fig. 1 the theoretical results for the real and imaginary parts of pp and pn amplitudes (a and c) computed at an energy of 100 MeV are shown as functions of the center-of-mass NN angle ϕ and compared with the experimental data. The calculations are performed using the EM potentials [3] with a LS cutoff ranging between 450 and 600 MeV. The experimental data are globally reproduced by the three potentials, with the only remarkable exception of the real part of the c_{pp} amplitude that is overestimated. It must be considered, however, that this is a

IOP Conf. Series: Journal of Physics: Conf. Series **981** (2018) 012002 doi:10.1088/1742-6596/981/1/012002

small quantity, i.e. two orders of magnitude smaller than the respective imaginary part, and it will only provide a very small contribution to the optical potential.

In Fig. 2 we show the results obtained at 100 MeV with the EGM potentials [4]. Also in this case, all three potentials are in overall good agreement with the experimental data with the only remarkable exception of the real part of the c_{pp} amplitude. In particular, they show very similar results and in many cases the yellow and torquoise bands are overlapped. Their trends are also very close to the ones shown in Fig. 1 for the EM potential and they display the same discrepancy in comparison with the experimental data for c_{pn} and around the peak of the imaginary part of c_{pp}.

Since Chiral Perturbation Theory (ChPT) is a low-momentum expansion of QCD, we expect that, as the energy is increased, larger discrepancies appear respect to empirical data. In Figs. 3 and 4 we present the results corresponding to the ones shown in Figs. 1 and 2 but at an energy of 200 MeV. As energy is increased, all potentials are still unable to reproduce the experimental data of the real part of the c_{pp} amplitude, but most of the chiral potentials give satisfactory results, in agreement with the data for all the other amplitudes, with one notable exception, i.e. cutoffs below 500 MeV. Based on these results, one can predict an unsatisfactory result for a nucleon-nucleus optical potential if EM-450 or EGM-450 are employed at energies well above 100 MeV.

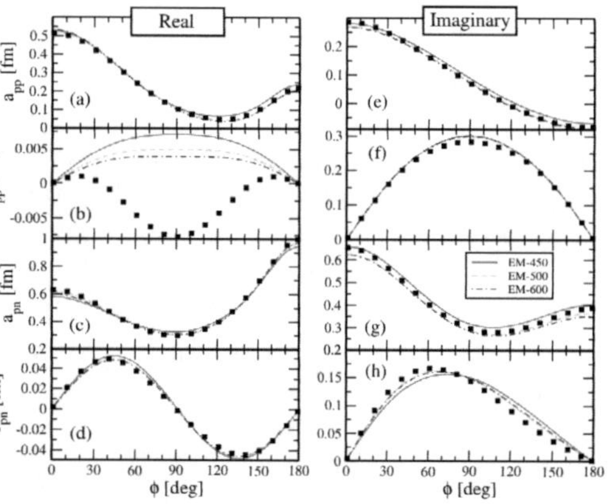

Figure 1. Real (left panel) and imaginary (right panel) parts of *pp* and *pn* Wolfenstein amplitudes (a and c) as functions of the center-of-mass NN angle ϕ. All the amplitudes are computed at 100 MeV using the EM potentials [3] with a LS cutoff ranging between 450 and 600 MeV. Data (black squares) are taken from Ref. [8].

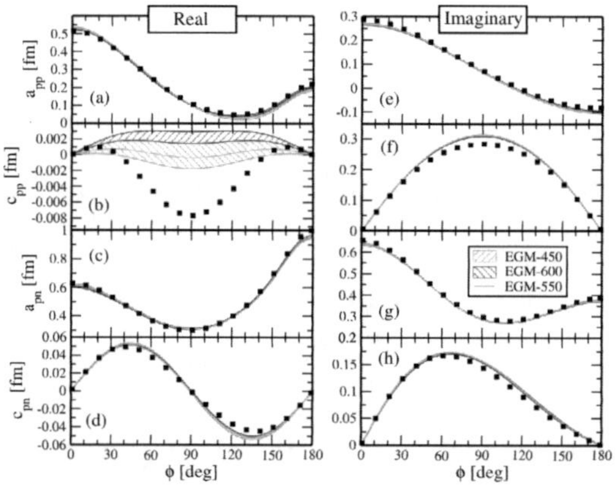

Figure 2. The same as in Fig. 1 using EGM potentials [4] with a LS cutoff ranging between 450 and 600 MeV. In two cases (Λ = 450 and 600 MeV) we show uncertainty bands produced by changing $\tilde{\Lambda}$ according to Eq. (34). Data (black squares) are taken from Ref. [8].

3.2. Proton elastic scattering on ^{16}O

In this section we present and discuss our numerical results for the NA elastic scattering observables calculated with the microscopic optical potential previously obtained. As a study case in our calculations we consider elastic proton scattering on ^{16}O.

In order to investigate and emphasize the differences between the different NN potentials and also on the basis of the results obtained for the NN amplitudes a and c, the scattering

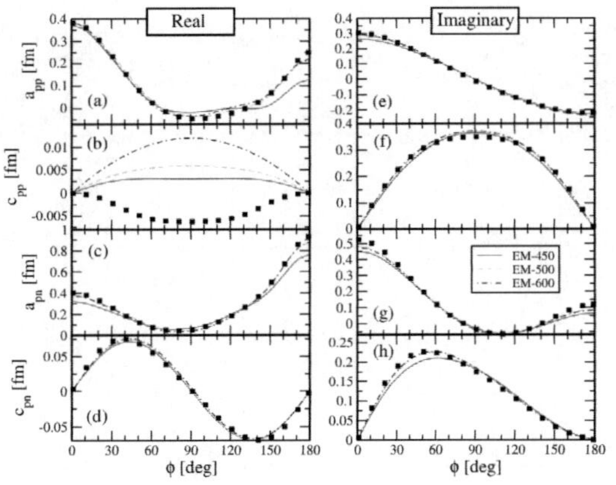

Figure 3. The same as in Fig. 1 but for an energy of 200 MeV. Data (black squares) are taken from Ref. [8].

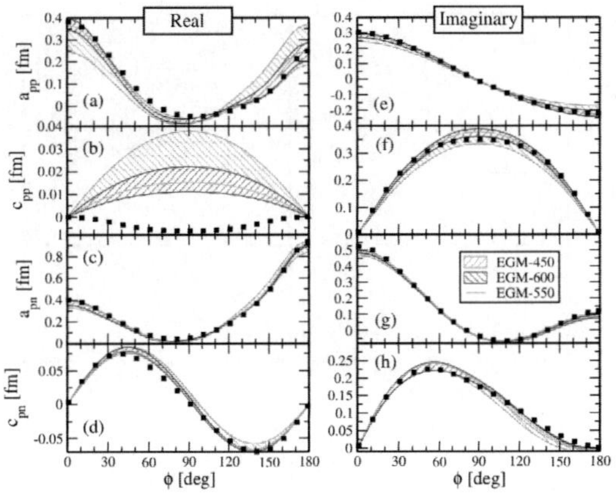

Figure 4. The same as is Fig. 2 but for an energy of 200 MeV. Data (black squares) are taken from Ref. [8].

observables have been calculated for two energies (100 and 200 MeV) for which experimental data are available.

With these calculations we intend to achieve the following goals: 1) to check the agreement of our theoretical predictions with the empirical data; 2) to study the limits of applicability of chiral potentials in terms of the proton energy; 3) to identify the best set of values for the LS and, eventually, SFR cutoffs.

In Figs. 5 and 6 we show the differential cross section ($d\sigma/d\Omega$), the analyzing power A_y, and the spin rotation Q for elastic proton scattering on ^{16}O as functions of the center-of-mass scattering angle θ with the above mentioned energies. In the left panels we show the results obtained with the EM potentials [3] while in the right panels we show the results obtained with the EGM potentials [4]. All potentials are denoted by the value of the LS cutoff.

In Fig. 5, at 100 MeV, all sets of potentials, regardless of cutoffs and theoretical approaches, give very similar results for all three observables, with the exception of A_y above 50 degrees, where all potentials overestimate the experimental data up to the maximum and then display an unrealistic downward trend, and Q around the maximum at 30 degrees. In particular, the experimental cross section is well reproduced by all potentials in the minimum region, between 30 and 35 degrees. Even if differences are rather small, potentials with the largest cutoff ($\Lambda = 600$ MeV) seem to provide the best description of A_y.

In Fig. 6 we plot the results obtained at 200 MeV. At this energy, it is clear that potentials obtained with the lower cutoffs (EM-450 and EGM-450) cannot be employed any further: in both cases, the differential cross sections are not satisfactorily reproduced and the behaviour of A_y and Q as a function of θ is in clear disagreement with the empirical one. On the other hand, the remaining sets of potentials well describe the experimental cross sections and the analyzing power A_y, that is reasonably described not only for small scattering angles but also for values larger than the minimum value up to about 45 degrees.

4. Conclusions

In this work we have obtained a new microscopic optical potential for elastic proton-nucleus scattering. Our optical potential has been derived as the first-order term within the spectator expansion of the nonrelativistic multiple scattering theory.

Microscopic optical potentials have been derived from two different versions of the chiral

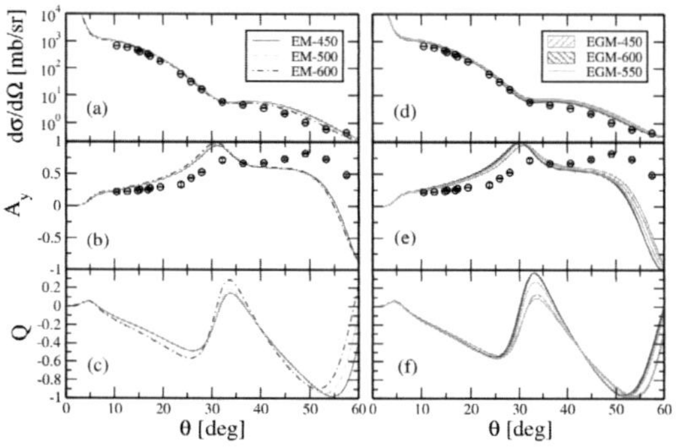

Figure 5. Scattering observables as a function of the center-of-mass scattering angle θ for elastic proton scattering on ^{16}O computed at 100 MeV (laboratory energy). On the left panel we employ the set of EM potentials [3] while in the right panel we show the EGM potentials [4]. Data are taken from Refs. [9, 10].

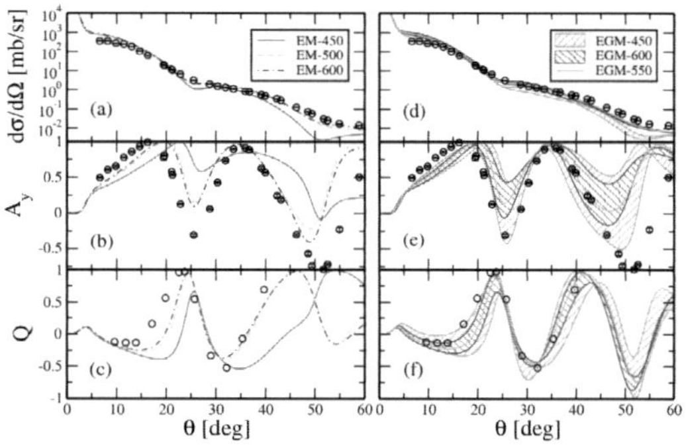

Figure 6. The same as is Fig. 5 but for an energy of 200 MeV. Data are taken from Refs. [9, 10].

potential at fourth order (N^{3}LO) based on the work of Entem and Machleidt (EM) [3] and Epelbaum, Glöckle, and Meißner (EGM) [4], which differ in the regularization scheme employed in the two-pion exchange term and in the choice of the cutoffs.

Results for pp and np amplitudes (a and c) obtained with different NN potentials have been presented and discussed. Since ChPT is a low-momentum expansion of QCD, the agreement of the chiral potential with the experimental data becomes, as expected, worse increasing the energy. While at 100 MeV all the NN potentials are able to reproduce the experimental amplitudes, with the only exception of the real part of c_{pp} amplitude, that is anyhow extremely small, at 200 MeV the set of potentials with lower cutoffs (450 MeV) fail to reproduce empirical data.

As case study for our investigation we have considered elastic proton scattering on ^{16}O. On the basis of all these results for ^{16}O we can draw two conclusions: 1) Potentials with lower cutoffs cannot reproduce experimental data at energies close to 200 MeV. 2) There is no appreciable difference in using 500 or 600 MeV as LS cutoffs, even if the EM-600 and EGM-600 potentials seem to have a slightly better agreement with empirical data, in particular looking at polarization observables.

References

[1] Kerman A K, McManus H and Thaler R M 1959 *Ann. of Phys.* **8** 551
[2] Vorabbi M, Finelli P and Giusti C 2016 *Phys. Rev.* C **93** 034619
[3] Entem D R and Machleidt R 2003 *Phys. Rev.* C **68** 041001
[4] Epelbaum E, Glöckle W and Meißner Ulf-G 2005 *Nucl. Phys.* A **747** 362
[5] Niksic T, Paar N, Vretenar D and Ring P 2014 *Comp. Phys. Comm.* **185** 1808
[6] Niksic T, Vretenar D, Finellip and Ring P 2002 *Phys. Rev. C* **66** 024306
[7] Chinn C R, Elster Ch, Thaler R M and Weppner S P 1995 *Phys. Rev. C* **52** 1992
[8] `http://nn-online.org/`
[9] `http://www.physics.umd.edu/enp/jjkelly/datatables.htm`
[10] `http://www.nndc.bnl.gov/exfor/exfor.htm`
[11] Chinn C R, Elster Ch and Thaler R M 1991 *Phys. Rev.* C **44** 1569
[12] Elster C, Liu L C and Thaler R M 1993 *J. of Phys.* G **19** 2123

IOP Conf. Series: Journal of Physics: Conf. Series **981** (2018) 012003 doi:10.1088/1742-6596/981/1/012003

A self-consistent equation of motion multiphonon method for odd mass nuclei

G De Gregorio[1,2], F Knapp[3], N Lo Iudice[1,2], P Veselý[4]

[1] Dipartimento di Fisica, Università di Napoli Federico II, Napoli, Italy
[2] Istituto Nazionale di Fisica Nucleare (INFN), Sezione di Napoli, Napoli, Italy
[3] Faculty of Mathematics and Physics, Charles University in Prague, Czech Republic
[4] Institute of Nuclear Physics, Academy of Sciences of the Czech Republic, 250 68 Řež, Czech Republic

E-mail: degregorio@na.infn.it

Abstract. An equation of motion phonon method, developed for even-even nuclear systems, is extended to odd nuclei and applied to $A = 17$ isobars. A calculation using a chiral potential is carried out in a space encompassing up to two phonons and, under some simplifying assumptions, three phonons. The low and high energy spectroscopic properties of ^{17}O and ^{17}F are thoroughly explored. The impact of the different phonon components on spectrum, transitions and dipole cross sections is discussed.

1. Introduction

Different methods have been developed to study the modifications of the single-particle states in odd nuclei induced by the core excitations. The basic mechanism is illustrated within the particle-vibration coupling (PVC) model [1, 2] in which a particle is coupled to the collective excitations of the core, commonly described in random-phase approximation (RPA). Recently, PVC calculations were performed within the framework of energy density functionals deduced from Skyrme forces [3–5] or relativistic meson-nucleon Lagrangians [6, 7] or from the theory of finite Fermi systems [8].

Calculations using NN + 3N chiral forces were performed within the context of self-consistent Green's function theory [9], no-core shell model (NCSM) [10] and coupled cluster [11–13].

An equation of motion phonon method (EMPM), proposed [14, 15] for even nuclei and applied mainly to study the dipole response in neutron rich nuclei [16–18], has been extended recently to odd nuclei [19]. It consists in constructing and solving iteratively a set of equations of motion to generate an orthonormal basis of states composed of a valence particle coupled to n-phonon states ($n = 1, 2, ...n...$), also generated within the EMPM, describing the excitations of a doubly magic core. The basis is then adopted to diagonalize the full eigenvalue problem. The whole procedure does not rely on any approximations and preserves the Pauli principle.

We report here on the formulation of the method and its application to ^{17}O and ^{17}F. The calculations are fully self-consistent and make use of a nucleon-nucleon (NN) chiral potential $V_\chi = NNLO_{opt}$ optimized so as to minimize the effects of the three-body forces [20].

IOP Conf. Series: Journal of Physics: Conf. Series **981** (2018) 012003 doi:10.1088/1742-6596/981/1/012003

2. Outline of the EMPM method for odd nuclei

We use the second quantization formalism and denote the particle creation (annihilation) operators by $a^\dagger_r$ (b_r). We label the particle anf hole states by p and h, respectively.

The main goal of the method is to generate a basis of n-phonon states $\mid \nu_n\rangle$, of energies E_{ν_n}, having the form

$$|\nu_n>= \sum_{p\alpha_n} C^{\nu_n}_{p\alpha_n} \mid (p \times \alpha_n)^v\rangle = \sum_{p\alpha_n} C^{\nu_n}_{p\alpha_n} \left\{a^\dagger_p \times \mid \alpha_n >\right\}^{\nu_n}, \quad (1)$$

where $a^\dagger_p$ creates a valence particle coupled to the n-phonon states

$$\mid \alpha_n >= \sum_{\lambda\alpha_{n-1}} C^{\alpha_n}_{\lambda\alpha_{n-1}} \left\{O^\dagger_\lambda \times \mid \alpha_{n-1}\rangle\right\} \quad (2)$$

and $O^\dagger_\lambda = \sum_{ph} c^\lambda_{ph}(a^\dagger_p \times b_h)^\lambda$ is the TDA phonon operator. The $\mid \alpha_n >$ are generated iteratively within the EMPM starting from $n = 1$ (TDA phonons) and form an orthonormal basis.

We start with constructing the equations of motion in the n-phonon subspace

$$\langle \alpha \parallel [b_p, H]^p \parallel \nu\rangle = \Big(E_\nu - E_\alpha\Big) X^\nu_{p\alpha}, \quad (3)$$

where

$$X^\nu_{p\alpha} = \langle \alpha \parallel b_p \parallel \nu\rangle = \sum_{p'\alpha'} D^\nu(p\alpha, p'\alpha')C^\nu_{p'\alpha'} \quad (4)$$

and $\mathcal{D}^\nu(p\alpha, p'\alpha') = \langle (p \times \alpha)^v \mid (p' \times \alpha')^v\rangle$ is the metric or overlap matrix. We have omitted the subscript n for simplicity.

After expanding the commutator and making use of the relation (4) we obtain the generalized eigenvalue equations

$$\sum_{p_1\alpha_1 p'\alpha'} \Big[\mathcal{A}^\nu(p\alpha, p_1\alpha_1) - E_\nu \delta_{pp_1}\delta_{\alpha\alpha_1}\Big]\mathcal{D}^\nu(p_1\alpha_1, p'\alpha')C^\nu_{p'\alpha'} = 0. \quad (5)$$

$\mathcal{A}^\nu$ is a matrix of the simple structure

$$\mathcal{A}^\nu(p\alpha, p'\alpha') = (\epsilon_p + E_\alpha)\delta_{pp'}\delta_{\alpha\alpha'} + \mathcal{V}^\nu_{p\alpha,p'\alpha'}, \quad (6)$$

where $\mathcal{V}^\nu$ is the particle-phonon potential given by

$$\mathcal{V}^\nu_{p\alpha,p'\alpha'} = \sum_{\sigma tq} [2j_\sigma + 1]^{1/2} W(\alpha\sigma v p'; \alpha' p) F^\sigma_{pp'tq} \langle n; \alpha' \parallel \Big[a^\dagger_t \times b_q\Big]^\sigma \parallel n; \alpha\rangle. \quad (7)$$

Here, $\langle n; \alpha' \parallel \Big[a^\dagger_t \times b_q\Big]^\sigma \parallel n; \alpha\rangle$ is the n-phonon density matrix, W the Racah coefficients, and F the Pandya transform of the two-body potential.

Eq. (5) represents the eigenvalue equation in the overcomplete basis $\left\{a^\dagger_p \times \mid \alpha >\right\}^\beta$. The redundant states are eliminated by the procedure outlined in [14, 21], based on the Cholesky decomposition method. It is therefore possible to turn the singular eigenvalue equation (5) into a non singular one yielding only the physical states.

Since recursive formulas hold for all quantities entering $\mathcal{A}$ and $\mathcal{D}$, the eigenvalue equations are solved iteratively and, thereby, yield a basis of orthonormal multiphonon states $\{|\nu_0 >, |\nu_1 >, \dots |\nu_n > \dots\}$.

Represented in such a basis, the Hamiltonian matrix is composed of a sequence of diagonal blocks, one for each n, mutually coupled by off-diagonal terms $\langle \nu_{n'} \mid H \mid \nu_n \rangle$. These are non vanishing only for $n' = n \pm 1, n \pm 2$ and are computed by means of recursive formulas. A matrix of such a simple structure can be easily diagonalized yielding eigenfunctions of the form

$$\mid \Psi_\nu \rangle = \sum_{\nu_n} C^{\nu}_{\nu_n} \mid \nu_n \rangle. \tag{8}$$

These states can be used to evaluate the transition amplitudes of a multipole operator $\mathcal{M}(\lambda\mu)$. If the initial and/or final states have dominant single particle character, we have

$$\langle \psi_{\nu'} \parallel \mathcal{M}(\lambda) \parallel \psi_\nu \rangle \simeq \mathcal{M}^{(\nu\nu')}_{00}(\lambda) + \mathcal{M}^{(\nu\nu')}_{01}(\lambda) + \mathcal{M}^{(\nu\nu')}_{10}(\lambda), \tag{9}$$

where $\mathcal{M}^{(\nu\nu')}_{00}(\lambda)$, $\mathcal{M}^{(\nu\nu')}_{01}(\lambda)$, and $\mathcal{M}^{(\nu\nu')}_{10}(\lambda)$ are, respectively, the particle-particle, particle-phonon, and phonon-particle transition amplitudes, given by

$$\mathcal{M}^{(\nu\nu')}_{00}(\lambda) = \sum_{pp'} C^{\nu}_{p} C^{\nu'}_{p'} \langle p' \parallel \mathcal{M}_\lambda \parallel p \rangle, \tag{10}$$

and

$$\mathcal{M}^{(\nu\nu')}_{01}(\lambda) = (-)^{v-v'} \mathcal{M}^{\nu\nu'}_{10}(\lambda) = \sum_{pk} C^{\nu}_{p} \mathcal{M}_{\lambda_k} P^{(\nu')}_{p\lambda_k}. \tag{11}$$

They are determined by the amplitudes of the transitions to the kth TDA state of spin $J_k = \lambda$

$$\mathcal{M}_{\lambda_k} = \langle k\lambda \parallel \mathcal{M}(\lambda) \parallel 0 \rangle = \frac{1}{[\lambda]^{1/2}} \sum_{ph} c^{(\lambda_k)}_{ph} \langle p \parallel \mathcal{M}_\lambda \parallel h \rangle \tag{12}$$

and by the weights $P^{(\nu')}_{p\lambda_k} = \sum_{\nu'_1} C^{\nu'}_{\nu'_1} X^{\nu'_1}_{p\lambda_k}$ which incorporate the joint contributions of the one-phonon components $\mid \nu'_1 \rangle$ of the final states $\mid \Psi_{\nu'} \rangle$ and of the λ-multipole particle-phonon configurations $\mid [p \times \lambda_k]^{\nu'_1} \rangle$ present in $\mid \nu'_1 \rangle$.

3. Calculations and results

The Hamiltonian we used is composed of an intrinsic kinetic operator T_{int} plus the optimized chiral two-body potential $V_\chi = NNLO_{opt}$ [20]. We generate a HF basis in a configuration space including up to the $N_{max} = 15$ harmonic oscillator major shell and derive the TDA phonons from a subset of HF states corresponding to $N = 12$.

The multiphonon particle-core basis is composed of all one-phonon states $\mid (p \times \alpha_1)^\nu \rangle$, the two-phonon $\mid (p \times \alpha_2)^\nu \rangle$ of energies $E_{\alpha_2} \leq 35$ MeV and the three-phonon $\mid (p \times \alpha_3)^\nu \rangle$ of energies $\epsilon_p + E_{\alpha_3} \leq 55$ MeV. The inclusion of the three-phonon particle-core states has required some approximations. We have neglected the phonon-phonon potential for the calculation of α_3 as well as the particle-core interaction $\mathcal{V}^{\nu}_{p\alpha_3, p'\alpha'_3}$ (diagonal approximation). We have also neglected the exchange terms between the odd particle and α_3 by putting $\mathcal{D}^{\nu}(p_1\alpha_3; p'\alpha'_3) = \delta_{p_1 p'} \delta_{\alpha_3 \alpha'_3}$.

The theoretical spectra of ^{17}O and ^{17}F are compared with experiments in Figs. 1 and 2, respectively. The one-phonon states alter appreciably the HF level scheme. They depress the lowest $5/2^+_1$ with respect to the other states thereby enhancing its distance from the other levels. This effect promotes the inversion between the $1/2^+_1$and $5/2^+_1$ in ^{17}F consistently with the experiments. Another important consequence is the occurrence of a large number of levels which enrich greatly the spectra. Unfortunately, these new states fall at too high energies ($\geq 11 MeV$) with respect to the corresponding experimental levels. The inclusion of the two-phonon states enhances the level density only at high energy and, therefore, leaves the low-lying

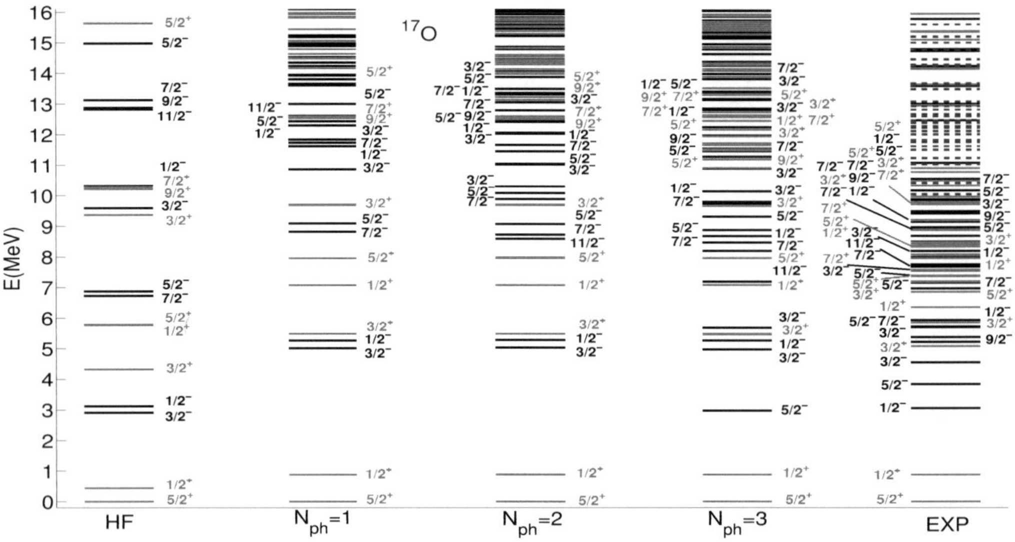

Figure 1. Theoretical versus experimental [22] spectra of ^{17}O. N_{ph} indicates the maximum phonon number. The dashed levels have unknown spin or parity or both.

sector unchanged. Only the three phonons, by pushing few negative parity states down in energy, enrich the low energy spectra but not sufficiently to reproduce the density of the experimental low-energy level schemes.

We use the intrinsic $E1$ operator referred to the CM coordinate which induces the effective charges $e_i = (N/A)e$ for protons and $e_i = (Z/A)e$ for neutrons. The E1 cross section is given by

$$\sigma_{int} = \int_{E_0}^{E} \sigma(\omega) d\omega = \frac{16\pi^3}{9\hbar c} \int_{E_0}^{E} \omega S(E1, \omega) d\omega, \tag{13}$$

where

$$S(E\lambda, \omega) = \sum_{\nu} B_\nu(E\lambda)\delta(\omega - \omega_\nu) \approx \sum_{\nu} B_\nu(E\lambda)\rho_\Delta(\omega - \omega_\nu) \tag{14}$$

is the strength function , ω is the energy variable, ω_ν the energy of the transition of multipolarity $E\lambda$ from the ground to the ν_{th} excited state of spin $J = \lambda$, and ρ_Δ is a Lorentzian of width Δ which replaces the δ function as a weight of the reduced transition probability

$$B_\nu(E\lambda) =\mid \langle \Psi_{\nu\lambda} \parallel \mathcal{M}(E\lambda\mu) \parallel \Psi_0 \rangle \mid^2 . \tag{15}$$

The one-phonon cross section in ^{17}O (Fig. 3(a)) is displaced upward by several MeV with respect to experiments and remains practically unchanged when we include the two phonons. It gets, instead, severely damped and down-shifted by the couplings to three phonons, though not sufficiently to reproduce the experimental main peak which remains ~ 2 MeV down. The unwanted secondary peak originates from a strong transition at ~ 20 MeV (Fig. 4(c)).

The phonon action in ^{17}F is analogous to the one exerted in ^{17}O. The cross section gets quenched and shifted mainly by the coupling to three phonons (Fig. 3(a)). Its behavior, however, is smoother than in ^{17}O. It exhibits a broad wiggly hump covering a wide energy range (20-40 MeV) arising from a huge numbers of closely packed small peaks. Unfortunately

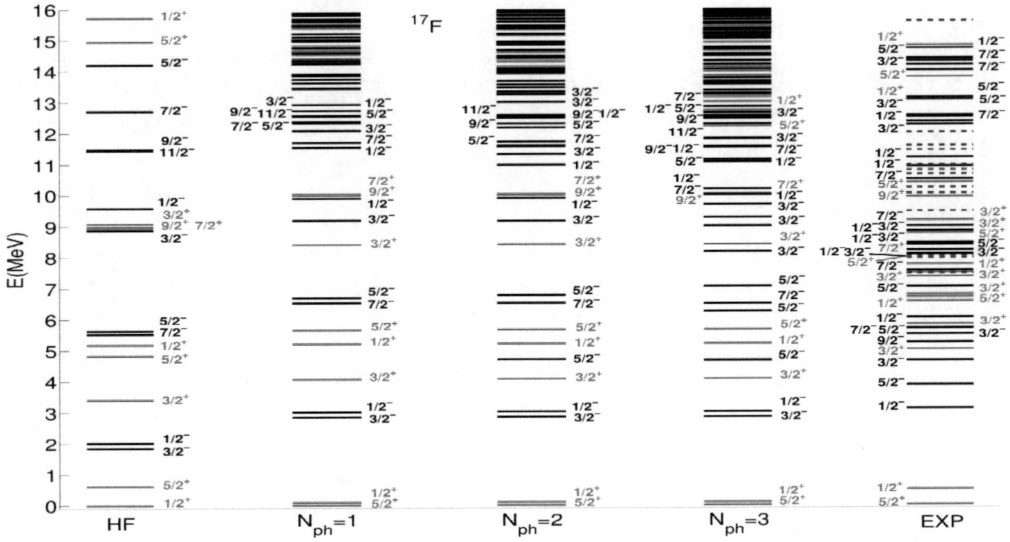

Figure 2. Level schemes of ^{17}F. The notations are the same as for ^{17}O

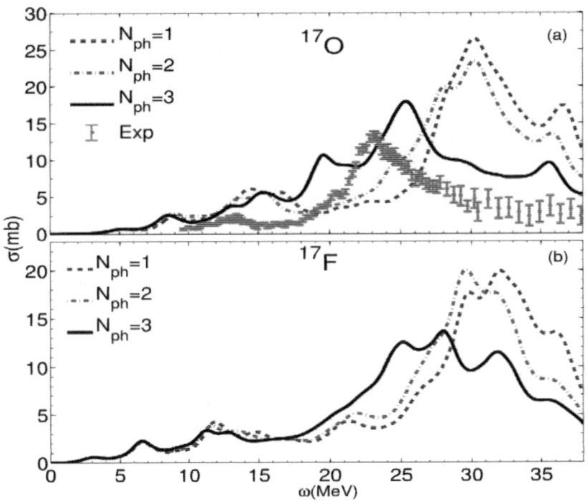

Figure 3. The theoretical E1 cross sections, computed in different multiphonon spaces, are compared in ^{17}F (b) and with the experimental ones [23] in and ^{17}O (a). A width of $\Delta = 2$ MeV have been used.

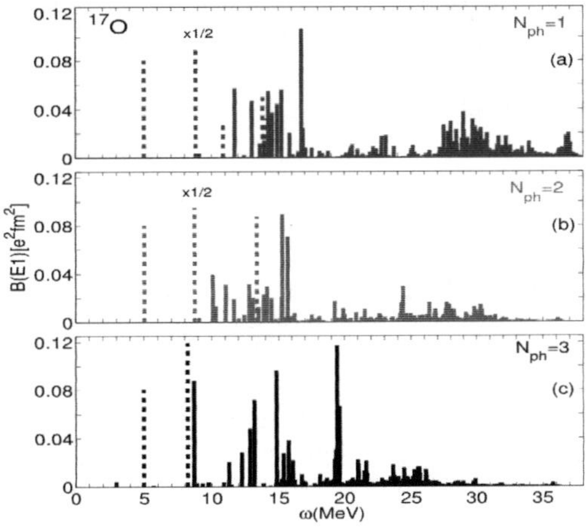

Figure 4. E1 strength distribution of ^{17}O computed in spaces including up to $N_{ph} = 1$ (a), $N_{ph} = 2$(b), and $N_{ph} = 3$(c) phonons.

no experimental data are available except for few low-lying transitions which we will discuss later.

The analysis of the $E1$ strength distribution in ^{17}O (Fig. 4) is helpful for clarifying the role of the different phonon subspaces. The coupling to two phonons induces an appreciable damping over the whole one-phonon spectrum, especially in the high-energy sector (Figs. 4 (a) and (b)). The three phonons deplete almost completely the high energy region and pack most of the strength in the range $5 - 25$ MeV. (Fig. 4(c)). This is the result of two actions. The three phonons shift the one-phonon energies downward and strengthen the amplitudes of several one-phonon states at the expenses of the two phonons.

The small peaks at low energies should correspond to the pygmy resonance. This is the case of the hump in the $13-15$ MeV interval. This arises almost entirely from the $7/2^-$ excitations and is likely to correspond to the pygmy resonance, which, according to the experimental analysis of Ref. [22], is due to the excitation of two $7/2^-$ states.

4. Conclusions

The multiphonon states enhance enormously the density of levels and compress the whole spectrum, consistently with the data. In particular, the one-phonon states improve the description of low-lying states and enhance the density of levels only at high energy. The low-energy sector is enriched only by the three phonons which push down in energy few one-phonon states, through their strong coupling. The three phonons exert also a crucial quenching action on the $E1$ transitions which reduces substantially the gap between the theoretical and experimental $E1$ cross sections. Sizable discrepancies still remain. The damping and energy shift, though appreciable, are not sufficient to reproduce completely peak, shape and magnitude of the cross section in ^{17}O.

A possible recipe for bridging the gap with experimental data may consist in improving the HF description of the single particle spectra which amounts to improve the nucleon-nucleon potential. In fact, as shown in Figs. 1 and 2, the levels or groups of levels above the Fermi surface are too far apart, especially as the energy increases, a common feature of HF spectra derived from NN interactions [24, 25]. These gaps are reduced substantially but not completely by the phonon coupling.

Acknowledgments

This work was partly supported by the Czech Science Foundation (project no. P203-13-07117S).

References

[1] Bohr A and Mottelson B R 1975 *Nuclear Structure, Vol. II* (W. A. Benjamin, New York)
[2] Mahaux C, Bortignon P F, Broglia R A and Dasso C H 1985 *Phys. Rep.* **120** 1
[3] Cao L G, Colò G, Sagawa H and Bortignon P 2014 *Phys Rev. C* 044314
[4] Niu Y F, Niu Z M, Colò G and Vigezzi E 2015 *Phys. Rev. Lett.* **114**(14) 142501
[5] Co' G, De Donno V, Anguiano M, Bernard R N and Lallena A M 2015 *Phys Rev. C* **92** 024314
[6] Litvinova E and Ring P 2006 *Phys Rev. C* **73** 044328
[7] Afanasjev A V and Litvinova E 2015 *Phys. Rev. C* **92**(4) 044317
[8] Gnezdilov N V, Borzov I N, Saperstein E E and Tolokonnikov S V 2014 *Phys. Rev. C* **89** 034304
[9] Cipollone A, Barbieri C and Navrátil P 2013 *Phys. Rev. Lett.* **111** 062501
[10] Barrett B R, Navrátil P and Vary J P 2013 *Progress in Particle and Nuclear Physics* **69** 131
[11] Hagen G, Papenbrock T and Hjorth-Jensen M 2010 *Phys. Rev. Lett.* **104**(18) 182501
[12] Jansen G R, Engel J, Hagen G, Navratil P and Signoracci A 2014 *Phys. Rev. Lett.* **113** 142502
[13] Hagen G, Papenbrock T, Hjorth-Jensen M and Dean D J 2014 *Rep. Prog. Phys.* **77** 096302
[14] Andreozzi F, Knapp F, Lo Iudice N, Porrino A and Kvasil J 2007 *Phys. Rev. C* **75**(4) 044312
[15] Bianco D, Knapp F, Lo Iudice N, Andreozzi F and Porrino A 2012 *Phys. Rev. C* **85**(1) 014313
[16] Knapp F, Lo Iudice N, Veselý P, Andreozzi F, De Gregorio G and Porrino A 2014 *Phys. Rev. C* **90** 014310
[17] Knapp F, Lo Iudice N, Veselý P, Andreozzi F, De Gregorio G and Porrino A 2015 *Phys Rev. C* **92** 054315
[18] De Gregorio G, Knapp F, Lo Iudice N and Vesely P 2016 *Physical Review C* **93**
[19] De Gregorio G, Knapp F, Lo Iudice N and Vesely P 2016 *Phys. Rev. C* **94**(6) 061301
[20] Ekström A, Baardsen G, Forssén C, Hagen G, Hjorth-Jensen M, Jansen G R, Machleidt R, Nazarewicz W, Papenbrock T, Sarich J and Wild S M 2013 *Phys. Rev. Lett.* **110**(19) 192502
[21] Andreozzi F, Knapp F, Lo Iudice N, Porrino A and Kvasil J 2008 *Physical Review C* **78** 054308
[22] Tilley D, Weller H and Cheves C 1993 *Nucl. Phys. A* **564** 1
[23] Jury J W, Berman B L, Faul D D, Meyer P and Woodworth J G 1980 *Phys Rev. C* **21** 503
[24] Bianco D, Knapp F, Lo Iudice N, Veselý P, Andreozzi F, De Gregorio G and Porrino A 2014 *J. Phys. G: Nucl. Part. Phys.* **41** 025109
[25] Roth R, Papakonstantinou Pand Paar N, Hergert H, Neff T and Feldmeier H 2006 *Phys Rev. C* **73** 044312

IOP Conf. Series: Journal of Physics: Conf. Series **981** (2018) 012004 doi:10.1088/1742-6596/981/1/012004

Direct reactions of weakly-bound nuclei within a one dimensional model

L Moschini[1 2 3], A Vitturi[1 2] and AM Moro[3]

[1]Dipartimento di Fisica e Astronomia "G. Galilei", Padova, Italy
[2]INFN, Sezione di Padova, Padova, Italy
[3]Departamento de Física Atómica, Molecular y Nuclear, Universidad de Sevilla, Apartado 1065, E-41080 Sevilla, Spain

E-mail: laura.moschini@pd.infn.it

Abstract. A line of research has been developed to describe structure and dynamics of weakly-bound systems with one or more valence particles. To simplify the problem we are assuming particles moving in one dimension and, despite the drastic assumption, the model encompasses many characteristics observed in experiments. Within this model we can describe, for example, one- and two-particle breakup and one- and two-particle transfer processes. We concentrate here in models involving weakly-bound nuclei with just one valence particle. Exact solutions obtained by directly solving the time-dependent Schroedinger equation can be compared with the results obtained with different approximation schemes (coupled-channels formalism, continuum discretization, etc). Our goal is to investigate the limitations of the models based on approximations, and in particular to understand the role of continuum in the reaction mechanism.

1. Introduction

In medium-light mass region of nuclear chart the drip-lines are characterized by the presence of the so-called halo nuclei. These systems can be described in terms of a core surrounded by one or more weakly-bound nucleons and present different features with respect to well-bound nuclei. While in a mean-field description bound systems do not fill all bound single-particle levels and their excitation basically promotes nucleons to empty bound levels, weakly-bound systems fully occupy all available bound states and so their excitation necessarily involve the promotion to the continuum with subsequent emission. For this reason the description of halo nuclei is more involved (in particular for more than two valence particles), even considering inert cores.

In this contribution we study direct reactions (i.e. elastic and inelastic scattering, transfer and breakup) involving nuclei with just one valence neutron. One could model the reaction following the time evolution of the wavefunction of this neutron: starting from an initial wavefunction $\Psi(\vec{r}, t_i)$ describing the system at $t_i = -\infty$, to obtain at $t_f = +\infty$ the final wavefunction $\Psi(\vec{r}, t_f)$. The projection of the final wavefunction on the different channels will lead to the probability for the different processes. The problem is traditionally solved in the so-called coupled-channels approach (or in its lower order approximations) in which the wavefunction of the system is expanded in a certain basis, and the solution of the problem moves to the study of the time evolution of the expansion coefficients. When this scheme is applied to weakly-bound systems, one has to deal with continuum: this involves a procedure of defining, discretizing and truncating the continuum part of the basis. The different discretization methods involve approximations and subsequent limitations that should be tested. Our proposal here is to test these approximations reducing the problem to one dimension. In this simplified case we can easily follow the time evolution for the exact total wavefunction $\Psi(x,t)$ and for the expansion coefficients, thus checking the validity of the different approximations connected to the discretization methods.

2. The model

Let us start by specifying the model Hamiltonian and the initial conditions.
The system is described by the Hamiltonian

$$\mathcal{H}(x,t) = -\frac{\hbar}{2\mu}\frac{\partial^2}{\partial x^2} + V_T(x) + V_P(x - x_P(t)), \tag{1}$$

where we include two potentials V_T and V_P, chosen with a Woods-Saxons shape and associated to target and to projectile respectively. In our "semiclassical" model the target is at rest and only the projectile potential moves according to a fixed trajectory[1]

$$x_P(t) = x_0 + \sqrt{\rho^2 + (vt)^2} - \rho, \tag{2}$$

that accounts for the target-projectile relative motion with asymtotic energy $E_P = 1/2\mu v^2$ and a distance of closest approach x_0. The variation of these two parameters can simulate different kinematical conditions due to different bombarding energies and impact parameters (partial waves), while the choice of the parameters of the two potentials accounts for the different masses of the colliding nuclei, the Q-values of the different final channels as well as the possibility of simulating weak-binding conditions.
By solving separately the time independent Schroedinger equation for each well we obtain two asymptotic sets of bound levels [1]; depending on the kind of reaction under study we will select one of these levels as the initial single-particle wavefunction $\Psi(x,t_i)$. For example, to describe a pick-up reaction we will choose as initial state a single-particle level in the target.

We can then proceed with the "exact" time evolution of the function $\Psi(x,t)$ by directly solving the time dependent Schroedinger equation [2, 3, 4]

$$i\hbar\frac{\partial}{\partial t}\Psi(x,t) = \mathcal{H}(x,t)\Psi(x,t). \tag{3}$$

At the end of time evolution we can evaluate the final probabilities for each reaction channel by projecting the final wavefunction onto the bound levels of each well (target states for elastic and inelastic scattering, projectile states for transfer channel)

$$\mathcal{P}_{elastic} = |\langle\Phi^T_{g.s.}(x)|\Psi(x,t_f)\rangle|^2, \tag{4}$$

$$\mathcal{P}_{inelastic} = |\langle\Phi^T_{i\neq g.s.}(x)|\Psi(x,t_f)\rangle|^2, \tag{5}$$

$$\mathcal{P}_{transfer} = |\langle\Phi^P_i(x)|\Psi(x,t_f)\rangle|^2; \tag{6}$$

we can also measure breakup probability either by direct subtraction

$$\mathcal{P}_{breakup} = 1 - \mathcal{P}_{elastic} - \mathcal{P}_{inelastic} - \mathcal{P}_{transfer}, \tag{7}$$

[1] According to the trajectory used, the projectile is changing its velocity v over the distance ρ, thus simulating the nuclear interaction with the target; in fact at $t = \pm\infty$ the trajectory tends to a uniform motion with constant velocity.

or by projection onto proper continuum states $\Phi(x, k)$ with momentum k

$$\mathcal{P}_{breakup} = \int |\langle \Psi(x, t_f)|\Phi(x, k)\rangle|^2 dk. \tag{8}$$

The same problem can then be solved along the popular framework used to describe a variety of reaction channels, namely the coupled-channels approach. To this end the system wavefunction is expressed as a combination of target and projectile states

$$\Psi(x, t) = \sum_{j=1}^{N_T} c_i^T(t)\Phi_j^T(x)e^{iE_j^T t/\hbar} + \sum_{j=1}^{N_P} c_i^P(t)\Phi_j^P(x)e^{iE_j^P t/\hbar}, \tag{9}$$

and, following reference [5], the solution of the problem is reduced to the time evolution of the coefficients $c_i(t)$ within the system of differential coupled equations

$$\begin{aligned} i\hbar\frac{\partial c_j^T}{\partial t} &= \sum c_k^T \langle \omega_j^T|V^P|\Phi_k^T\rangle + \sum c_k^P \langle \omega_j^T|V^T|\Phi_k^P\rangle, \\ i\hbar\frac{\partial c_j^P}{\partial t} &= \sum c_k^T \langle \omega_j^P|V^P|\Phi_k^T\rangle + \sum c_k^P \langle \omega_j^P|V^T|\Phi_k^P\rangle. \end{aligned} \tag{10}$$

Since the two bases are not orthonormal we introduce a dual basis $\omega_j^{T,P}$ associated to each well, which is time dependent and is defined in terms of overlaps between the two bases states. In this way states of a dual basis are orthonormal to states of the potential basis

$$\langle \Psi_m^I|\omega_n^J\rangle = \delta_{I,J}\delta_{n,m}. \tag{11}$$

In order to describe breakup processes we also include continuum states, following different discretization procedures. At the end, the probability to populate the different final channels is measured as

$$P_j^{(T,P)} = |c_j^{(T,P)}(t_f)|^2. \tag{12}$$

Note that this has to be calculated only for $t_f = +\infty$ since, due to non-orthonormality of the basis, the total probability is not conserved during the collision. Only at the end, when overlaps between target and projectile states vanish, the total probability gets unitary again.

3. Results

By comparing the "exact" and coupled-channels results, we can check the necessary approximations (such as bases truncation, continuum discretization, ...) which are particularly relevant in case of weakly-bound systems. For this reason we have compared results obtained for well-bound and weakly-bound systems in a variety of situations [6].

We report here two specific cases. For the well-bound case we choose a situation in which the initial state is bound by an energy of -7.39 MeV, as it is presented in figure 1. The projectile asymptotic energy is fixed at 1 MeV. In figure 2 we show the potentials (lower panel) and the final squared wavefunction (upper panel), obtained with different methods, at the end of the time evolution.

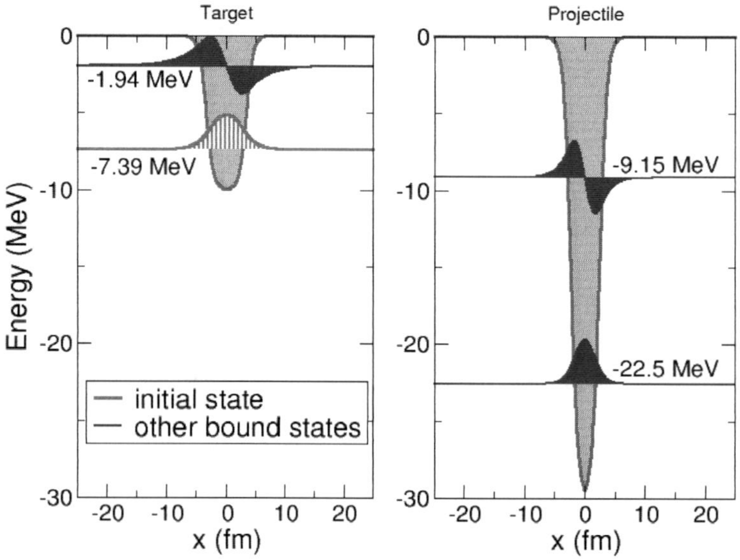

Figure 1: The target (left) and projectile (right) potentials with their bound states and respective binding energies. In red is higlighted the initial state of the reaction in the bound case.

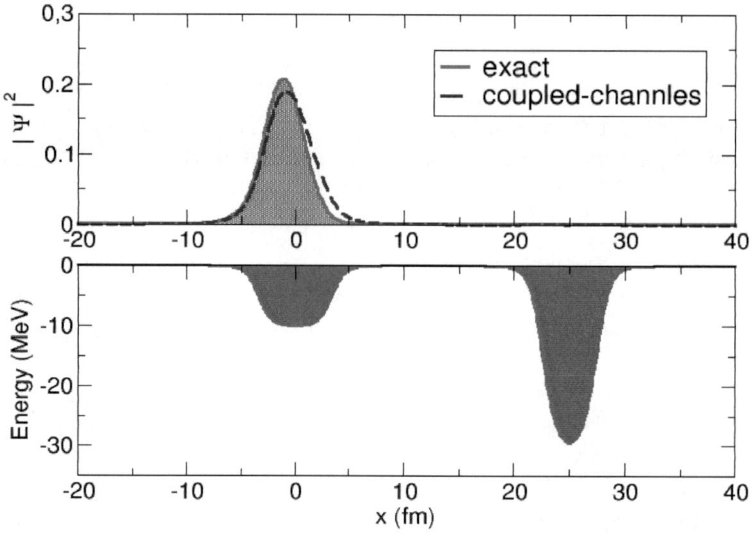

Figure 2: Lower panel: the target and projectile potentials at the end of the time evolution. Upper panel: the final squared "exact" wavefunction (red) and the final squared wavefunction obtained with the coupled-channel method (blue).

IOP Conf. Series: Journal of Physics: Conf. Series **981** (2018) 012004 doi:10.1088/1742-6596/981/1/012004

The results obtained by solving the time-dependent Schroedinger equation and those obtained in the coupled-channels formalism without including the continuum in the bases are listed in table 1. The two methods agree quite well, which means that in this bound case the continuum does not have a crucial role in the reaction mechanism.

Table 1: Results for the different reaction channels for the well-bound case obtained by solving the time dependent Schroedinger equation (*"Exact"*) and the coupled-channels formalism without including the continuum (*"Coupled-channels - no continuum"*).

	Exact	Coupled-channels - no continuum
Elastic	83%	84%
Inelastic	14%	16%
Transfer (g.s)	0.6%	0.002%
Transfer (1)	0.03%	0.2%
Breakup	2.4%	-

By following the time evolution of the probability calculated using equation (12), presented in figure 3, we can notice the non-conservation of total probability during collision, but we can also see the convergence to the final value after the collision, restoring the unitarity of the problem. This is due to non-orthonormality of the separated target and projectile bases, as it was mentioned in the previous Section 2.

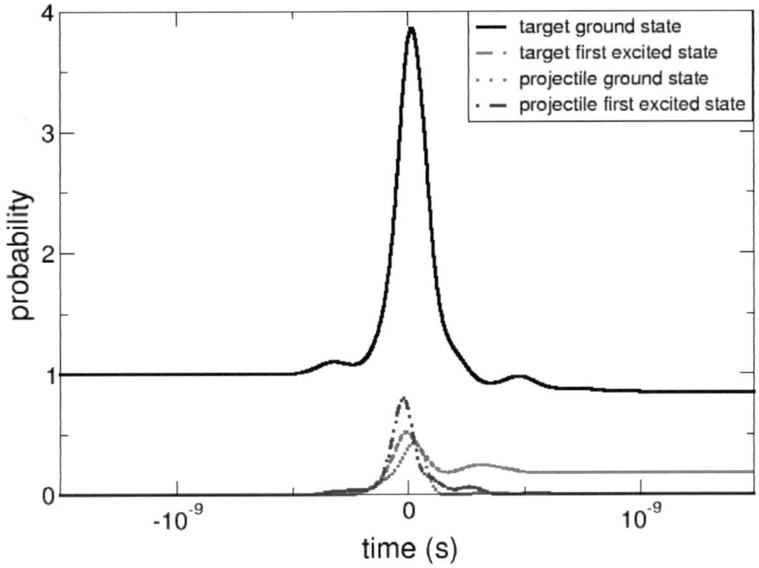

Figure 3: Time evolution of the probability for the coupled-channels method in the bound case.

IOP Conf. Series: Journal of Physics: Conf. Series **981** (2018) 012004 doi:10.1088/1742-6596/981/1/012004

To simulate the drip-line conditions we consider now a weakly-bound case as the one illustrated in figure 4. As initial state we choose a particle moving in the target weakly-bound ground state with energy −0.28 MeV. The projectile asymptotic energy is fixed at 5 MeV. In figure 5 we present the final "exact" wavefunction for this weakly-bound case.

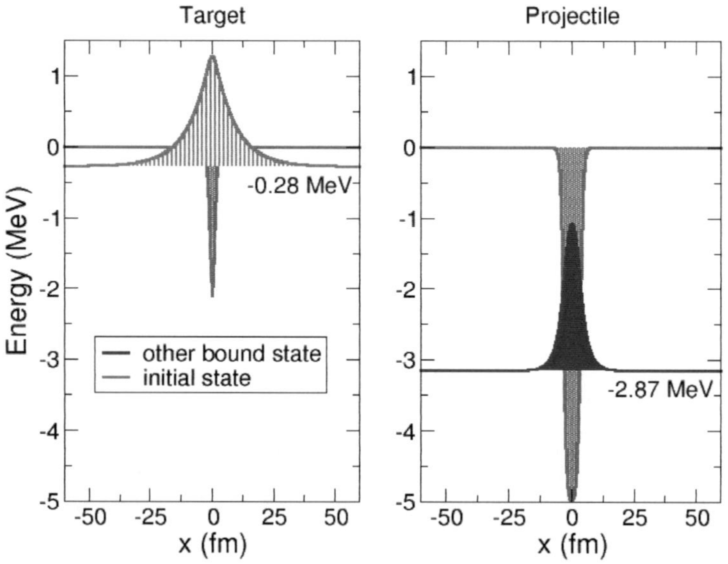

Figure 4: The target (left) and projectile (right) potentials with their bound states and respective binding energies. The squared initial wavefunction is higlighted in red.

In table 2 we show the final probabilities for each reaction channel, obtained applying different methods. First of all, by following the "exact" time evolution, we find that the breakup probability is 74%, and in fact looking at figure 5 the wavefunction is spread in the continuum. Therefore, the coupled-channels solution with only bound states does not give a satisfactory response: in this case the inclusion of the continuum is mandatory.
By adding the continuum part to the target basis we get close to the "exact" results. In this specific case we have defined the continuum states in a range of radius 40 fm centered on the target potential (note that target functions can not overlap with projectile states at the end of the time evolution because of model definition), and we have discretized the basis by imposing boundary conditions generated by an infinite square well.

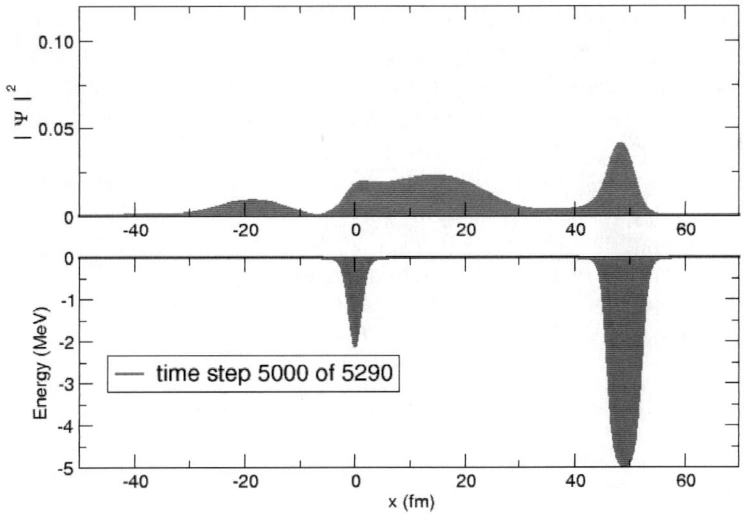

Figure 5: Lower panel: the position of the target and projectile potentials at the end of the time evolution. Upper panel: the final squared "exact" wavefunction.

Table 2: Results for the population of the different reaction channels in the weakly-bound case obtained by solving the time dependent Schroedinger equation (*"Exact"*), the coupled-channels formalism without including the continuum (*"Coupled-channels - no continuum"*), and the coupled-channels formalism including the lowest fifty continuum states (*"Coupled-channels - target continuum"*).

	Exact	Coupled-channels - no continuum	Coupled-channels - target continuum
Elastic	21%	95%	21%
Transfer	5%	5%	0.04%
Breakup	74%	-	79%

It is important to note that continuum waves can be obtained in other different ways: through the use of pseudostates constructed by diagonalizing the potentials in different bases (harmonic oscillator, transformed harmonic oscillator, ...), or grouping oscillating waves into bins in a CDCC-like method; the procedure choice affects the results.
The duration of the time evolution is also fundamental not to compromise the correct solution. If it is too short, the overlap between target and projectile bases is not zero when we calculate the final probabilities. On the contrary, if the system evolution is too long, the continuum component of the final wavefunction could leave the potentials area and go outside the range where we defined the bases.

IOP Conf. Series: Journal of Physics: Conf. Series **981** (2018) 012004 doi:10.1088/1742-6596/981/1/012004

4. Conclusions

We have developed a one dimensional model to describe direct reactions involving halo nuclei with one valence neutron. The approach consists in following the time evolution of the system total wavefunction by solving the time dependent Schroedinger equation, and the time evolution along a coupled-channels approach. By comparing the two results we can clarify the validity of the necessary approximations: basis truncation and continuum discretization. The latter point is particularly relevant in the case of weakly-bound systems where the inclusion of the continuum is practically unavoidable.

References

[1] Moschini L, Pérez-Bernal F and Vitturi A 2016 *J. Phys. G: Nucl. Part. Phys.* **43** 045112
[2] Vitturi A and Moschini L 2015 *J. Phys.: Conf. Ser.* **590** 012007
[3] Vitturi A, Moschini L, Hagino K and Moro AM 2015 *AIP Conf. Proc.* **1681** 060001
[4] Bonche P, Koonin S and Negele JW 1976 *Phys. Rev. C* **13** 1226
[5] Esbensen H, Broglia RA and Winther A 1983 *Ann. Phys.* **146** 149-173
[6] Moschini L, Moro AM and Vitturi A, in preparation

Ab initio optical potentials and nucleon scattering on medium mass nuclei

A. Idini[1], C. Barbieri[1], P. Navrátil[2]
[1]Department of Physics, University of Surrey, Guildford, GU2 7XH, UK
[2]TRIUMF, 4004 Wesbrook Mall, Vancouver, British Columbia, V6T 2A3, Canada

Abstract. We show first results for the elastic scattering of neutrons off oxygen and calcium isotopes obtained from *ab initio* optical potentials. The potential is derived using self–consistent Green's function theory (SCGF) with the saturating chiral interaction NNLO_{sat}. Calculations are compared to available scattering data and show that it is possible to reproduce low energy scattering observables in medium mass nuclei from first principles.

1. Introduction

Recent years have seen considerable advances in the theory of optical potentials. Non locality effects have been shown to be crucial for describing three-body processes [1, 2], the importance of both scattering and bound states in the coupling to breakup channels has been explored [3], and global dispersive optical potentials have been developed [4].

The greatest challenge remains, however, the one of describing the nuclear structure and scattering consistently, from the same theory. Many-body Green's function methods are particularly suited to attempt this for medium and large nuclei since their central quantity, the self-energy, is naturally linked to the Feshbach theory of optical potentials [5, 6]. In particular, the particle part of the self-energy is equivalent to the original formulation of Feshbach, while its hole part describes the structure of the target [7]. Nuclear field theory is one of the first (semi phenomenological) attempts to build such a theory for atomic nuclei [8, 9] and it has been extended to nuclear transfer reactions [10, 11]. Another incarnation of Green's function related theories is the dispersive optical model (DOM) [12], which is a data driven formulation of global (local and non local) potentials constructed as the best possible parameterization of the microscopic self-energy [13, 4]. Finally, in the last years the nuclear structure method (NSM) followed by the authors of Ref. [14] obtained good reproduction of ^{40}Ca scattering based on the Gogny D1S interaction.

For transfer reactions, such as (d, p), it would be particularly important to have an optical potential that is deduced consistently from the same Hamiltonian used in the proton-neutron channel [2]. To do so, one needs *realistic* nuclear interactions and *ab initio* calculations of elastic nucleon-nucleus scattering. The no-core shell model with continuum (NCSMC) has been successful to calculate scattering and transfer reactions for light targets [15, 16, 17]. On the other hand, the self consistent Green's function (SCGF) formalism [18, 19] can calculate the *ab initio* optical potential directly even for havier nuclei. This approach has been used to calculate phase shifts [20] and to investigate analytical properties of DOMs [21]. However, these calculations were limited to two-body forces and a direct comparison to the experiment has been hindered by the lack of realistic interactions capable to reproduce accurately nuclear radii. Coupled cluster theory has also been recently employed [22], in combination with a Green function approach, to calculate phase shifts for ^{16}O.

Three-body interactions have been recently implemented for SCGF in [23, 24, 25]. Moreover, the introduction of the NNLO$_{\rm sat}$ interaction [26] has allowed a good reproduction of nuclear saturation and, hence, of radii and binding energies across the oxygen [27] and calcium chains [28]. Although this interaction has limitations regarding the symmetry energy in neutron rich nuclei, we are now in the position to make a meaningful comparison of first principle approaches to scattering data. Here, we perform state of the art SCGF calculations to test the quality of current *ab initio* methods and of the NNLO$_{\rm sat}$ Hamiltonian in predicting elastic scattering.

2. The microscopic optical potential

The irreducible self-energy, $\Sigma^\star(\omega)$, has the following general spectral representation,

$$\Sigma^\star_{\alpha\beta}(E) = \Sigma^{(\infty)}_{\alpha\beta} + \sum_{i,j} \mathbf{M}^\dagger_{\alpha,i} \left[\frac{1}{E - (\mathbf{K}^> + \mathbf{C}) + i\eta} \right]_{i,j} \mathbf{M}_{j,\beta} + \sum_{r,s} \mathbf{N}_{\alpha,r} \left[\frac{1}{E - (\mathbf{K}^< + \mathbf{D}) - i\eta} \right]_{r,s} \mathbf{N}^\dagger_{s,\beta} \,, \tag{1}$$

where α and β label the single particle quantum numbers and $\Sigma^{(\infty)}$ is the correlated and energy independent mean field. We perform calculations with the third order algebraic diagrammatic construction [ADC(3)] method, where the matrices $\mathbf{M}$ ($\mathbf{N}$) couple single particle states to intermediate 2p1h (2h1p) configurations, $\mathbf{C}$ ($\mathbf{D}$) are interaction matrices among these configurations and $\mathbf{K}$ are their unperturbed energies [29, 30].

We use a spherical harmonic oscillator basis consisting of $N_{\rm max}$+1 oscillator shells, so the optical potential for a given partial wave (l, j) is expressed in terms of the oscillator radial functions $R_{n,l}(r)$ as

$$\Sigma^{\star\, l,j}(r, r'; E) = \sum_{n,n'} R_{n,l}(r)\, \Sigma^{\star\, l,j}_{n,n'}(E)\, R_{n',l}(r') \,, \tag{2}$$

which is non local and energy–dependent. We solve the corresponding scattering problem in the full one-body space (so that, even if the many–body structure is described in a $N_{\rm max}$ oscillator space, the kinetic energy is treated exactly, without truncations) and account for the non locality and l, j dependence of Eq. (2). For each partial wave and parity, the phase shifts $\delta(E)$ are obtained as function of the projectile energy, from where the differential cross section is calculated. We will show results for incident energies in the laboratory frame, except for Fig. 6 below.

3. Results

In the following, we consider the volume integrals of the real (J_V) and imaginary (J_W) parts of the self-energy (i.e., the optical potential):

$$J_V(E) = 4\pi \int \mathrm{d}r r^2 \int \mathrm{d}r' r'^2 \sum_{l,j} \Re e\{\Sigma^{\star\, l,j}(r, r'; E)\} \,, \tag{3}$$

$$J_W(E) = 4\pi \int \mathrm{d}r r^2 \int \mathrm{d}r' r'^2 \sum_{l,j} \Im m\{\Sigma^{\star\, l,j}(r, r'; E)\} \,, \tag{4}$$

since these are strongly constrained by experimental data [6].

Fig. 1 shows the volume integrals of the neutron-^{16}O potential for different model space truncations. Both the part of the self-energy below the Fermi surface (which describes the

IOP Conf. Series: Journal of Physics: Conf. Series **981** (2018) 012005 doi:10.1088/1742-6596/981/1/012005

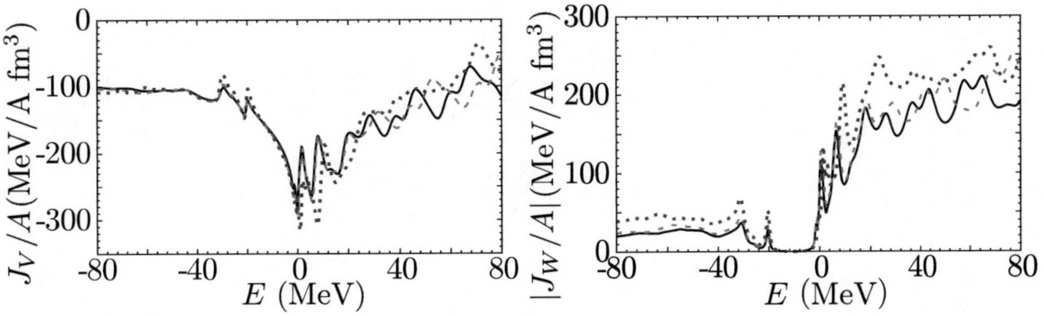

Figure 1. Volume integrals of the real (*left*) and imaginary (*right*) parts of the neutron-^{16}O optical potential calculated for different numbers oscillator shells in the model space: $N_{\max} = 7$ (dotted), 11 (dashed) and 13 (solid lines). Note that $\Im m\{\Sigma^\star(E{=}E_F)\}{=}0$, so $J_W(E{=}E_F){=}0$, where E_F is the Fermi energy. Thus, the potential for particle (holes) states is above (below) the gap in the J_W plot.

structure of the target) and the resonant structures for scattering at low energy are substantially converged already for $N_{\max}{=}11$. The oscillations seen at higher energies ($E >$10 MeV) are an artefact of using a discretized model space and keep changing with $N_{\max}$. They can fade away for an infinite space, or by exploiting an appropriate basis with continuum degrees of freedom.

Fig. 2 shows $|J_W|$ for neutron scattering on closed sub-shell Ca isotopes. The gap at the Fermi surface, where $\Im m\Sigma^\star(E){=}0$, shifts to higher energies and eventually crosses the continuum threshold with increasing neutron number. Compared to previous calculations using the Argonne v_{18} and N^{3}LO(500) interactions [21], the NNLO$_{\rm sat}$ predicts an increased level density in the proximity of the Fermi energy, as expected for a correct nuclear saturation.

In Fig. 3, the neutron $s_{1/2}$ and $d_{3/2}$ phase shifts for ^{16}O are shown for $N_{\max}{=}11$ and 13. In the $s_{1/2}$ case, the resonance at $E \approx$5 MeV changes by $\approx$ 0.5 MeV between the two spaces. Note that this state is dominated by 2p1h components and thus it can still be affected by many–body truncations. The wiggles computed for energies $E >$8 MeV are due to similar but very narrow resonances. Again, these are likely to be sensitive to the discretisation of the model space and drift when increasing the number of oscillator shells. On the contrary, in the $d_{3/2}$ case, the principal sharp resonance is dominated by pure single particle components, therefore it is well converged respect to the model space truncation $N_{\max}$. We calculate its energy to be $\approx$ 1.15 MeV in the c.o.m. frame, while the experimental value is 0.94 MeV.

Fig. 4 shows the phase shifts for other representative partial waves. The calculated $p_{1/2}$ and $p_{3/2}$ have both a sub-threshold bound state, although the $p_{3/2}$ is experimentally only observed in the continuum. We calculate a narrow $f_{7/2}$ resonance at 3.5 MeV close to the experimental one at 3.77 MeV (laboratory energy) [31]. Note however that there are other $f_{7/2}$ narrower resonances that are also seen experimentally lying at 1.65 and 3.01 MeV. In general, we find that NNLO$_{sat}$ predicts the location of dominant quasiparticle and holes states with a (conservative) accuracy of <2 MeV for this nucleus.

Wavefunctions can also be calculated by diagonalizing the self–energy in the continuum. Fig. 5, shows overlap wavefunctions for the addition and removal of a $p_{3/2}$ neutron to and from ^{16}O, as predicted by NNLO$_{sat}$ and SCGF.

Finally, Fig. 6 compares the differential cross section for the elastic scattering of neutrons off ^{40}Ca with the experiment at 13.56 MeV c.o.m. energy, with $N_{\max}{=}11$. Minima in the cross section are reproduced reasonably well, confirming the correct prediction of matter radii, but there appears to be a general lack of absorption. This may be due to either missing doorway configurations (3p2h and beyond) or to the (still crude) model space. Note that proton scattering

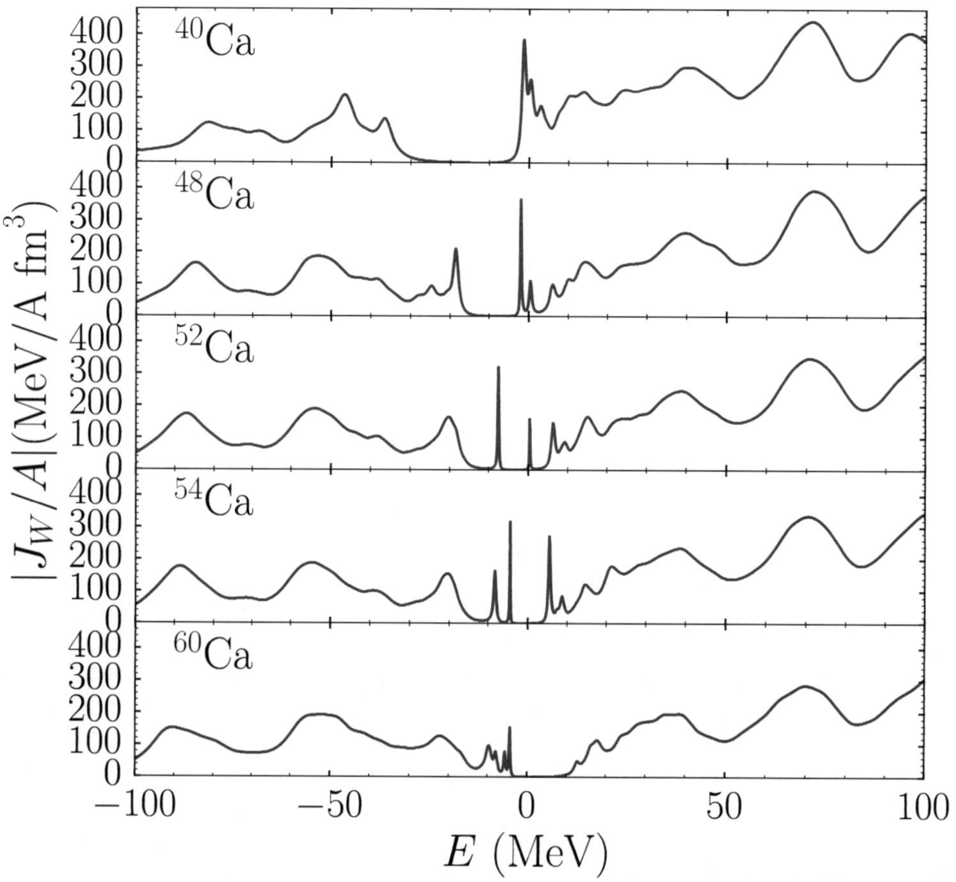

Figure 2. Volume integral of the imaginary part of the neutron optical potentials, $J_W(E)$, for targets at the subshell closure of calcium isotopes: ^{40}Ca, ^{48}Ca, ^{52}Ca, ^{54}Ca, ^{60}Ca, calculated at N_{max}=11.

on ^{40}Ca was also computed in Ref. [32].

4. Conclusions

Even with the limitations of a (non optimal) oscillator basis, we found that most important features of optical potentials are well reproduced. In the long term, it will be necessary to properly account for the continuum in calculating the self-energy and to improve the realistic nuclear interactions. Nevertheless, it is clear from the present results that reliable *ab initio* calculations of optical potentials are now a goal within reach.

Acknowledgments

A.I. is supported by the Royal Society and Newton Fund through the Newton International Fellowship scheme. This work was also supported by the United Kingdom Science and Technology Facilities Council (STFC) under Grant No. ST/L005816/1 and by the NSERC

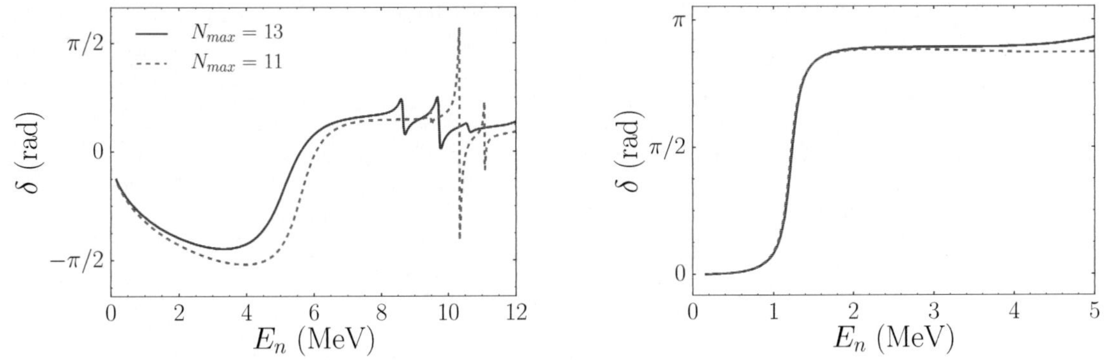

Figure 3. Nuclear phase shifts, $\delta(E)$, for neutrons scattering off ^{16}O as a function of the incident neutron energy. *Left panel:* $s_{1/2}$ partial wave phase shifts for $N_{\max}$=11 (dashed) and 13 (solid). All oscillations seen at energies above 8 MeV are calculated narrow resonances (originating form poles of the self-energy) and go through a full shift of π. *Right panel:* $d_{3/2}$ partial wave phase shifts for for $N_{\max}$=11 (dashed) and 13 (solid).

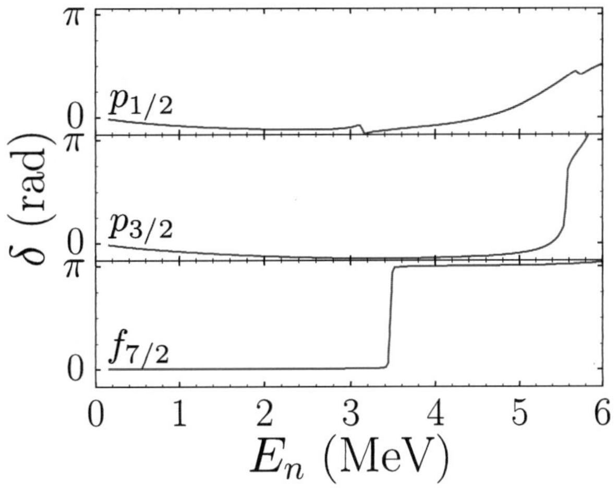

Figure 4. Nuclear phase shifts, $\delta(E)$, for scattering off ^{16}O as a function of the incident neutron energy for $p_{1/2}$, $p_{3/2}$, and $f_{7/2}$ partial waves. Experimental phase shifts from [31] (not shown on the figure) are qualitatively reproduced (e.g. the broad resonance of $p_{1/2}$) and the energy of narrow resonances is generally reproduced within 1 MeV. See also discussion the text.

Grant No. SAPIN-2016-00033. TRIUMF receives federal funding via a contribution agreement with the National Research Council of Canada. Calculations were performed by using HPC resources from the DiRAC Complexity system at the University of Leicester (BIS National E-infrastructure capital grant No. ST/K000373/1 and STFC grant No. ST/K0003259/1).

IOP Conf. Series: Journal of Physics: Conf. Series **981** (2018) 012005 doi:10.1088/1742-6596/981/1/012005

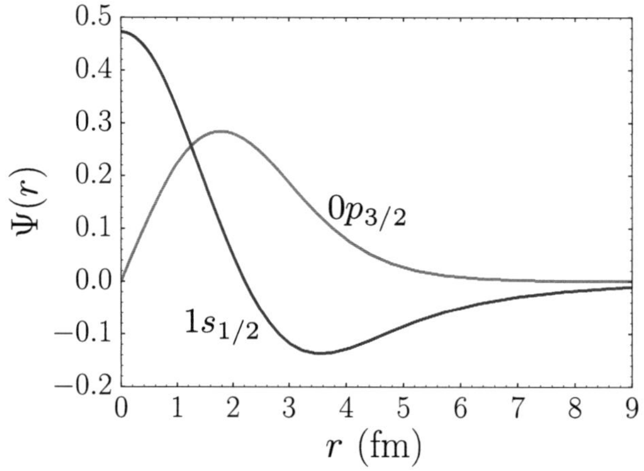

Figure 5. Radial nuclear overlap wave function for the neutron $s_{1/2}$ and $p_{3/2}$ states, obtained by diagonalizing the self energy in the exact continuum, without considering center of mass correction. The $0p_{3/2}$ hole state at −20.704 MeV (solid red line), is compared to the $1s_{1/2}$ particle state at −3.095 MeV. The spectroscopic factors are 0.80 and 0.86 respectively.

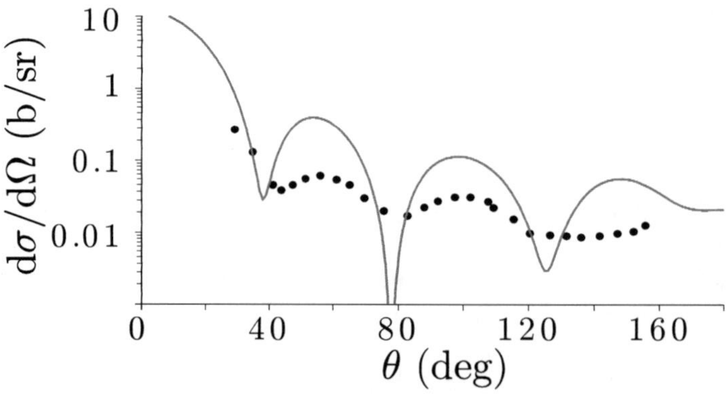

Figure 6. Plot of differential cross section for neutron elastic scattering on ^{40}Ca at 13.56 MeV of *center of mass energy* compared with experimental data from [33].

References

[1] Timofeyuk N K and Johnson R C 2013 *Phys. Rev. Lett.* **110**(11) 112501 URL http://link.aps.org/doi/10.1103/PhysRevLett.110.112501
[2] Bailey G W, Timofeyuk N K and Tostevin J A 2016 *Phys. Rev. Lett.* **117**(16) 162502 URL http://link.aps.org/doi/10.1103/PhysRevLett.117.162502
[3] Fernández-García J P, Zadro M, Di Pietro A, Figuera P, Fisichella M, Goryunov O, Lattuada M, Marchetta C, Moro A M, Musumarra A, Ostashko V, Pellegriti M G, Scuderi V, Strano E and Torresi D 2015 *Phys. Rev. C* **92**(5) 054602 URL http://link.aps.org/doi/10.1103/PhysRevC.92.054602
[4] Dickhoff W H, Charity R J and Mahzoon M H 2017 *J. Phys. G* **44** 033001 URL http://stacks.iop.org/0954-3899/44/i=3/a=033001
[5] Feshbach H 1958 *Annals of Physics* **5** 357 – 390 ISSN 0003-4916 URL http://www.sciencedirect.com/science/article/pii/0003491658900071
[6] Jeukenne J P, Lejeune A and Mahaux C 1977 *Phys. Rev. C* **16**(1) 80–96 URL http://link.aps.org/doi/10.1103/PhysRevC.16.80
[7] Escher J and Jennings B K 2002 *Phys. Rev. C* **66** 034313
[8] Mahaux C, Bortignon P F, Broglia R A and Dasso C H 1985 *Phys. Rep.* **120** 1
[9] Idini A, Barranco F and Vigezzi E 2012 *Phys. Rev. C* **85** 014331 URL http://link.aps.org/doi/10.1103/PhysRevC.85.014331
[10] Idini A, Potel G, Barranco F, Vigezzi E and Broglia R A 2015 *Phys. Rev. C* **92**(3) 031304 URL http://link.aps.org/doi/10.1103/PhysRevC.92.031304
[11] Broglia R A, Bortignon P F, Barranco F, Vigezzi E, Idini A and Potel G 2016 *Physica Scripta* **91** 063012 URL http://stacks.iop.org/1402-4896/91/i=6/a=063012
[12] Johnson C H and Mahaux C 1988 *Phys. Rev. C* **38** 2589
[13] Charity R J, Sobotka L G and Dickhoff W H 2006 *Phys. Rev. Lett.* **97**(16) 162503 URL http://link.aps.org/doi/10.1103/PhysRevLett.97.162503
[14] Blanchon G, Dupuis M, Arellano H F and Vinh Mau N 2015 *Phys. Rev. C* **91**(1) 014612 URL http://link.aps.org/doi/10.1103/PhysRevC.91.014612
[15] Navrátil P, Roth R and Quaglioni S 2010 *Phys. Rev. C* **82**(3) 034609 URL http://link.aps.org/doi/10.1103/PhysRevC.82.034609
[16] Baroni S, Navrátil P and Quaglioni S 2013 *Phys. Rev. Lett.* **110**(2) 022505 URL http://link.aps.org/doi/10.1103/PhysRevLett.110.022505
[17] Raimondi F, Hupin G, Navrátil P and Quaglioni S 2016 *Phys. Rev. C* **93**(5) 054606 URL http://link.aps.org/doi/10.1103/PhysRevC.93.054606
[18] Dickhoff W and Barbieri C 2004 *Progress in Particle and Nuclear Physics* **52** 377 – 496 ISSN 0146-6410 URL http://www.sciencedirect.com/science/article/pii/S0146641004000535
[19] Somà V, Duguet T and Barbieri C 2011 *Phys. Rev. C* **84**(6) 064317 URL http://link.aps.org/doi/10.1103/PhysRevC.84.064317
[20] Barbieri C and Jennings B K 2005 *Phys. Rev. C* **72** 014613
[21] Waldecker S, Barbieri C and Dickhoff W H 2011 *Phys. Rev. C* **84** 034616
[22] Rotureau J, Danielewicz P, Hagen G, Nunes F and Papenbrock T 2016 *arXiv:1611.04554 [nucl-th]*
[23] Carbone A, Cipollone A, Barbieri C, Rios A and Polls A 2013 *Phys. Rev. C* **88**(5) 054326 URL http://link.aps.org/doi/10.1103/PhysRevC.88.054326
[24] Cipollone A, Barbieri C and Navrátil P 2013 *Phys. Rev. Lett.* **111** 062501
[25] Cipollone A, Barbieri C and Navrátil P 2015 *Phys. Rev. C* **92**(1) 014306 URL http://link.aps.org/doi/10.1103/PhysRevC.92.014306
[26] Ekström A, Jansen G, Wendt K, Hagen G, Papenbrock T, Carlsson B, Forssén C, Hjorth-Jensen M, Navrátil P and Nazarewicz W 2015 *Phys. Rev. C* **91** 051301
[27] Lapoux V, Somà V, Barbieri C, Hergert H, Holt J D and Stroberg S R 2016 *Phys. Rev. Lett.* **117**(5) 052501 URL http://link.aps.org/doi/10.1103/PhysRevLett.117.052501
[28] Garcia Ruiz R F and *et al* 2016 *Nat Phys* **12** 594–598 URL http://dx.doi.org/10.1038/nphys3645
[29] Schirmer J, Cederbaum L S and Walter O 1983 *Phys. Rev. A* **28**(3) 1237 URL http://link.aps.org/doi/10.1103/PhysRevA.28.1237
[30] Barbieri C and Carbone A 2016 *arXiv preprint arXiv:1611.03923*
[31] Lister D and Sayres A 1966 *Phys. Rev.* **143** 745
[32] Hagen G and Michel N 2012 *Phys. Rev. C* **86**(2) 021602 URL http://link.aps.org/doi/10.1103/PhysRevC.86.021602
[33] Honoré G M, Tornow W, Howell C R, Pedroni R S, Byrd R C, Walter R L and Delaroche J P 1986 *Phys. Rev. C* **33**(4) 1129–1140 URL http://link.aps.org/doi/10.1103/PhysRevC.33.1129

IOP Conf. Series: Journal of Physics: Conf. Series **981** (2018) 012006 doi:10.1088/1742-6596/981/1/012006

^{7}Li(d,p)^{8}Li transfer reaction in the NCSM/RGM approach

F Raimondi[1,2,a], G Hupin[3,4,5,b], P Navrátil[2,c] and S Quaglioni[4,d]

[1]Department of Physics, Faculty of Engineering and Physical Sciences, University of Surrey, Guildford GU2 7XH, United Kingdom.
[2]TRIUMF, 4004 Wesbrook Mall, Vancouver BC, V6T 2A3, Canada.
[3]CEA, DAM, DIF, F-91297 Arpajon, France.
[4]Lawrence Livermore National Laboratory, P.O Box 808, L-414, Livermore, California 94551, USA
[5]Institut de Physique Nucléaire, IN2P3-CNRS, Université Paris-Sud, F-91406 Orsay Cedex, France.

E-mail:
[a]f.raimondi@surrey.ac.uk
[b]hupin@ipno.in2p3.fr
[c]navratil@triumf.ca
[d]quaglioni1@llnl.gov

Abstract. Recently, we applied an *ab initio* method, the no-core shell model combined with the resonating group method, to the transfer reactions with light p-shell nuclei as targets and deuteron as the projectile. In particular, we studied the elastic scattering of deuterium on ^{7}Li and the ^{7}Li(d,p)^{8}Li transfer reaction starting from a realistic two-nucleon interaction. In this contribution, we review of our main results on the ^{7}Li(d,p)^{8}Li transfer reaction, and we extend the study of the relevant reaction channels, by showing the dominant resonant phase shifts of the scattering matrix. We assess also the impact of the polarization effects of the deuteron below the breakup on the positive-parity resonant states in the reaction. For this purpose, we perform an analysis of the convergence trend of the phase and eigenphase shifts, with respect to the number of deuteron pseudostates included in the model space.

1. Introduction

The stripping of a proton (p) or a neutron (n) from the deuteron (d) projectile is the simplest nuclear transfer reaction. The deuteron is the first stable isotope synthesized after the Big Bang; henceforth, (d,p) and (d,n) are among the key reactions in the formation of the light elements in the primordial nucleosynthesis. Because of its simplicity, the transfer of a nucleon, stripped from the deuteron, into a target nucleus is an efficient process implemented in the experimental studies of the ground-state (g.s.) and excited states of nuclei, of their energies, spin, and parity properties [1].

The aim of an *ab initio* description of the deuteron-induced transfer reaction is to connect the dynamics of the scattering to the first principles, which in low-energy nuclear reaction are the nucleons, considered as the relevant degrees of freedom, interacting though realistic forces. Such a description has been successfully carried out for light target nuclei in the framework of two *ab initio* many-body methods: the no-core shell model [2] combined with the resonating group method (NCSM/RGM) [3], and the no-core shell model with continuum (NCSMC) [4, 5], which are both suited for realistic nucleon-nucleon (NN) and three-nucleon (3N) forces. The former is a cluster approach applied to the A-nucleon wave function, which is partitioned in a a nucleons projectile and a (A-a) nucleons target, allowing for the description of scattering states. The latter is built by treating on the same footing the cluster states of the NCSM/RGM method, and the A-nucleons wave-functions computed in the NCSM approach. The NCSM/RGM and NCSMC approaches have been first applied to the elastic (d-α) [6, 7] and ^{3}H$(d,n)^4$He and ^{3}He$(d,p)^4$He transfer reactions [8, 9].

In a recent work [10], we extended the scope of the NCSM/RGM description of the deuteron-induced transfer reactions to p-shell nuclei. We studied the ^{7}Li(d,p)^{8}Li transfer reaction, whose excitation function below the deuteron-breakup energy is adopted as a calibration tool for the measurement of the radiative proton capture on ^{7}Be [11]. We could then compute the energy, spin and parity of the resonances above the d+^{7}Li threshold in the ^{9}Be spectrum. Even in the energy range below the deuteron breakup, the internal excitations of the deuteron (i.e. polarization and virtual breakup) are expected to influence the shape of the cross section. Their correct description poses a challenge to the approaches that do not include explicitly the breakup of the deuteron. In the NCSM/RGM, the formalism for the description of the three-body continuum has been fully worked out with the introduction of the three-cluster wavefunction ansatz [12], but it is not yet implemented in the transfer reaction calculations. In the present approach, the virtual breakup of the two-nucleon projectile is approximated by discretizing the continuum, obtained by considering the positive-energy states (pseudostates) of the NCSM spectrum.

In this contribution, we revisit the main equations of the NCSM/RGM formalism, and our results for the ^{7}Li(d,p)^{8}Li transfer reaction of Ref. [10], then we discuss the binary-cluster channels that contribute to the resonances above the p+^{8}Li threshold. Moreover, we assess our treatment of the deuteron continuum, via an analysis of the convergence of the resonant phase and eigenphase shifts in terms of the number of deuteron pseudostates included in the model space basis.

2. Basic equations of the NCSM/RGM formalism

A general presentation of the equations of the NCSM/RGM formalism can be found in Ref. [3]. The idea of the cluster wave function, typical of the RGM approach, is exploited by building the following expansion over the antisymmetrized binary-cluster channel states $|\Phi^{J^\pi T}_{\nu r}\rangle$,

$$|\Psi^{J^\pi T}\rangle \;=\; \sum_\nu \int dr\, r^2 \frac{g^{J^\pi T}_\nu(r)}{r}\, \hat{\mathcal{A}}_\nu\, |\Phi^{J^\pi T}_{\nu r}\rangle \,, \tag{1}$$

with the coefficient of the expansion $g^{J^\pi T}_\nu(r)$ being the unknown amplitudes of the relative motion between the clusters. The quantum numbers J, π, and T, are the total angular momentum, parity and isospin.

Due to the introduction of the auxiliary variable r in Eq. (1), formally distinct from the inter-cluster relative coordinate $\vec{r}_{A-a,a} = r_{A-a,a}\hat{r}_{A-a,a}$, the antisymmetrization operator $\hat{\mathcal{A}}_\nu$ acts only on the channel states. For the case of the binary cluster, the operator $\hat{\mathcal{A}}_\nu$ acts on the product state,

$$|\Phi^{J^\pi T}_{\nu r}\rangle \;=\; \Big[\big(|A_t\,\alpha_t I_t^{\pi_t} T_t\rangle \,\big|A_p\,\alpha_p I_p^{\pi_p} T_p\big\rangle\big)^{(sT)}\, Y_\ell\left(\hat{r}_{A-a,a}\right)\Big]^{(J^\pi T)} \frac{\delta(r - r_{A-a,a})}{r r_{A-a,a}}, \tag{2}$$

where $\left|A_{t(p)}\,\alpha_{t(p)} I_{t(p)}^{\pi_{t(p)}} T_{t(p)}\right\rangle$ are translational-invariant eigenstates of the target (projectile). Each channel is identified by a index $\nu = \{A_t\alpha_t I_t^{\pi_t} T_t; A_p\alpha_p I_p^{\pi_p} T_p; s\ell\}$ collecting the total spin

IOP Conf. Series: Journal of Physics: Conf. Series **981** (2018) 012006 doi:10.1088/1742-6596/981/1/012006

s, the relative orbital angular momentum ℓ, and the quantum numbers of each cluster: angular momentum I, parity π, isospin T and energy α.

The target and projectile states in the binary-cluster channel are in turn obtained from the intrinsic Hamiltonian $\hat{H} = \hat{T}_{int} + \hat{V}$, composed by the internal kinetic energy $\hat{T}_{int}$ and nuclear interaction $\hat{V}$, by solving the following eigenvalue problem,

$$\hat{H}\,|A\,\alpha I^{\pi}T\rangle = E_{\alpha}\,|A\,\alpha I^{\pi}T\rangle\,. \tag{3}$$

In the NCSM approach [2] the eigenstates of Eq. (3) are obtained by diagonalizing $\hat{H}$ in a model space spanned by a complete harmonic oscillator (HO) basis. The size of the basis is fixed by the maximum number $N_{\rm max}$ of HO quanta, while the same HO frequency Ω is used for both clusters in the binary-cluster channel.

Note that the RGM cluster ansatz of Eq. (1) is linked to the NCSM method consistently, that is the same microscopic Hamiltonian of Eq. (3) is used. The basis states $\hat{\mathcal{A}}_{\nu}\,|\Phi^{J^{\pi}T}_{\nu r}\rangle$ of Eq. (1) give rise to a coupled-channel set of equations for the unknowns $g^{J^{\pi}T}_{\nu}(r)$,

$$\sum_{\nu}\int dr\,r^2\left[\mathcal{H}^{J^{\pi}T}_{\nu'\nu}(r',r) - E\,\mathcal{N}^{J^{\pi}T}_{\nu'\nu}(r',r)\right]\frac{g^{J^{\pi}T}_{\nu}(r)}{r} = 0\,, \tag{4}$$

with the norm and Hamiltonian kernel given by,

$$\mathcal{N}^{J^{\pi}T}_{\nu'\nu}(r',r) = \left\langle \Phi^{J^{\pi}T}_{\nu'r'}\right|\hat{\mathcal{A}}_{\nu'}\hat{\mathcal{A}}_{\nu}\left|\Phi^{J^{\pi}T}_{\nu r}\right\rangle\,, \tag{5}$$

and

$$\mathcal{H}^{J^{\pi}T}_{\nu'\nu}(r',r) = \left\langle \Phi^{J^{\pi}T}_{\nu'r'}\right|\hat{\mathcal{A}}_{\nu'}\hat{H}\hat{\mathcal{A}}_{\nu}\left|\Phi^{J^{\pi}T}_{\nu r}\right\rangle\,, \tag{6}$$

respectively.

In our implementation of the deuteron-induced reactions, two different mass partitions are taken into account in the asymptotic states: the one with the deuteron (A-2,2) and the one with the scattered nucleon (A-1,1). As a consequence, two types of kernels appear in the matrices (5) and (6). The diagonal kernels correspond to the elastic channel of the reaction, where the deuteron is present as a scattered particle in the asymptotic state or as a cluster in the intermediate composite nucleus, that is

$$\begin{aligned}\mathcal{H}^{J^{\pi}T}_{\nu'\nu}(r',r) &= \left\langle \Phi^{J^{\pi}T}_{\nu'r'}\right|\hat{\mathcal{A}}_{(A-2,2)}H\hat{\mathcal{A}}_{(A-2,2)}\left|\Phi^{J^{\pi}T}_{\nu r}\right\rangle = \left\langle \Phi^{J^{\pi}T}_{\nu'r'}\right|H\hat{\mathcal{A}}^2_{(A-2,2)}\left|\Phi^{J^{\pi}T}_{\nu r}\right\rangle \\ &= \left[T_{\rm rel}(r') + \bar{V}_C(r') + E^{I'_1T'_1}_{\alpha'_1} + E^{I'_2T'_2}_{\alpha'_2}\right]\mathcal{N}^{J^{\pi}T}_{\nu'\nu}(r',r) + \mathcal{V}^{J^{\pi}T}_{\nu'\nu}(r',r),\end{aligned} \tag{7}$$

where $T_{\rm rel}(r')$ and $\bar{V}_C(r')$ are the relative kinetic energy and average Coulomb interaction, respectively; $E^{I'_1T'_1}_{\alpha'_1}$ and $E^{I'_2T'_2}_{\alpha'_2}$ are NCSM energy eigenvalues for the two clusters, and $\mathcal{V}^{J^{\pi}T}_{\nu'\nu}$ is the potential kernel (see Eq. (B2) of Ref. [10]). The off-diagonal kernels couple different mass partitions and correspond to the transfer process: the complete list of the expressions of the diagonal and coupling kernels, both for the norm and the Hamiltonian, can be found in Refs. [6], [8] and [10]. Once the norm and Hamiltonian kernels have been computed, the set of coupled integral-differential equations in (4) is solved on a Lagrange mesh within the microscopic R-matrix method [13].

3. Results on ^{7}Li$(d,p)^8$Li transfer reaction and ^{9}Be energy spectrum

For the description of the ^{9}Be spectrum above the d+^{7}Li threshold, and of the ^{7}Li$(d,p)^8$Li transfer reaction, we start from specific choices of the nuclear interaction and model space:

- *Interaction*: The chiral N^{3}LO NN potential of Ref. [14], evolved through a similarity renormalization group (SRG) transformation with evolution parameter Λ=2.02 fm^{-1}. We do not include 3N forces.
- *HO basis*: The nuclear wave function is expanded in the HO basis, with frequency of $\hbar\Omega = 20$ MeV and two truncations corresponding to a total number of excitations above the $2\hbar\Omega$ minimum-energy configuration of $N_{\rm max} = 6$ and 8. To match the corresponding absolute number of HO quanta, we described the deuteron in $N_{\rm max} = 8$, and 10 model spaces, respectively.
- *Model space*: Two ^{7}Li states ($\frac{3}{2}^-$ g.s. and $\frac{1}{2}^-$ first excited state) and four ^{8}Li states (2^+ g.s. and 1^+, 3^+, 0^+ excited states) are taken as the clusters in the binary products of Eq. (2). For the deuteron, we included the g.s. and up to 4 pseudostates in the $^3S_1$-$^3D_1$ channel. Tables 1 and 2 give the energies of ^{7}Li and ^{8}Li states, and deuteron g.s. and pseudostates, respectively.

Table 1. Ground-state and excitation energies of ^{7}Li and ^{8}Li calculated within the NCSM with $N_{\rm max}$= 6, 8 and 10 in the HO basis and HO frequency $\hbar\Omega$=20 MeV, compared to the experiment. The values in the last column have been adjusted in order to reproduce the Q-value of the ^{7}Li(d,p)^{8}Li reaction, as explained in Section 3.2.

Nucleus	State J^π	E (MeV) $N_{\rm max}$ 6	8	10	Exp	Threshold ^{7}Li(d,p)^{8}Li
^{7}Li	$\frac{3}{2}^-$	-36.20	-38.01	-38.94	-39.25	-38.01
	$\frac{1}{2}^-$	-35.80	-37.64	-38.60	-38.77	-37.53
^{8}Li	2^+	-37.60	-39.66	-40.75	-41.28	-40.04
	1^+	-36.36	-38.47	-39.63	-40.30	-39.06
	3^+	-34.76	-36.78	-37.86	-39.02	-37.78
	0^+	-33.75	-36.16	-37.56		-36.83

3.1. $^7Li(d,p)^8Li$ *scattering (eigen)phase shifts*

A first insight on the dynamics of ^{7}Li$(d,p)^8$Li transfer reaction can be obtained from the different $(J^\pi T)$ components of the scattering matrix. The different binary-cluster states of Eq. (1), having good angular momentum, parity and isospin, contribute to different partial waves in the scattering matrix. The impact of different partial waves can be inferred from the eigenphase shifts, while the importance of a specific binary-cluster channel state within a partial wave is given by the phase shifts.

In Fig. 1(a) and (b) we show a selection of the computed $T = \frac{1}{2}$ eigenphase shifts for negative- and positive-parity states, respectively. These partial waves are responsible for the shape and strength of the ^{7}Li$(d,p)^8$Li cross section, as shown in Section 3.2.

The positive-parity eigenphase shifts show a resonant shape just above the p+^{8}Li threshold. In particular the dominant eigenphase shift carries the $\left(\frac{5}{2}^+\frac{1}{2}\right)$ quantum numbers. The

Table 2. Ground-state and pseudostate energies of the deuteron calculated within the NCSM, with $N_{\rm max}$ = 8, 10 and 12 basis space, and HO frequency $\hbar\Omega$=20 MeV. In the calculations we included up to 4 pseudostates in the $^3S_1$-$^3D_1$ channel.

	E (MeV)		
	$N_{\rm max}$= 8	$N_{\rm max}$= 10	$N_{\rm max}$= 12
g.s.	-1.96	-2.12	-2.13
1*	9.91	8.36	6.93
2*	15.22	12.82	11.06
3*	33.24	26.6	22. 80
4*	40.20	33.23	28.45

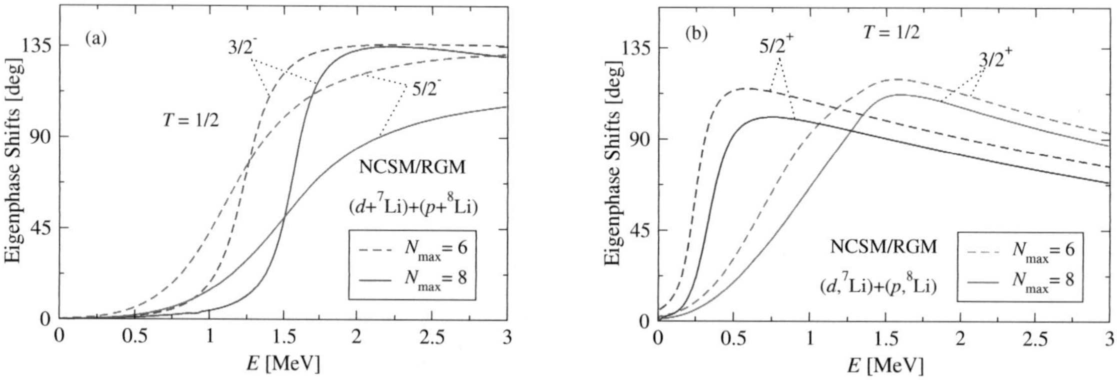

Figure 1. Calculated (a) negative and (b) positive-parity eigenphase shifts within the coupled (d,^{7}Li)+(p,^{8}Li) NCSM/RGM basis, as a function of the relative kinetic energy in the c.m. frame and with respect to the p+^{8}Li threshold. The SRG-N^{3}LO NN potential with Λ=2.02 fm^{-1}, and the HO frequency of $\hbar\Omega$=20 MeV were used.

deformation and virtual breakup of the deuteron can be important even at low energies, therefore they may influence these resonances located just above the reaction threshold. In this respect, we analyse the influence of the deuteron pseudostates in Fig. 2, where the dependence of the $J^\pi T = \frac{3}{2}^+\frac{1}{2}$ and $\frac{5}{2}^+\frac{1}{2}$ eigenphase shifts on the number of deuteron states in the $^3S_1$-$^3D_1$ channel is shown. Both resonances are significantly enhanced by the inclusion of the deuteron pseudostates. We cannot claim that the solid line curves in Fig. 2, corresponding to 4 pseudostates included in the model space, are fully converged. However, considering the $\frac{3}{2}^+$ eigenphase shift in Fig. 2(a), the relative difference between the resonance positions computed with 3 and 4 pseudostates is within 5%, whereas the position of the resonance in the calculation without the deuteron continuum is shifted by 50% with respect to our best attempt. Therefore, we are not expecting significant changes of the shape of the eigenphase shifts, by adding more pseudostates in the calculation.

Figure 3(a) further compares the two main phase shifts, labeled with the notation $^{2s}\ell_{J^\pi}$, contributing to the $J^\pi T = \frac{5}{2}^+\frac{1}{2}$ resonant state. In the d-^{7}Li mass partition (solid red line), the ^{7}Li g.s. is coupled to the deuteron in the relative P-wave motion, while the g.s. of ^{8}Li has an S-wave coupling with the proton in the p-^{8}Li channel (black solid line), giving rise to a phase shift with a clear resonant behavior. The dependence of the p-^{8}Li $^6S_{5/2^+}$ channel on the number of deuteron pseudostates is shown in Fig. 3(b). Again, we see that the inclusion of

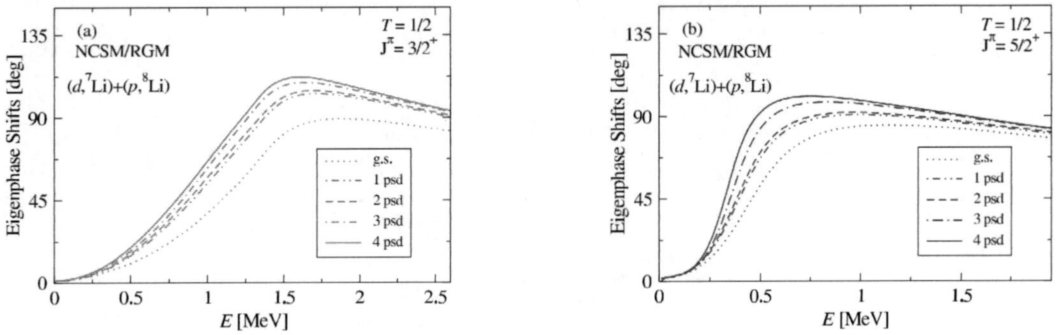

Figure 2. Trend of convergence with respect to the deuteron pseudostates of Table 2, for the calculated (a) $J^\pi T = \frac{3}{2}^+\frac{1}{2}$ and (b) $J^\pi T = \frac{5}{2}^+\frac{1}{2}$ eigenphase shifts within the coupled $(d,^7\text{Li})+(p,^8\text{Li})$ NCSM/RGM basis, as a function of the relative kinetic energy in the c.m. frame and with respect to the $p+^8$Li threshold. The SRG-N^{3}LO NN potential with Λ=2.02 fm^{-1}, and the HO frequency of $\hbar\Omega$=20 MeV were used.

the pseudostates in the model space changes drastically the position of the resonance, which is shifted by ~600 KeV towards the $p+^8$Li threshold when 4 pseudostates are taken into account on top of the deuteron ground state.

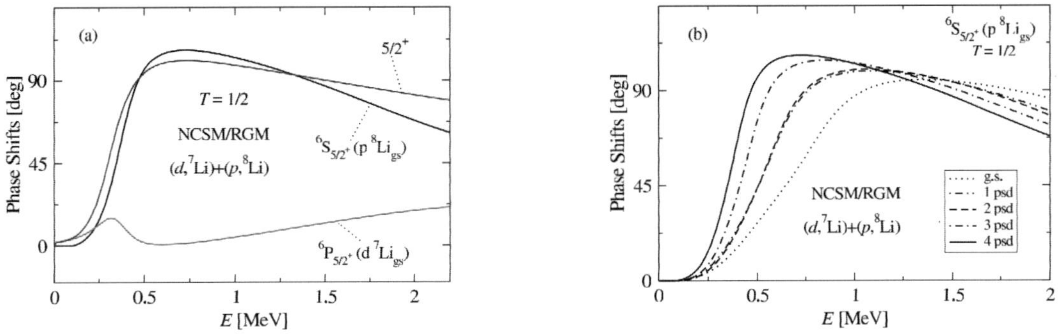

Figure 3. (a) Eigenphase shifts for $J^\pi T = \frac{5}{2}^+\frac{1}{2}$ (solid blue line) compared to the d-^{7}Li (solid red line) and p-^{8}Li (solid black line) elastic phase shifts contributing to the same $J^\pi T$ scattering amplitude through a P- and S-wave respectively. (b) Trend of convergence with respect to the deuteron pseudostates of Table 2, for the p-^{8}Li $^6S_{5/2^+}$ elastic phase shift. All results were obtained within the coupled $(d,^7\text{Li})+(p,^8\text{Li})$ NCSM/RGM basis, and are plotted as a function of the relative kinetic energy in the c.m. frame, with respect to the $p+^8$Li threshold.

The non-resonant behavior of the channel states in the d-^{7}Li mass partition is a general feature of our calculations when both binary-cluster channels are coupled together. For instance, the $^6P_{5/2^+}$ channel, which is a strong resonance in the uncoupled calculation with only d-^{7}Li mass partition, appears to be quenched in Fig. 3(a) (see Ref. [10] for a thorough discussion).

In Figs. 4(a) and (b) we show the negative- and positive-parity phase shifts corresponding to the channel states that contribute to the $J^\pi T$ states of Figs. 1(a) and (b), respectively. They all belong to the p-^{8}Li mass partition and none of them, except for $^5F_{5/2^-}$, shows a clear resonant behavior. We note that in the negative-parity partial waves, the dominant channels contain an

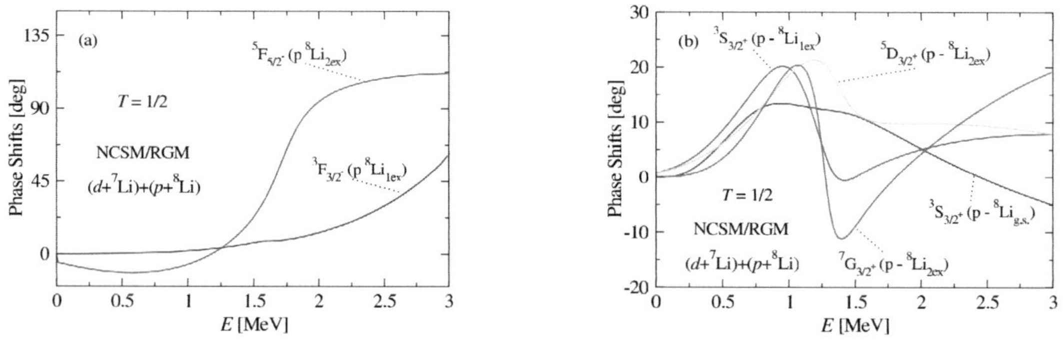

Figure 4. Calculated (a) negative and (b) positive-parity phase shifts contributing to the partial waves shown in Figs. 1, with the corresponding description of the binary-channel states in terms of the cluster state, total spin s and relative-motion angular momentum ℓ of the binary-channel, according to the notation $^{2s}\ell_{J^\pi}$ (see Eq. (1)). The SRG-N^{3}LO NN potential with Λ=2.02 fm^{-1}, and the HO frequency of $\hbar\Omega$=20 MeV were used.

excited state of ^{8}Li; also the $J^\pi T = \frac{3}{2}^+ \frac{1}{2}$ eigenphase shift contains contributions from the ^{8}Li excited states, and consequently it is shifted at higher energy compared to the $J^\pi T = \frac{5}{2}^+ \frac{1}{2}$ eigenphase shift. From this survey on the phase shifts, we conclude that the partial wave $\frac{5}{2}^+ \frac{1}{2}$, which corresponds the first resonant state above the p+^{8}Li threshold, is the only one showing a significant contribute of the $d-^7$Li channels, which are otherwise suppressed in the calculations coupling both the mass partitions.

3.2. ^{7}*Li(d,p)*^{8}*Li integrated cross section*

The shape of the excitation functions of the ^{7}Li(d,p)^{8}Li transfer reaction above the d+^{7}Li and p+^{8}Li thresholds, at about 16 MeV with respect to the ^{9}Be g.s., reflects the pattern found in the (eigen)phase shifts of Figs. 1, 3 and 4.

In Fig. 5, we compare the calculated ^{7}Li(d,p)^{8}Li integrated cross section to the experimental data of Refs. [15, 16, 17, 18] for deuteron energies in the laboratory frame up to about 2.3 MeV, that is the energy range below the breakup threshold of the deuteron. The experimental data displayed in Fig. 5 are concentrated around the deuteron kinetic energy of 0.78 MeV, where the cross section exhibits a resonance of width $\Gamma \approx 0.2$ MeV. The position of this peak, with recommended value of 0.147$\pm$0.011 b, is measured routinely to calibrate the ^{7}Be targets used in experimental studies of the ^{7}Be(p,γ)^{8}B radiative capture [11].

The present computational limit with respect to the size of the HO basis, is given by $N_{\max} = 8$, corresponding to the solid line in Fig. 5. We observe that the first resonant peak of the cross section approaches the experimental E $\sim$ 0.78 MeV resonance, when we increase the size of the HO basis from $N_{\max} = 6$ to 8. Nevertheless, the position of this first peak is overestimated by about 0.33 MeV, which is likely due to the fact that our wave function is not yet converged. The fact that we overestimate the position of the first peak by about 0.33 MeV is related to the underestimation of the Q-value of the reaction, as it is computed from the values of binding energies in Table 1: The experimental Q-value is -0.192 MeV, whereas the energies of the g.s. in our calculation give a Q-value of -0.556 and -0.465 MeV for $N_{\max}$= 6 and 8, respectively.

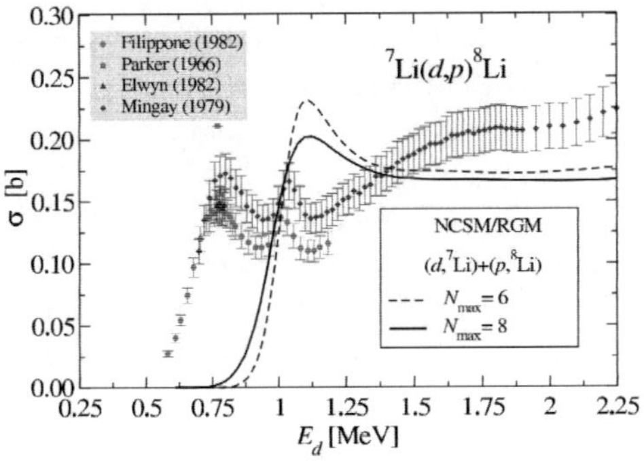

Figure 5. ^{7}Li(d,p)^{8}Li integrated cross section for deuteron laboratory energies up to 2.25 MeV computed within the NCSM/RGM approach at $N_{\rm max} = 6$ (thin-dashed line) and 8 (solid line) compared to the experimental data from Refs. [15, 16, 17, 18] (symbols).

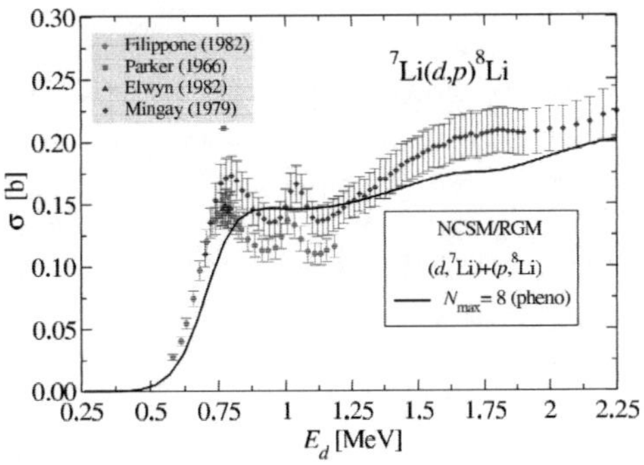

Figure 6. Same as Fig. 5, but for the integrated cross section computed in the NCSM/RGM phenomenology approach, in which the experimental energy difference between d+^{7}Li and p+^{8}Li thresholds is taken as input in the calculation (see text for details).

The wrong threshold affects the NCSM/RGM calculation and its impact on the computed cross section can be seen in Fig. 6, where the NCSM energies for the d, ^{7}Li and ^{8}Li clusters are correct in order to reproduced the d+^{7}Li and p+^{8}Li thresholds with a desired level of accuracy. The values of the clusters energies used to obtain the curve in Fig. 6 are shown in Table 1. This way to proceed, denoted as 'NCSM/RGM phenomenology', brings the calculated total cross section in fairly good agreement with the measured one in Fig. 6, with the position of the first peak slightly overestimated and the trend of the cross section qualitatively reproduced, except the second peak at about 1 MeV above the d+^{7}Li threshold. The lack of this peak in the calculated cross section could be due to the missing ^{8}Be(α-α)-n mass partition in the model space.

An interesting issue regarding the resonant peak at $\sim$ 0.78 MeV concerns the determination of its spin and parity. With the reasonable assumption that the first peak in our integrated cross section corresponds to the first experimental resonance, we can contribute to decide between conflicting spin-parity assignments derived from different phenomenological R-matrix analyses [19, 20]. As illustrated in the eigenphase shifts of Fig. 1 and phase shifts of Fig. 3, our calculation supports a $\frac{5}{2}^+$ spin-parity assignment, as suggested by the analysis in Ref. [20]. In general, by studying the contribution of the different partial waves to the total cross section (see Fig. 9 of Ref. [10]), we saw that the positive-parity phase shifts reach the maximum of their gradient at lower energies than the negative-parity ones.

4. Conclusion

We revisited the application of the NCSM/RGM approach to the ^{7}Li(d,p)^{8}Li transfer reaction [10]. This study is the first application of an *ab initio* method to the deuteron-induced transfer reaction with p-shell ($A > 4$) targets.

We found that the interplay between deuteron-^{7}Li and proton-^{8}Li channels explains some features of the ^{9}Be spectrum in the energy region where the two thresholds are open. In this work, we strengthen the conclusion that the first resonant peak in the integrated cross section of the reaction, detected at deuteron energy of 0.78 MeV, contains a significant contribution from both the mass partition channels. The other important resonances are instead dominated by the proton-^{8}Li channels, with a corresponding quenching of the deuteron-^{7}Li ones.

The study of the evolution of the phase and eigenphase shifts curves with respect to the number of the deuteron pseudostates, has shown the crucial effect of the continuum given by the deuteron polarization and its virtual breakup. While the trend of convergence in terms of the pseudostates suggests that the inclusion of more pseudostates could still affect quantitatively the position of the resonances, the main conclusions of our study are confirmed: in particular, the discussion on the experimental spin-parity assignments of the 0.78 MeV resonance, which is used as a calibration for the target thickness in the proton-capture experiments on ^{7}Be. We found that our calculations support a spin-parity assignment of $J^\pi = \frac{5}{2}^+$ for this resonance, suggesting a reaction mechanism dominated by the coupling of the P-wave $d-^7$Li incoming channel to the S-wave in the $p-^8$Li exit channel.

Acknowledgments
TRIUMF receives federal funding via a contribution agreement with the National Research Council of Canada. This work was supported in part by NSERC under Grant No. 401945-2011, by LLNL under Contract DE- AC52- 07NA27344, and by the U.S. Department of Energy, Office of Science, Office of Nuclear Physics, under Work Proposal Number SCW1158. Computing support came from the LLNL institutional Computing Grand Challenge Program and from an INCITE Award on the Titan supercomputer of the Oak Ridge Leadership Computing Facility (OLCF) at ORNL.

References

[1] Jones K L 2013 *Physica Scripta* **2013** 014020 URL http://stacks.iop.org/1402-4896/2013/i=T152/a=014020
[2] Navrátil P, Quaglioni S, Stetcu I and Barrett B R 2009 *Journal of Physics G: Nucl. and Part. Phys.* **36** 083101 URL http://stacks.iop.org/0954-3899/36/i=8/a=083101
[3] Quaglioni S and Navrátil P 2009 *Phys. Rev. C* **79**(4) 044606 URL http://link.aps.org/doi/10.1103/PhysRevC.79.044606
[4] Baroni S, Navrátil P and Quaglioni S 2013 *Phys. Rev. Lett.* **110**(2) 022505 URL http://link.aps.org/doi/10.1103/PhysRevLett.110.022505
[5] Baroni S, Navrátil P and Quaglioni S 2013 *Phys. Rev. C* **87**(3) 034326 URL http://link.aps.org/doi/10.1103/PhysRevC.87.034326
[6] Navrátil P and Quaglioni S 2011 *Phys. Rev. C* **83**(4) 044609 URL http://link.aps.org/doi/10.1103/PhysRevC.83.044609
[7] Hupin G, Quaglioni S and Navrátil P 2015 *Phys. Rev. Lett.* **114**(21) 212502 URL http://link.aps.org/doi/10.1103/PhysRevLett.114.212502
[8] Quaglioni S, Navrátil P, Roth R and Horiuchi W 2012 *Journal of Physics: Conference Series* **402** 012037 URL http://stacks.iop.org/1742-6596/402/i=1/a=012037
[9] Navrátil P and Quaglioni S 2012 *Phys. Rev. Lett.* **108**(4) 042503 URL http://link.aps.org/doi/10.1103/PhysRevLett.108.042503
[10] Raimondi F, Hupin G, Navrátil P and Quaglioni S 2016 *Phys. Rev. C* **93**(5) 054606 URL http://link.aps.org/doi/10.1103/PhysRevC.93.054606
[11] Adelberger E G and *et al* 1998 *Rev. Mod. Phys.* **70**(4) 1265–1291 URL http://link.aps.org/doi/10.1103/RevModPhys.70.1265
[12] Romero-Redondo C, Quaglioni S, Navrátil P and Hupin G 2014 *Phys. Rev. Lett.* **113**(3) 032503 URL http://link.aps.org/doi/10.1103/PhysRevLett.113.032503
[13] Baye D and Heenen P H 1986 *Journal of Physics A: Mathematical and General* **19** 2041 URL http://stacks.iop.org/0305-4470/19/i=11/a=013
[14] Entem D R and Machleidt R 2003 *Phys. Rev. C* **68**(4) 041001 URL http://link.aps.org/doi/10.1103/PhysRevC.68.041001
[15] Parker P D 1966 *Phys. Rev.* **150**(3) 851–856 URL http://link.aps.org/doi/10.1103/PhysRev.150.851
[16] Mingay D 1979 *South African Journal of Physics* **2**(3) 107
[17] Elwyn A J, Holland R E, Davids C N and Ray W 1982 *Phys. Rev. C* **25**(5) 2168–2173 URL http://link.aps.org/doi/10.1103/PhysRevC.25.2168

[18] Filippone B W, Elwyn A J, Ray W and Koetke D D 1982 *Phys. Rev. C* **25**(5) 2174–2180 URL `http://link.aps.org/doi/10.1103/PhysRevC.25.2174`
[19] Decharge D, Surget G, Bruno G and Decharge M Y 1972 *Le Journal de Physique* **33**(5-6) 485
[20] Friedland E and Venter I 1971 *Z. Physik* **243** 126

IOP Conf. Series: Journal of Physics: Conf. Series **981** (2018) 012007 doi:10.1088/1742-6596/981/1/012007

Effects of the coupling of quasiparticle and collective vibrations on the properties of ^{120}Sn

Enrico Vigezzi
INFN Sezione di Milano, Milano, Italy
E-mail: vigezzi@mi.infn.it

Abstract. Assuming quasiparticles and collective vibrations as fundamental modes of excitation and taking into account their interplay within the framework of Nuclear Field Theory, it is possible to give an accurate and comprehensive description of the low-energy spectrum of the superfluid nucleus ^{120}Sn.

1. Introduction

A relevant issue in the field of nuclear structure is whether it is possible to obtain a quantitative and comprehensive description of low-energy nuclear spectroscopic properties assuming quasiparticles and collective vibrations as fundamental modes of excitation, and considering the many-body processes associated with their interweaving in order to renormalise the properties obtained in the mean field approximation. This question was addressed (see [1] and refs. therein) in the framework of Nuclear Field Theory for the case of the superfluid nucleus ^{120}Sn. It was shown that theory can indeed account, within a 10% average error level, for a set of structure and reaction data which provide a rather complete characterization of ^{120}Sn. In the present contribution more details will be given concerning the influence of the mean field on the final results and concerning the possibility of determining an optimised set of bare single-particle energies in the valence shell, instead of taking them from a Hartree-Fock field obtained with an effective interaction fitted on global nuclear properties. After an outline of the formalism, I will discuss the renormalization of the pairing gap, which is enhanced by the pairing interaction induced by the exchange of collective phonons, and the renormalization of quasiparticles, leading to the fragmentation of the quasiparticle strength. Theoretical results will be compared to data, and the stability of the results will be addressed, varying the deformation of the lowest quadrupole vibration - which is the main factor determining the strength of the particle-vibration coupling - around its empirical value.

2. Theoretical framework

In this section the theoretical framework will be briefly summarised. See refs. [1, 2] for more details. The calculation will be divided in two steps. Starting from a fixed set of single particle levels, the HFB equations in the pairing $^1S_0$ channel are solved using the Argonne v_{14} bare interaction. In order to obtain convergence a large set of levels must be included in the calculation, up to a cutoff of about 1 GeV, due to the effect of the hard core in the interaction. In this way a set of quasiparticles energies and of quasiparticles amplitudes is produced. An

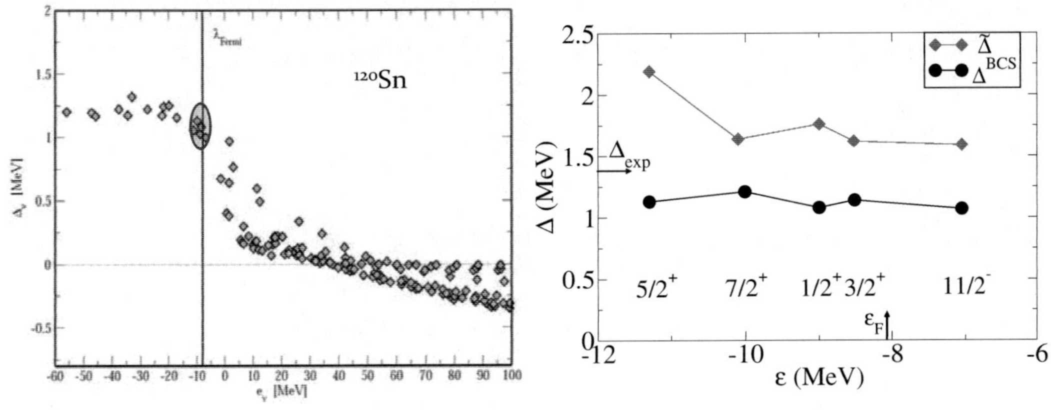

Figure 1. (left) Diagonal part of the pairing gaps, resulting from the HFB calculation performed with the Argonne interaction and the SLy4 mean field. The circle encloses the gaps associated with the five valence orbitals lying close to the Fermi energy. They are shown in detail in the right part of the figure, together with the gaps $\tilde{\Delta}$ obtained after the inclusion of the renormalization processes. Also indicated are the Fermi energy and the experimental value of the pairing gap derived from the odd-even mass difference, $\Delta_{exp} \approx 1.4$ MeV.

example is shown in the left panel of Fig. 1, which shows the diagonal part of the pairing gap associated with quasiparticles up to excitation energies of 100 MeV, obtained from a calculation performed with the Hartree-Fock mean field computed with the SLy4 effective interaction. Note the change of sign of the gap, which is associated with the presence of the v_{14} repulsive core. The pairing gaps associated with the five valence orbitals with quantum numbers $a = s_{1/2}, d_{3/2}, d_{5/2}, g_{7/2}$ and $h_{11/2}$ and lying close to the Fermi energy in ^{120}Sn are shown in more detail in the right panel of Fig. 1. We observe that the pairing gap close to the Fermi energy displays a weak state-dependence. In fact, as it will be discussed below, the pairing properties close to the Fermi energy can be accurately reproduced with a much simpler BCS calculation performed with a monopole force with an appropriate pairing strength.

Many-body effects are included in the second step of the calculation, in which only the five lowest valence quasiparticles are included, neglecting the effects of the particle-vibration coupling on the radial wave functions. Starting from the quasi particle energies E_a and quasi particle amplitudes u_a, v_a obtained with the bare pairing interaction, one solves the energy-dependent eigenvalue problem, fragmenting the original quasiparticle strength into a set of states $a(n)$ and obtaining a set of renormalized quasi particle energies $\tilde{E}_{a(n)}$ and associated amplitudes $\tilde{u}_{a(n)}, \tilde{v}_{a(n)}$:

$$\begin{pmatrix} (\epsilon_a - \epsilon_F) + \tilde{\Sigma}^{11}_{a(n)} & \tilde{\Sigma}^{12}_{a(n)} \\ \tilde{\Sigma}^{12}_{a(n)} & -(\epsilon_a - \epsilon_F) + \tilde{\Sigma}^{22}_{a(n)} \end{pmatrix} \begin{pmatrix} \tilde{u}_{a(n)} \\ \tilde{v}_{a(n)} \end{pmatrix} = \tilde{E}_{a(n)} \begin{pmatrix} \tilde{u}_{a(n)} \\ \tilde{v}_{a(n)} \end{pmatrix}, \tag{1}$$

where ϵ_a denotes the HF single-particle energy, while the new normal self-energies are given by

$$\begin{aligned} \tilde{\Sigma}^{11}_{a(n)} &= u_a^2 \Sigma^{11pho}_{a(n)} + v_a^2 \Sigma^{22pho}_{a(n)} - 2u_a v_a \Sigma^{12pho}_{a(n)} \\ \tilde{\Sigma}^{22}_{a(n)} &= u_a^2 \Sigma^{22pho}_{a(n)} + v_a^2 \Sigma^{11pho}_{a(n)} + 2u_a v_a \Sigma^{12pho}_{a(n)}. \end{aligned} \tag{2}$$

For a given eigenvalue $\tilde{E}_{a(n)} > 0$ there exists a corresponding eigenvalue $\tilde{E}_{a(-n)} = -\tilde{E}_{a(n)}$.

IOP Conf. Series: Journal of Physics: Conf. Series **981** (2018) 012007 doi:10.1088/1742-6596/981/1/012007

The quantity $\Sigma^{11pho}_{a(n)}$ is given by

$$\Sigma^{11pho}_{a(n)} = \sum_{b,m,\lambda,\nu} \frac{V^2(ab(m)\lambda\nu)}{\tilde{E}_{a(n)} - \tilde{E}_{b(m)} - \hbar\omega_{\lambda\nu}} + \sum_{b,m,\lambda,\nu} \frac{W^2(ab(m)\lambda\nu)}{\tilde{E}_{a(n)} + \tilde{E}_{b(m)} + \hbar\omega_{\lambda\nu}}, \quad (3)$$

and $\Sigma^{22pho}_{a(n)} = -\Sigma^{11pho}_{a(-n)}$.

The quasiparticle-phonon matrix elements connecting the quasiparticle a with the quasiparticle-phonon configuration $b(m)\lambda\nu$ are given by

$$V(ab(m)\lambda\nu) = h(ab\lambda\nu)(u_a\tilde{u}_{b(m)} - v_a\tilde{v}_{b(m)}) \quad ; \quad W(ab(m)\lambda\nu) = h(ab\lambda\nu)(u_a\tilde{v}_{b(m)} + v_a\tilde{u}_{b(m)}), \quad (4)$$

where

$$h(ab\lambda\nu) = -(-1)^{j_a - j_b}\beta_{\lambda\nu} < a|r_1\frac{\partial U}{\partial r_1}|b >< j_b||Y_\lambda||j_a > \left[\frac{1}{(2j_a+1)(2\lambda+1)}\right]^{1/2} \quad (5)$$

is the basic particle-phonon vertex [3]. This vertex contains a finite range form factor and is suitable for the coupling to collective surface modes. It avoids convergence ¡ problems related to the coupling with uncorrelated particle-hole excitations, which arise when the coupling is calculated selfconsistently using zero-range Skyrme forces.

One can separate the abnormal self-energy into two terms, writing

$$\tilde{\Sigma}^{12}_{a(n)} = \Delta^{BCS}_a + \tilde{\Sigma}^{12pho}_{a(n)}. \quad (6)$$

The first term, Δ^{BCS}_a, is the pairing gap associated with the bare interaction which was obtained in the first step of the calculation. The second term is equal to

$$\tilde{\Sigma}^{12pho}_{a(n)} = - \sum_{b,m,\lambda,\nu} \frac{(2j_b+1)}{2} V_{ind}(a(n)b(m))\tilde{u}_{b(m)}\tilde{v}_{b(m)} \quad (7)$$

and contains the induced pairing interaction:

$$V_{ind}(a(n)b(m)) = \sum_{\lambda,\nu} \frac{2h^2(ab\lambda\nu)}{(2j_b+1)} \times \left[\frac{1}{\tilde{E}_{a(n)} - \tilde{E}_{b(m)} - \hbar\omega_{\lambda\nu}} - \frac{1}{\tilde{E}_{a(n)} + \tilde{E}_{b(m)} + \hbar\omega_{\lambda\nu}}\right]. \quad (8)$$

The renormalized pairing gap $\tilde{\Delta}_{a(n)}$ is obtained multiplying $\tilde{\Sigma}^{12}_{a(n)}$ by the $Z-$factor (which is close to the fraction of the quasiparticle strength remaining in the $n-$fragment, although it does not coincide with it)

$$Z_{a(n)} = \left(1 - \frac{\tilde{\Sigma}^{11}_{a(n)} - \tilde{\Sigma}^{11}_{a(-n)}}{2\tilde{E}_{a(n)}}\right)^{-1}. \quad (9)$$

The renormalised pairing gaps obtained with the Argonne interaction and the SLy4 mean field are shown in the right panel of Fig. 1. The induced interaction increases the value of the gap, and prevails over the reduction caused by the $Z-$factor.

The properties of the phonons are not explicitly renormalized. They will instead be obtained from QRPA calculations with separable forces, using coupling constants reproducing the empirical energy $\hbar\omega_{\lambda_1}$ and deformation parameter β_{λ_1} for $\lambda = 2, 3$.

IOP Conf. Series: Journal of Physics: Conf. Series **981** (2018) 012007 doi:10.1088/1742-6596/981/1/012007

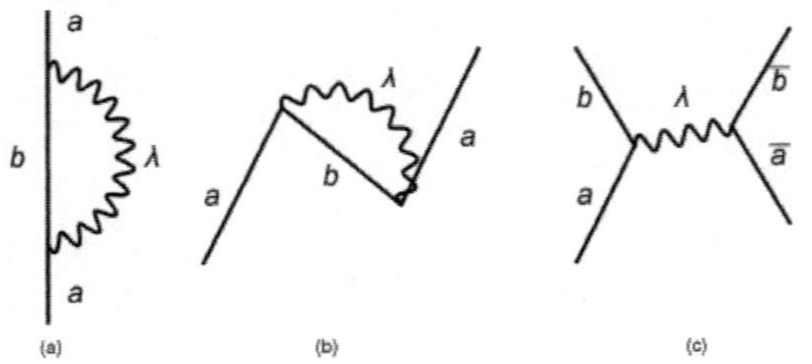

Figure 2. Basic diagrams taken into account in the present study, which renormalize the normal and abnormal self-energies obtained in mean field calculations: polarization (a), correlation (b), induced pairing interaction (c) processes.

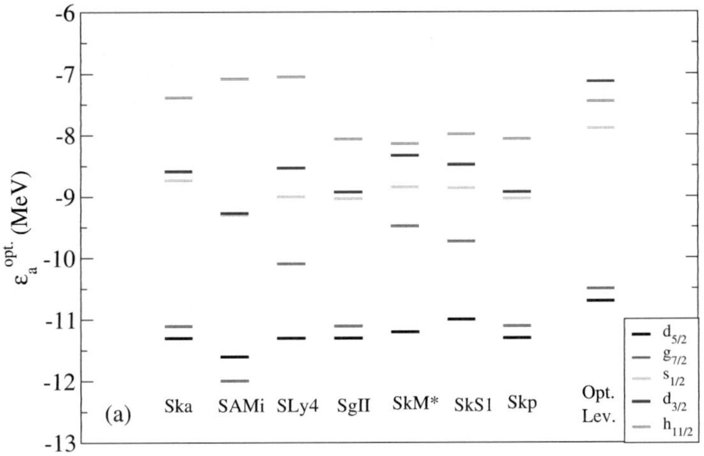

Figure 3. Energies of the five valence single-particle energies associated with the Skyrme forces discussed in the text (ordered according to increasing effective mass). The last column shows the optimised energies discussed in text.

3. Pairing gaps and effective masses

In this section, the dependence of the pairing gaps on the effective mass associated with the mean field will be discussed. First of all let us study the case of the bare interaction. The pairing gaps calculated with the Argonne interaction with the mean field generated by seven Skyrme forces characterized by different values of m_k, the effective mass in nuclear matter at saturation density, are shown in Fig. 4(a) by green diamonds. One observes an overall increase of the gap with increasing m_k superposed to significant fluctuations.

The fluctuations in the gaps are associated with the details of the energy distribution of the five valence orbitals, shown in Fig. 3, while the overall dependence on m_k is determined by the whole single-particle spectrum. The two effects can be disentangled by solving the pairing gap equations using always the same single-particle energies for the five valence orbitals. In this

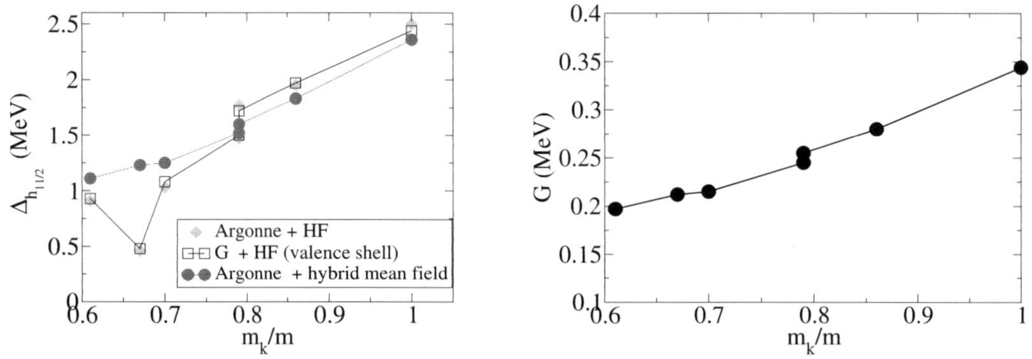

Figure 4. (a) Bare pairing gap of the $h_{11/2}$ orbital, as a function of the effective mass associated with the mean fields generated with the effective forces shown in Fig. 3 (cf. text). (b) Dependence of the monopolar pairing coupling constant on the effective mass.

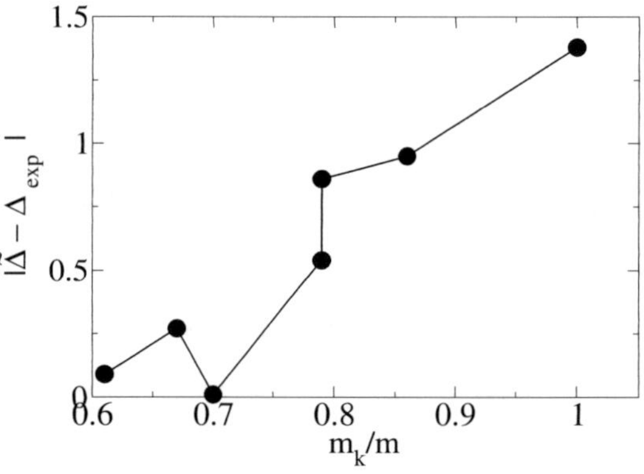

Figure 5. Absolute difference between the experimental gap and the theoretical value of the renormalised gap, calculated as a function of the effective mass associated with the mean field.

case the pairing gap dependence becomes much smoother (red dots in Fig. 4(a), obtained using the valence levels of the SGII interaction). The results of these calculations can be accurately reproduced by a BCS calculation with a monopole interaction acting only in the valence shell, using the values of the pairing constant G reported in Fig. 4(b) as a function of m_k. It is remarkable that, once the function $G(m_k)$ has been determined, the gaps shown in Fig 4(a) by open squares, obtained by simple BCS calculations using only the corresponding five valence levels, can reproduce in detail the pairing gaps obtained by the calculation with the Argonne interaction using different mean fields (green diamonds in Fig. 4(a)).

The values of the renormalised gaps $\tilde{\Delta}$ are shown Fig. 5. The comparison with the experimental value favours effective masses in the range $m_k \approx 0.6m - 0.8m$. It should be remarked that the above discussion is oversimplified concerning at least three effects neglected here, which act on the pairing gap in different way. First, the momentum dependence associated with the Skyrme interaction is not realistic, and tends to make the gaps too low [4]. On the

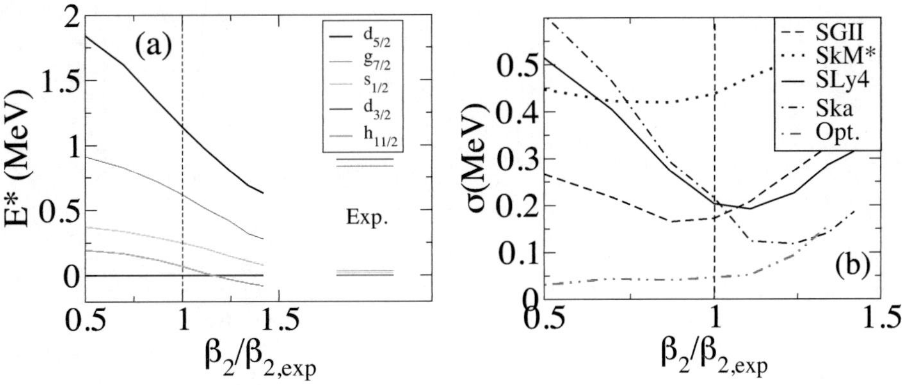

Figure 6. (a) The experimental excitation energies of the lowest valence levels (referred to the energy of the $3/2^+$ level) are compared to the theoretical renormalised quasiparticle energies, computed as a function of the coupling strength, starting from SLy4 single-particle energies. (b) The mean square root deviation between the experimental and the theoretical spectrum shown in (a) (solid line) is compared to the deviations obtained starting from with other three Skyrme forces and with the optimised single-particle levels shown in Fig. 7 (a) and discussed in the text (red dash-dotted line).

other hand, renormalization effects due to spin modes have not been included, and as a result gaps are about too large [2, 5]. Finally, the inclusion of three-body force, rarely considered until now, would tend to make the gap smaller, according to the results of ref. [6]. The corrections associated with these effects are difficult to quantify, but are of the order of 300 keV.

4. Renormalization processes and optimised levels

The quasiparticle spectrum obtained with the SLy4 mean field and the Argonne interaction is shown in Fig. 6(a) as a function of the ratio between the quadrupole deformation β_2 used in the calculation and the experimental value $\beta_{2,exp} = 0.13$, in order to assess the stability of the results. The mean square root deviation σ between theory and experiment is shown in Fig. 6(b), also for other forces. Consistent results are obtained for the SGII, SLy4 and Ska interactions, which lead to $\sigma \approx 200$ keV and to a minimum close to β_{exp} (although larger value of β_2 are equally acceptable in the case of SGII). It has been shown in ref. [1], that, using the SLy4 mean field, good quantitative agreement (within 10-20%) between theory and experiment can be obtained for a variety of observables, including electromagnetic transitions, multiplet splittings, one- and two-nucleon transfer absolute cross sections. On the other hand, it was observed that the $5/2^+$ strength distribution originating from the strongly fragmented $d_{5/2}$ orbital was not well reproduced, and that this was related to the specific position of the orbitals in the SLy4 mean field. Better results were obtained shifting the energy of the $d_{5/2}$ by 600 keV towards the Fermi energy.

We have already noticed that the results obtained with the Argonne interaction and a given mean field associated with an effective mass m_k can be reproduced using a monopole pairing interaction with the value of the pairing coupling strength appropriate to m_k. We note that in such calculations, the essential ingredients are the values of the single-particle energies of the five valence orbitals; the change in the mean field leads to relatively small differences in the matrix elements. In the following, we study the possibility to determine optimized values of the single-particle energies, and show results obtained with matrix elements calculated with

IOP Conf. Series: Journal of Physics: Conf. Series **981** (2018) 012007 doi:10.1088/1742-6596/981/1/012007

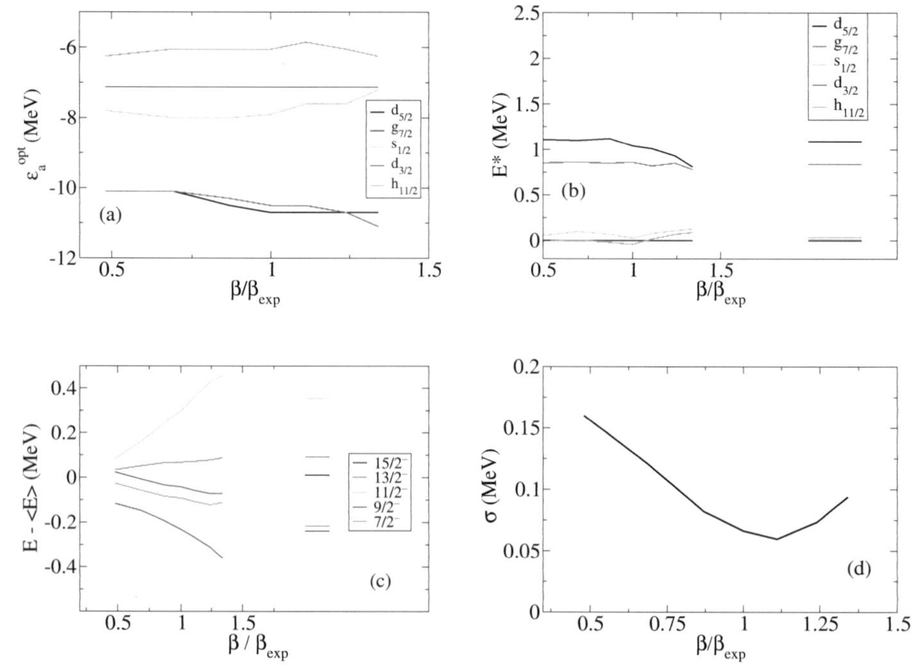

Figure 7. (a)Energies of the optimised five valence single-particle energies ϵ_a^{opt} as a function of the quadrupole deformation parameter. (b) The experimental excitation energies of the lowest valence levels (referred to the energy of the $3/2^+$ level) are compared to the theoretical spectrum obtained starting from the optimized single-particle levels, as a function of the quadrupole deformation parameter. (c) The experimental energies of the members of the $h_{11/2}\otimes 2^+$ multiplet are compared with the theoretical values, calculated as a function of the quadrupole parameter. The associated mean square deviation is shown in (d).

single-particle wave functions generated with the SLy4 interaction.

The optimised single-particle energies have been determined minimising the mean square root difference between the experimental quasi particle spectrum and the theoretical renormalized energies $\tilde{E}(\bar{n})$, where $\bar{n}$ indicates the fragment carrying the largest quasiparticle strength for given quantum number. The resulting optimised levels have been calculated as function of β/β_{exp} and are shown in Fig. 7(a). The spectrum obtained for $\beta = \beta_{exp}$ is compared in Fig. 3 with those associated with the different Skyrme forces representative of different values of the effective mass. The fit does not determine the absolute position of the levels, because the quasiparticle spectrum is unchanged by shifting all the single-particle energies and the Fermi energy ϵ_F by the same amount. One can fix the absolute position of the energy levels by comparing the experimental value of the neutron separation energy in ^{120}Sn ($S_n = 9.1$ MeV) and ^{121}Sn (6.2 MeV) with the theoretical value, $S_n \approx -\epsilon_F \pm \varepsilon_0$, where ε_0 denotes the values of the lowest calculated quasiparticle. It can be seen that all the considered Skyrme forces lead to a more bound scheme. The distance between the $g_{7/2}$ and the $d_{5/2}$ optimised levels is smaller than in most Skyrme spectra, in particular than in the case of SLy4. This confirms the analysis of the $5/2^+$ strength function mentioned above.

The renormalised quasiparticle spectrum is shown in Fig. 7(b). The deviation from experiment is smaller than 50 keV and is, by construction, smaller than that obtained starting

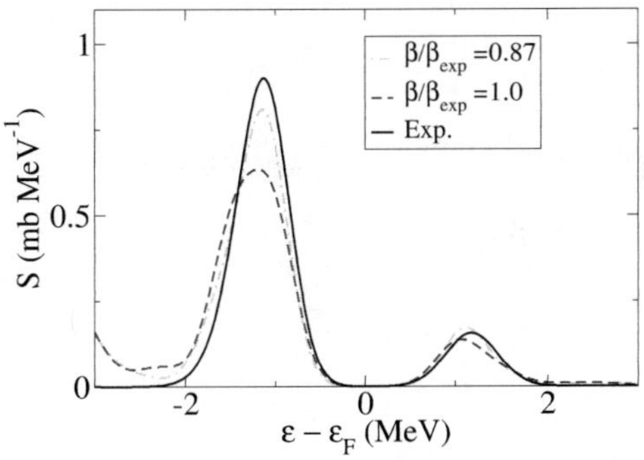

Figure 8. The experimental occupation factors associated with the $5/2^+$ levels extracted from pickup and stripping one-transfer reaction are convoluted with a gaussian with $\sigma = 0.25$ MeV in order to obtain the experimental strength function shown by the black curve. The other curves show the theoretical strength functions calculated with the optimised energy levels for two values of the quadrupole deformation parameter.

from the HF orbitals (Fig. 6(b)). It is noticed that relatively good accuracy is obtained with the Ska and the SGII mean fields, which have a low effective mass and $g_{7/2}$ and $d_{5/2}$ orbitals which lie close to each other. Also shown are the calculated levels belonging to the $h_{11/2} \otimes 2^+$ multiplet with the associated deviation between theory and experiment (Fig. 7(c),(d)). Finally, in Fig. 8 the experimental and the theoretical $5/2^+$ strength function are compared. In this case, the best results are obtained with a value of the deformation parameter somewhat smaller than the experimental one.

5. Conclusions

The possibility of using optimised energy levels constitutes an interesting alternative to the use of single-particle levels from Hartree Fock calculations with effective interactions fitted on empirical bulk properties, which contain correlations which may lead in principle to double counting with respect to those explicitly considered in the many-body renormalization processes. In this respect, it seems anyway reassuring that the results obtained in the present analysis are in most cases quantitatively similar to those obtained with effective interactions associated with reasonable effective masses, lying in the range $m_k = 0.6m - 0.8m$.

6. Acknowledgment

This work has been carried out in collaboration with F. Barranco, R.A. Broglia, A. Idini, and G. Potel. Funding from the European Union Horizon 2020 research and innovation program under Grant Agreement No. 654002 is acknowledged.

References

[1] Idini A, Potel G, Barranco F, Vigezzi E and Broglia RA 2015 Phys. Rev. C **92** 031304
[2] Idini A, Barranco F and Vigezzi E 2012 Phys. Rev. C **85** 014331
[3] Bohr A and Mottelson BR 1975 *Nuclear Structure*, Vol. II, Benjamin
[4] Lesinski T, Duguet, Bennaceur K and Meyer J 2009 Eur. Phys. J A **40** 121
[5] Gori G, Ramponi F, Barranco F, Bortignon PF, Broglia RA Colò G and Vigezzi E 2005 Phys. Rev. C **72** 011302(R)
[6] Duguet T, Lesinski T, Hebeler K and Schwenk A 2010 Mod. Phys. Lett. **25** 1989

Bridging the gap: effective interactions from microscopic dynamics

Omar Benhar

INFN and Dipartimento di Fisica, Sapienza University, I-00185 Rome, Italy

E-mail: omar.benhar@roma1.infn.it

Abstract. I discuss the derivation of an effective interaction, obtained from a realistic nuclear Hamiltonian using the formalism of Correlated Basis Functions (CBF) and the cluster expansion technique. Unlike the bare nucleon-nucleon potential, this effective interaction is well behaved, and suitable for carrying out perturbative calculations of a variety of equilibrium and non-equilibrium properties of nuclear matter using the basis of eigenstates of the non interacting system. The robustness and accuracy of the proposed approach are thoroughly examined, comparing the results of its application to the Femi hard-sphere system to the predictions of low-density expansions. Selected properties of nuclear matter, at both zero and nonzero temperatures, are reported and discussed.

1. Introduction

Approaches based on effective interactions are widely used to study the properties of strongly interacting many-body systems, when the bare interaction between the constituents cannot be treated in perturbation theory using the basis states describing non interacting particles [1, 2].

Effective interactions specifically designed to reproduce the bulk properties of nuclear matter (see, e.g., Refs. [3, 4]), while being remarkably successful in a number of instances, fail to provide a quantitative account of nucleon-nucleon scattering—both in free space and in the nuclear medium—whose understanding is needed for the description of important non-equilibrium properties [5, 6]. For example, the results reported in Ref. [5] clearly show that the determination of the shear viscosity and thermal conductivity of pure neutron matter, relevant to many astrophysical applications [7, 8], requires effective interactions derived from *ab initio* microscopic approaches, capable of explaining the observed nucleon-nucleon scattering data in the zero-density limit [5].

The authors of Refs. [9, 10] have developed a procedure to determine the effective interaction in nuclear matter using the Correlated Basis Function (CBF) formalism and the cluster expansion technique. While this scheme has been thoroughly tested through comparison between its results and those obtained from G-matrix perturbation theory in pure neutron matter [5], the analysis of a somewhat simpler many-body system, several properties of which can be accurately calculated and expressed in analytic form, provides further insight into the validity and robustness of the underlying assumptions.

The results of studies of the Fermi hard-sphere system strongly suggest that the CBF effective interaction approach, provides accurate estimates of fundamental quantities other than the ground-state energy, such as the self-energy determining the two-point Green's function. In this context, the capability to carry out perturbative calculations beyond the lowest order—without incurring in the severe difficulties arising from the use of a non orthogonal basis—has been shown to play a critical role [11, 12].

Nuclear applications also cover a number of quantities of great astrophysical interest, such as the ground state energy and the quasiparticle spectrum of uniform matter at fixed baryon density and large neutron excess, at both zero and nonzero temperatures.

In Section 2, the derivation of the CBF effective interaction is outlined exploiting the pedagogical example of point-like fermions interacting through the hard-sphere potential, while Sections 3 and 4 are devoted to the results of numerical calculations of a variety of properties of the hard-sphere system and uniform nuclear matter, respectively. Finally, in Section 5 I summarize the present status of the CBF effective interaction approach, and outline the prospects for future developments.

2. Formalism

This section introduces the main elements entering the derivarion of the CBF effective interactions, in the simple case in which the bare potential only depends on the separation distance between interacting particles.

2.1. Definition of the CBF effective interaction

The nucleon-nucleon interaction is known to be strongly repulsive at short distances, as clearly indicated by the saturation of the charge-density distributions measured by elastic electron-nucleus scattering [13]. As a consequence, the Fermi hard-sphere system, i.e. a uniform system of point-like spin one-half particles interacting through the potential

$$v(r) = \begin{cases} \infty & r < a \\ 0 & r > a \end{cases} , \tag{1}$$

has been long recognised as a valuable model for investigating concepts and approximations employed to study the properties of nuclear systems. Setting the degeneracy of the momentum eigenstates to $\nu = 2$ and 4, the hard-sphere system is meant to model pure neutron matter (PNM) and isospin-symmetric nuclear matter (SNM), respectively. Note that in both instances the system can be described in terms of a single parameter $c = k_F a$, the Fermi momentum k_F being simply related to the density ρ through $k_F = (6\pi^2\rho/\nu)^{1/3}$.

Within the CBF approach, the *correlated* states of the hard-sphere system are obtained from the non interacting Fermi gas (FG) states through the transformation

$$|n\rangle = \frac{F|n_{FG}\rangle}{\langle n_{FG}|F^\dagger F|n_{FG}\rangle^{1/2}} , \tag{2}$$

where the operator F, embodying the correlation structure induced by the interaction potential, is written in the form

$$F = \prod_{j>i} f(r_{ij}) , \tag{3}$$

with

$$f(r_{ij} \leq a) = 0 \quad , \quad \lim_{r_{ij}\to\infty} f(r_{ij}) = 1 , \tag{4}$$

$r_{ij} = |\mathbf{r}_i - \mathbf{r}_j|$ being the interparticle distance.

In principle, the shape of the two-body correlation function, $f(r_{ij})$, at $r_{ij} > a$ can be determined from functional minimisation of the expectation value of the hamiltonian

$$H = T + V = \sum_i \frac{k_i^2}{2m} + \sum_{j>i} v(r_{ij}) , \tag{5}$$

in the correlated ground state. In the above equation, $k_i = |\mathbf{k}_i|$, m denotes the particle mass and v is the potential of Eq. (1).

The effective interaction

$$V_{\rm eff} = \sum_{j>i} v_{\rm eff}(r_{ij}) \ , \tag{6}$$

is defined by the relation

$$\frac{1}{N}\langle H\rangle = \frac{1}{N}\frac{\langle 0|H|0\rangle}{\langle 0|0\rangle} = \frac{3k_F^2}{10m} + \frac{1}{N}\langle 0_{FG}|V_{\rm eff}|0_{FG}\rangle \ , \tag{7}$$

where N denotes the number of particles. Note that the above equation shows that, unlike he G-matrix, the CBF effective interaction is defined *not* in operator form, but rather in terms of its expectation value in the Fermi gas ground state.

The calculations discussed in the following Sections are largely based on the assumption—that will be tested comparing the results to those obtained from different many-body approaches—that perturbative calculations involving matrix elements of $V_{\rm eff}$ between Fermi gas states provide accurate estimates of *all* properties of the Fermi hard-sphere system.

2.2. Cluster expansion technique

The calculation of matrix elements of any many-body operator between correlated states involves largely irreducible $3N$-dimensional integrations. This problem, that becomes quickly intractable with increasing N, can be circumvented expanding the matrix element in a series, the terms of which represent the contributions of subsystems (clusters) involving an increasing number of particles [14]. Since correlations are short ranged, at not too high density the cluster expansion is expected to be rapidly convergent.

The effective interaction of Ref. [10] is derived expanding the left hand side of Eq. (7), and keeping the two-body cluster contribution only. The resulting expression is

$$\frac{1}{N}\langle H\rangle = \frac{3k_F^2}{10m} + (\Delta E)_2 \ , \tag{8}$$

with

$$(\Delta E)_2 = \frac{\rho}{2}\int d^3r \ \frac{1}{m}\left[\boldsymbol{\nabla} f(r)\right]^2\left[1 - \frac{1}{\nu}\ell^2(k_F r)\right] , \tag{9}$$

where the Fermi gas density-matrix, referred to as the Slater function, is defined as $\ell(x) = 3\left(\sin x - x\cos x\right)/x^3$. The form of $v_{\rm eff}$ follows immediately from Eqs. (6)-(9), implying

$$v_{\rm eff}(r) = \frac{1}{m}\left[\boldsymbol{\nabla} f(r)\right]^2 \ . \tag{10}$$

As pointed out above, the shape of $f(r)$ can be obtained from the functional minimisation of the Hamiltonian expectation value in the correlated ground state. Within the two-body cluster approximation, this procedure yields the Euler-Lagrange equation

$$g''(r) - g(r)\left[\frac{\Phi''(r)}{\Phi(r)} + m\lambda\right] = 0 \ , \tag{11}$$

where

$$g(r) = f(r)\Phi(r) \quad , \quad \Phi(r) \equiv r\sqrt{1 - \frac{1}{\nu}\ell^2(k_F r)} \ . \tag{12}$$

Equation (11) is solved with the boundary conditions $f(a) = 0$ and $f(d) = 1$, the additional constraint $f'(d) = 0$ being fulfilled through the introduction of the Lagrange multiplier λ. The correlation range is determined requiring that the ground-state energy per particle obtained at first order in the effective interaction be equal to the corresponding result obtained from accurate many-body calculations, carried out within the Fermi Hyper-Netted Chain (FHNC) approach [15, 16] or using Quantum Monte Carlo techniques [17] using the bare Hamiltonian.

3. The Fermi hard-spheres system

In this section, several equilibrium and non-equilibrium properties of the Fermi hard-sphere systems obtained using the CBF effective interaction are compared to the predictions of different many-body approaches. All calculations have been carried out setting the hard-sphere radius to $a = 1$ fm and the particle mass to $m = 1\ \text{fm}^{-1}$.

3.1. Equilibrium properties

The expansion of the ground-state energy of the quantum mechanical hard-sphere system in powers of the dimensionless parameter c, defined in Section 2.1, was first discussed in the 1950s by Huang and Yang [18], who were able to derive its terms up to order c^2.

More recent calculations have been carried out exploiting the formalism developed to describe the scattering process involving two particles interacting through a strongly repulsive potential [19]. The main element of this approach is the replacement of the bare interaction with the t-matrix, which amounts to including the contribution of the infinite series of ladder diagrams. This technique, which in general allows to achieve a fast convergence of perturbative calculations, becomes essential when dealing with the hard-core interaction of Eq. (1).

The resulting expressions of the ground-state energy per particle are

$$E_0 = \frac{k_F^2}{2m}\left[\frac{3}{5} + \frac{2}{\pi}c + \frac{12}{35\pi^2}(11 - 2\ln 2)\,c^2 + 0.78c^3 + \frac{32}{9\pi^3}\left(4\pi - 3\sqrt{3}\right)c^4 \ln c\right] , \tag{13}$$

for degeneracy $\nu = 4$ and

$$E_0 = \frac{k_F^2}{2m}\left[\frac{3}{5} + \frac{2}{3\pi}c + \frac{4}{35\pi^2}(11 - 2\log 2)\,c^2 + 0.230c^3\right] , \tag{14}$$

for $\nu = 2$.

The ground-state energy is conveniently parametrized by the dimensionless quantity ζ, defined through the equation

$$E_0 = \frac{3k_F^2}{10m}(1 + \zeta) . \tag{15}$$

Figures 1 and 2, taken form Ref. [12], show the c-dependence of ζ—providing a measure of the deviation of E_0 from the energy of the non interacting Fermi gas—corresponding to $\nu = 2$ and $\nu = 4$, respectively. The results of Diffusion Monte Carlo (DMC) calculations, used as baseline for the determination of the effective interaction, are compared those obtained from the Variational Monte Carlo (VMC) and FHNC approaches, the latter being only available at $c \geq 0.5$. For reference, the predictions of the low-density expansions of Ref. [19] are also displayed, by the solid lines.

It clearly appears that the VMC and FHNC results are very close to one another, thus showing that at $c \geq 0.5$ the FHNC approximation does provide an upper bound to the ground-state energy. The accuracy of the variational result is measured by the difference between the VMC—or, equivalently, FHNC—values of ζ and those obtained from DMC. In the case

IOP Conf. Series: Journal of Physics: Conf. Series **981** (2018) 012008 doi:10.1088/1742-6596/981/1/012008

of degeneracy ν=2, illustrated in Fig. 1, this difference ranges between ∼2% and ∼9% at $0.2 \leq c \leq 1$. Note that a 9% difference in ζ translates in a difference of less that 3% in the ground-state energy E_0. The low-density expansion turns out to be also quite accurate, its predictions being within 5% of the DMC results at $c < 0.5$. The results corresponding to $\nu = 4$, shown in Figure 2, exhibit the same pattern.

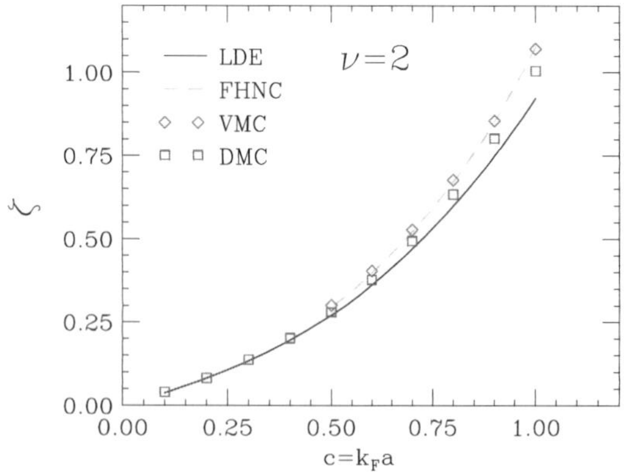

Figure 1. c-dependence of the quantity ζ, defined by Eq. (15), for degeneracy ν=2. The solid and dashed lines show the results obtained from the low-density expansion and the variational FHNC approach, respectively. The VMC and DMC results are represented by diamonds and squares.

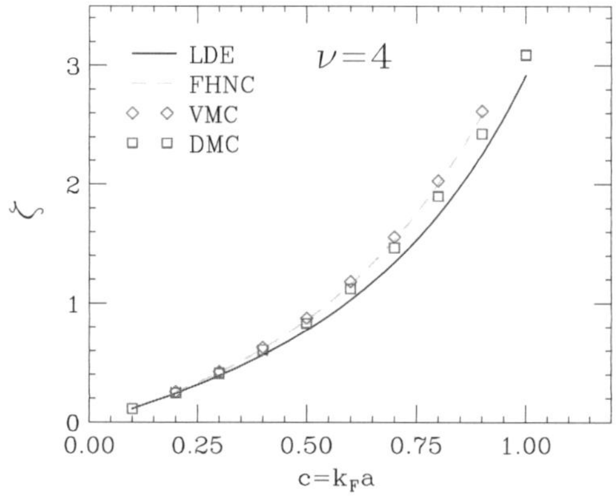

Figure 2. Same as in Fig. 1, but for degeneracy $\nu = 4$.

The effective interaction obtained from the procedure described in Section 2 has been employed to carry out a second-order perturbative calculation of the self-energy $\Sigma(k, E)$, which has been in turn used as an input to obtain the quantities describing single-particle properties.

Figure 3 shows the energy spectrum of the hard-sphere system with $\nu = 2$, defined as [11]

$$e(k) = \frac{k^2}{2m} + \text{Re }\Sigma[k, e(k)] \ , \tag{16}$$

obtained at first- and second-order in the effective interaction. The former provides the energy-independent correction corresponding to the Hatree-Fock approximation, while the latter brings about an explicit energy dependence. The momentum-dependence of $e(k)$ is illustrated for two values of density, corresponding to $c = 0.3$ and 0.6. The c-dependence of the corresponding effective mass, defined as

$$m^\star(k) = \left(\frac{1}{k}\frac{de}{dk} \right)^{-1} \ , \tag{17}$$

with

$$\frac{de}{dk} = \left[\frac{k}{m} + \frac{\partial}{\partial k}\text{Re }\Sigma(k, E) \right] \times \left[1 - \frac{\partial}{\partial E}\text{Re }\Sigma(k, E) \right]^{-1}_{E=e(k)} \ , \tag{18}$$

evaluated at $|\mathbf{k}| = k_F$, is displayed in Fig. 4. From Figs. 3 and 4, it clearly appears that the appearance of the energy dependence of the self-energy, while resulting in a small modifications of the spectrum, dramatically affects bot magnitude and density dependence of $m^\star(k_F)$. For comparison, Fig. 4 also shows the effective mass predicted by the low-density expansion

$$\frac{m^\star(k_F)}{m} = 1 + \frac{8}{15\pi^2}(7\ln 2 - 1)c^2 \; . \tag{19}$$

obtained by Galitskii back in the 1950s [20].

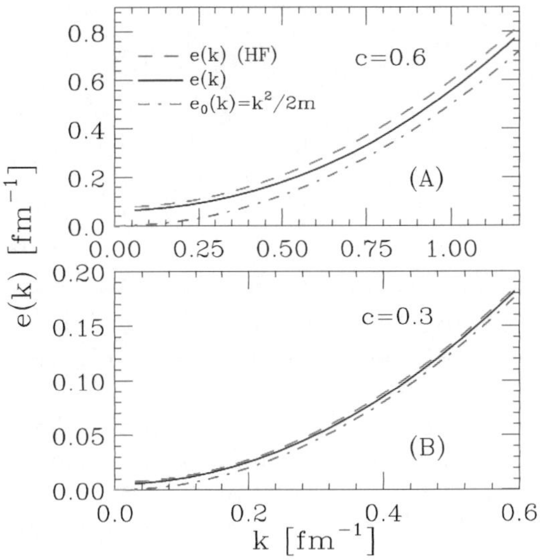

Figure 3. Single-particle energy spectrum of the fermion hard-sphere system of degeneracy ν=2 at $c = 0.3$ [panel (B)] and 0.6 [panel(A)]. The dashed and solid lines correspond to the first order (*i.e.* Hartree-Fock) and second order perturbative calculation of the self-energy $\Sigma(k, E)$, respectively. For comparison, the dot-dash lines show the kinetic energy spectrum.

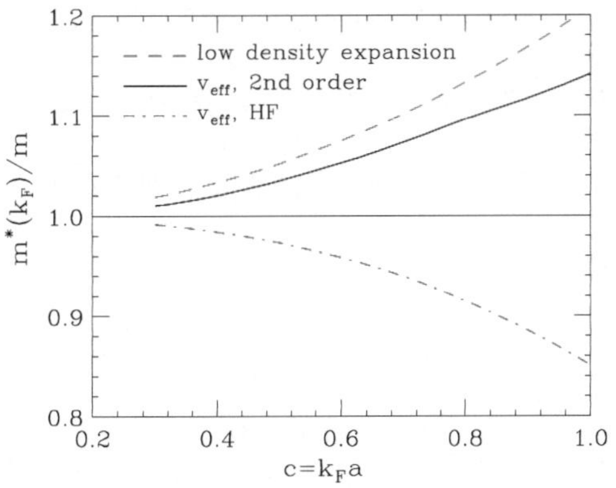

Figure 4. c-dependence of the ratio $m^\star(k_F)/m$ for the hard-sphere system of degeneracy ν=2. The dot-dash and solid lines represent the results of calculations carried out using the first- and second-order approximations to the self-energy $\Sigma(k, E)$. For comparison, the dashed line shows the results computed using the low-density expansion of Eq. (19).

The calculated self-energy can be also used to obtain the two-point Green's function, defined as

$$G(k, E) = \frac{Z_k}{E - e(k) + i\Gamma_k} \; , \tag{20}$$

with $e(k)$ given by Eq. (16). The spectroscopic factor Z_k, yielding the normalization of the quasiparticle state of momentum k, and the corresponding width Γ_k can be obtained from the self-energy using the relations

$$Z_k = \left[1 - \frac{\partial}{\partial E}\mathrm{Re}\Sigma(k, E)\right]^{-1}_{E=e(k)} \quad , \quad \Gamma_k = Z_k \mathrm{Im}\Sigma(k, E) \; . \tag{21}$$

The momentum distribution

$$n(k) = \frac{1}{2\pi i}\int_C d\omega \; G(k, \omega) \; , \tag{22}$$

where the integration on the complex vaiable ω is carried out along a closed contour in the upper half-plane Im $\omega > 0$, exhibits a discontinuity at $k = k_F$, given by

$$n(k_F - \eta) - n(k_F + \eta) = Z = Z_{k_F} \ , \tag{23}$$

with $\eta = 0^+$. At second order in the effective interaction, $n(k)$ can be conveniently written in the form

$$n(k) = n_<(k) + n_>(k) \ , \tag{24}$$

with $n_<(k > k_F) = n_>(k < k_F) = 0$, and

$$n_<(k < k_F) = 1 + \left[\frac{\partial}{\partial E} \text{Re}\Sigma_\text{p}(k, E) \right]_{E=k^2/2m} \ , \tag{25}$$

$$n_>(k > k_F) = - \left[\frac{\partial}{\partial E} \text{Re}\Sigma_\text{c}(k, E) \right]_{E=k^2/2m} \ , \tag{26}$$

where $\Sigma_\text{p}(k, E)$ and $\Sigma_\text{c}(k, E)$ denote the polarization and correlation contributions to the second order self-energy, respectively (see, e.g., Ref. [21]). Note that the above equations imply that within the Hartree-Fock approximation $n(k) = \theta(k_F - k)$, and $Z = 1$.

Figure 5 shows a comparison between the momentum distribution of the Fermi hard-sphere system with degeneracy $\nu = 4$, evaluated at $c = 0.55$, and the momentum distribution of SNM at equilibrium density, corresponding to Fermi momentum $k_F = 1.33$ fm^{-1}, obtained by the authors of Ref. [22] using non-orthogonal CBF perturbation theory at second order. The emerging picture suggests that nucleons in nuclear matter behave like hard spheres of radius $a = 0.55/1.33 \approx 0.4$ fm.

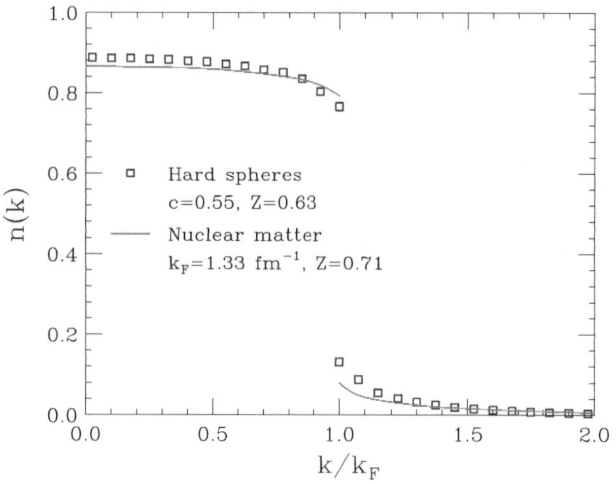

Figure 5. Comparison between the momentum distribution of the Fermi hard-sphere system obtained using the CBF effective interaction (squares) and that of SNM at equilibrium density reported in Ref. [22] (solid line).

3.2. Collision probability and transport properties

Within the approach based on Landau's theory of normal Fermi liquids [23], originally developed by Abrikosov and Khalatnikov [24, 25], the shear viscosity and thermal conductivity coefficients—denoted η and κ, respectively—are determined from the momentum and energy fluxes obtained from the kinetic equation for the distribution function

$$\frac{\partial n_\mathbf{k}}{\partial t} + \frac{\partial n_\mathbf{k}}{\partial \mathbf{r}} \cdot \frac{\partial e_\mathbf{k}}{\partial \mathbf{k}} - \frac{\partial n_\mathbf{k}}{\partial \mathbf{k}} \cdot \frac{\partial e_\mathbf{k}}{\partial \mathbf{r}} = I\left[n_\mathbf{k}\right] \ , \tag{27}$$

where $e_{\mathbf{k}}$ denotes the energy of a quasiparticle carrying momentum $\mathbf{k}$, and $I[n_{\mathbf{k}}]$ is the collision integral, the definition of which involves the in-medium scattering probability W.

In general, the scattering probability depends on the initial and final momenta of the particles participating in the process. In the low-temperature limit, however, the system is strongly degenerate, and only quasiparticles occupying states in the vicinity of the Fermi surface can be involved in interactions. As a consequence, the magnitudes of their momenta can be all set equal to the Fermi momentum, and W reduces to a function of the two angular variables θ and ϕ only. The former is the angle between the initial momenta, whereas the latter is the angle between the planes specified by the initial and final momenta, respectively.

Besides the collision probability, that can be obtained in Born approximation using the CBF effective interaction, the calculation of the shear viscosity and thermal conductivity involves the effective mass, defined by Eq. (17). The results corresponding to the case of degeneracy $\nu = 2$ are reported in Fig. 6 and 7, showing the c-dependence of the T-independent quantities ηT^2 and κT, where T denotes the temperature.

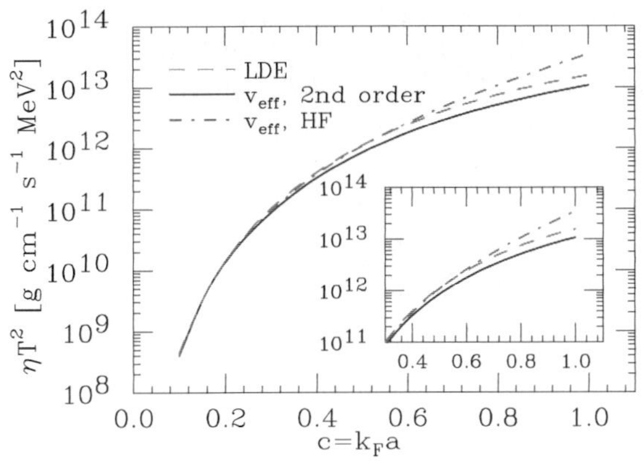

Figure 6. c-dependence of the quantity ηT^2, where η is the shear viscosity coefficient of the Fermi hard-sphere system of degeneracy $\nu = 2$. The dot-dash and solid lines represent the results of calculations carried out using the first and second order approximations for the effective mass, respectively, while the dashed line shows the results of the low-density expansion of Ref. [26], including terms of order up to c.

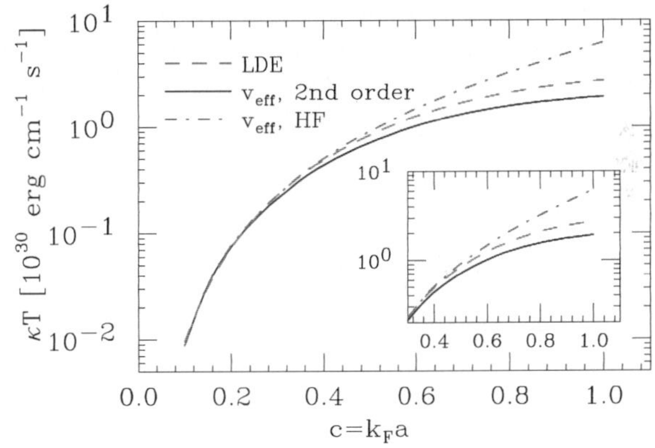

Figure 7. c-dependence of the quantity κT, where κ is the thermal conductivity of the fermion hard-sphere system of degeneracy $\nu = 2$. The dot-dash and solid lines represent the results of calculations carried out using the first and second order approximations for the effective mass, respectively, while the dashed line shows the results of the low-density expansion of Ref. [26], including terms of order up to c.

The most striking feature emerging from the figures is the sizable effect of the second order contributions to the effective mass, highlighted in the insets. It clearly appears that these corrections lead to sharp increase of $m^{\star}$, which in turn implies a decrease of the shear viscosity and thermal conductivity coefficients.

4. Nuclear matter at zero and nonzero temperature

The extension of the approach based on the CBF effective interaction to describe the properties of uniform nuclear matter does not involve any conceptual problems, although the complex operatorial structure of the nuclear interaction, reflected by the correlation functions, and the occurrence of strong three-body forces entail non trivial computational difficulties.

IOP Conf. Series: Journal of Physics: Conf. Series **981** (2018) 012008 doi:10.1088/1742-6596/981/1/012008

The procedure developed by the authors of Ref. [27, 28], taking into account the contributions of three-nucleon clusters, allows to use a realistic nuclear Hamiltonian, comprising two- and three-nucleon potentials. The results reported in this paper have been obtained with the nucleon-nucleon potential referred to as Argonne v_6' (AV6P) [29]—embodying the spin-isospin dependence of the nuclear forces, as well as the presence of non-spherically-symmetric interactions—and the Urbana IX (UIX) three-nucleon potential [30].

The resulting effective interaction—derived using as baseline the FHNC variational estimates of the ground state energy of PNM and SNM—can be used to evaluate the energy per nucleon of neutron reach and spin polarized matter at fixed baryon density

$$\rho = \sum_\lambda \rho_\lambda = \rho \sum_\lambda x_\lambda \ , \tag{28}$$

where $\lambda = 1, 2, 3, 4$ labels spin-up protons, spin-down protons, spin-up neutrons and spin-down neutrons, respectively, the corresponding densities being $\rho_\lambda = x_\lambda \rho$. In SNM $x_1 = x_2 = x_3 = x_4 = 1/4$, while in PNM $x_1 = x_2 = 0$ and $x_3 = x_4 = 1/2$.

At first order in the CBF effective interaction, the energy per baryon can be written in the form

$$\frac{E}{A} = \frac{3}{5} \sum_\lambda x_\lambda \frac{{k_{F_\lambda}}^2}{2m} + \frac{\rho}{2} \sum_{\lambda\mu} x_\lambda x_\mu \int d^3r \left[v^{\rm eff,d}_{\lambda\mu}(r) - v^{\rm eff,e}_{\lambda\mu}(r) \ell(k_{F_\lambda} r) \ell(k_{F_\mu} r) \right] \tag{29}$$

where $v^{\rm eff,d}_{\lambda\mu}(r)$ and $v^{\rm eff,e}_{\lambda\mu}(r)$ denote the direct and exchange matrix elements of the effective interaction in spin-isospin space, $k_{F_\lambda} = (6\pi^2 \rho_\lambda)^{1/3}$ is the Fermi momentum of the nucleons of type λ and

$$\rho_\lambda \ell(k_{F_\lambda} r) \equiv \frac{1}{V} \sum_{\mathbf{k}} e^{i\mathbf{k}\cdot\mathbf{r}} n_\lambda(\mathbf{k}) \ , \tag{30}$$

where $n_\lambda(\mathbf{k}) = \theta(k_{F_\lambda} - |\mathbf{k}|)$ is the zero-temperature Fermi distribution and V is the normalization volume.

The solid lines of Figures 8 and 9 illustrate the density dependence of the energy per nucleon of PNM and SNM, respectively, obtained from Eq. (29) using the CBF effective interaction. The shaded regions show the FHNC results obtained using the bare Hamiltonian, with the associated theoretical uncertainty arising from the treatment of the kinetic energy [14]. For comparison, the results of a calculation carried out using the Auxiliary Field Diffusion Monte Carlo (AFDMC) technique are also displayed. It clearly appears that the FHNC variational estimates provide a very accurate upper bound to the ground state energy. Note, however, that while the empirical equilibrium density of SNM is well reproduced, the corresponding energy is underestimated.

The main advantage of the effective interaction approach is the possibility of carrying out calculations of a variety of nuclear matter properties of astrophysical interest using perturbation theory and the basis of eigenstates of the non interacting system. Of great importance, in this context, is the ground state energy of matter at fixed baryon density and arbitrary proton fraction, $x_p = x_1 + x_2$, shown in Fig. 10.

To the extent to which thermal effects do not lead to modifications of the underlying nuclear dynamics, the approach described in this paper can be readily generalized to treat nuclear matter at nonzero temperature, by replacing the $T = 0$ Fermi distribution appearing in Eq. (30) with the corresponding distribution at temperature $T > 0$

$$n_\lambda(k, T) = \{1 + \mathrm{e}^{\beta[e_\lambda(k) - \mu_\lambda]}\}^{-1} \ , \tag{31}$$

IOP Conf. Series: Journal of Physics: Conf. Series **981** (2018) 012008 doi:10.1088/1742-6596/981/1/012008

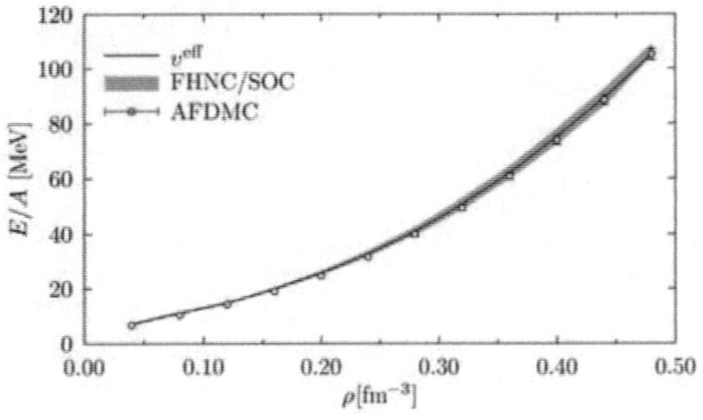

Figure 8. Density dependence of the energy per nucleon of PNM. The solid lines display results obtained using Eqs. (29)-(30) and the CBF effective interaction.The variational FHNC results are represented by the shaded regions, accounting for the uncertainty arising from the treatment of the kinetic energy, while the open circles correspond to the results obtained using the AFDMC technique.

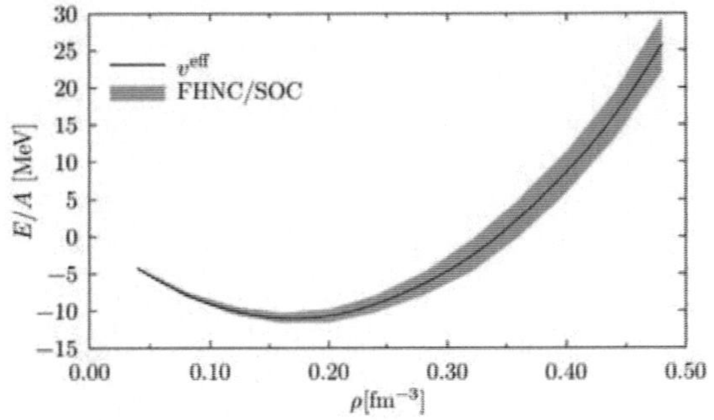

Figure 9. Density dependence of the energy per nucleon of SNM. The solid lines display results obtained using Eqs. (29)-(30) and the CBF effective interaction.The variational FHNC results are represented by the shaded regions, accounting for the uncertainty arising from the treatment of the kinetic energy.

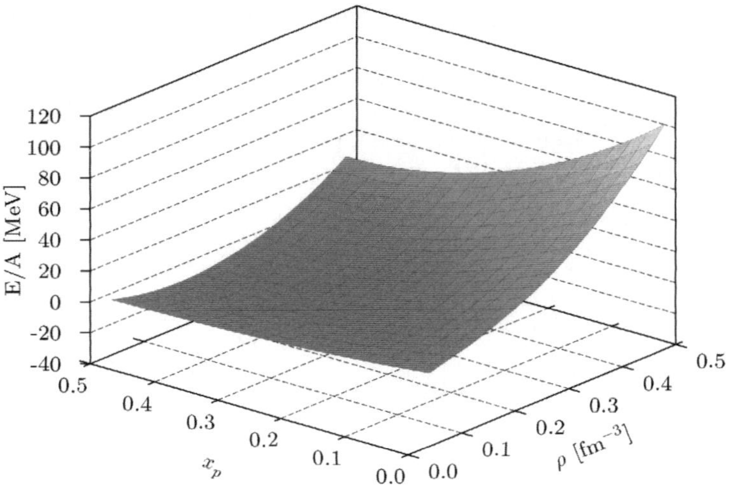

Figure 10. Energy per nucleon of uniform nuclear matter, computed as a function of baryon density, ρ, and proton fraction, $x_p = x_1 + x_2$, using Eqs. (29)-(30) and the CBF effective interaction.

whwre $e_\lambda(k)$ is the energy of a nucleon of type λ carrying momentum k, and the chemical potential μ_λ is determined by the constraint

$$\frac{1}{V}\sum_{\mathbf{k}\lambda} n_\lambda(k,T) = \rho_\lambda \ . \tag{32}$$

As an example, Figs. 11 and 12 show the temperature- and density-dependence of the Gibbs free energy per baryon of PNM and SNM, respectively, defined as

$$\frac{F}{N} = \frac{E_0 - TS}{N} \ , \tag{33}$$

where S denotes the entropy.

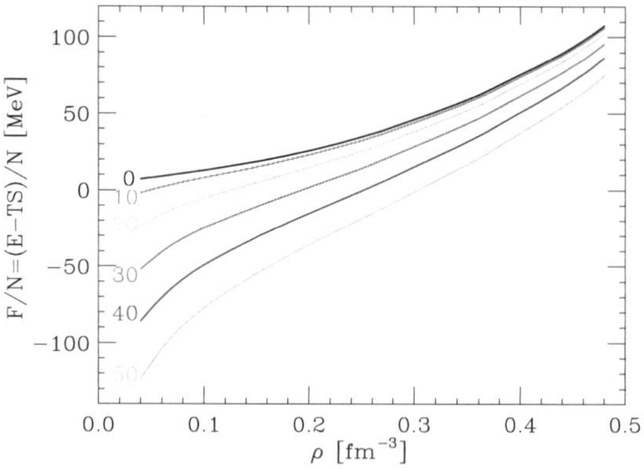

Figure 11. Density dependence of the Gibbs free energy per nucleon of PNM. The lines are labelled according to the temperature, T, expressed in units of MeV.

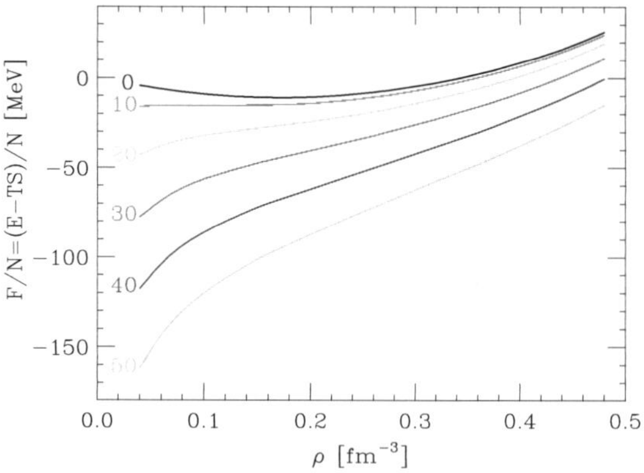

Figure 12. Same as in Fig. 11, but for SNM.

5. Summary and outlook

The results reported in this paper provide convincing evidence that the effective interaction obtained from a microscopic Hamiltonian using the (CBF) formalism and the cluster expansion technique is a powerful tool to carry out consistent calculations of a variety of properties of strongly interacting many-body systems, ranging from the ground-state energy to the quasiparticle spectrum, the in-medium collision probability and the transport coefficients.

The outcome of systematic studies of the Fermi hard-sphere system is quite encouraging, and suggest that the CBF effective interaction approach can be employed to describe a variety of nuclear matter properties, exploiting the flexibility of perturbation theory in the Fermi gas basis.

Future applications to neutron star matter will include the calculations of the shear and bulk viscosity coefficients, the superfluid gaps and the neutrino emission and absorption rates.

Acknowledgements

The results reported in this paper have been obtained in collaboration with Alesandro Lovato, Angela Mecca and Artur Polls.

References

[1] Fetter A L and Walecka J D 1971 *Quantum Theory of Many-Particle Systems* (New York: McGraw-Hill)
[2] Ring P and Schuck P 1988 *The Nuclear Many-Body Problem* (Berlin-Heidelberg: Springer)
[3] Rikovska Stone J, Miller J C, Koncewicz R, Stevenson P D and Strayer M R 2003 *Phys. Rev. C* **68** 034324
[4] Chabanat E, Bonche P, Haensel P, Meyer J and Schaeffer R 1997 *Nucl. Phys. A* **627** 710
[5] Benhar O, Polls A, Valli M and Vidaña I 2010 *Phys. Rev. C* **81** 024305
[6] Zhang H F, Lombardo U and Zuo W 2010 *Phys. Rev. C* **82** 015805
[7] Andersson N, Comer G and Glampedakis K 2005 *Nucl. Phys. A* **763** 212
[8] Shen G and Reddy S 2014 *Phys. Rev. C* **89** 032802
[9] Cowell S and Pandharipande V 2006 *Phys. Rev. C* **73** 025801
[10] Benhar O and Valli M 2007 *Phys. Rev. Lett.* **99** 232501
[11] Mecca A, Lovato A, Benhar O and Polls A 2015 *Phys. Rev. C* **91** 034325
[12] Mecca A, Lovato A, Benhar O and Polls A 2016 *Phys. Rev. C* **93** 035802
[13] Frois B and Papanicolas C 1987 *Annual Review of Nuclear and Particle Science* **37** 133
[14] Clark J W 1979 *Prog. Part. Nucl. Phys.* **2** 89
[15] Fantoni S and Rosati S 1974 *Nuovo Cimento* **20** 179
[16] Fantoni S and Rosati S 1975 *Nuovo Cimento* **25** 593

IOP Conf. Series: Journal of Physics: Conf. Series **981** (2018) 012008 doi:10.1088/1742-6596/981/1/012008

[17] Kalos M H 1971 *Monte Carlo Methods in Quantum Problems* (Netherlands: Springer)
[18] Huang K and Yang C Y 1957 *Phys. Rev.* **105** 767
[19] Bishop R F 1973 *Ann. Phys. (NY)* **77** 106
[20] Galitskii V M 1958 *Sov. Phys. JETP* **7** 104
[21] Jeukenne J P, Lejeune A and Mahaux C 1976 *Phys. Rep.* **83** 104
[22] Fantoni S and Pandharipande V R 1984 *Nucl. Phys. A* **427** 473
[23] Baym G and Pethick C 1991 *Landau Fermi-Liquid Theory* (New York: John Wiley & Sons)
[24] Abrikosov A A and Khalatnikov I L 1957 *Sov. Phys. JETP* **5** 887
[25] Abrikosov A A and Khalatnikov I L 1959 *Rep. Prog. Phys.* **22** 329
[26] Rainwater J C and Mohling F 1976 *J. Low Temp. Phys.* **23** 519
[27] Lovato A, Losa C, and Benhar O 2013 *Nucl. Phys. A* **901** 22
[28] Lovato A, Benhar O, Gandolfi S, and Losa C 2002 *Phys. Rev. C* **89** 182501
[29] Wiringa R B and Pieper S C 2014 *Phys. Rev. Lett.* **89** 025804
[30] Pudliner B S, Pandharipande V R, Carlson J, and Wiringa R B 1995 *Phys. Rev. Lett.* **74** 4396

TNPI2016 - XV Conference on Theoretical Nuclear Physics in Italy IOP Publishing
IOP Conf. Series: Journal of Physics: Conf. Series **981** (2018) 012009 doi:10.1088/1742-6596/981/1/012009

Nuclear matter calculations with chiral interactions

Domenico Logoteta[1], **Ignazio Bombaci**[2,1] **and Alejandro Kievsky**[1]

[1] INFN, Sezione di Pisa, Largo Bruno Pontecorvo 3, I-56127 Pisa, Italy
[2] Dipartimento di Fisica, Universitá di Pisa, Largo Bruno Pontecorvo 3, I-56127 Pisa, Italy

E-mail: domenico.logoteta@pi.infn.it, ignazio.bombaci@unipi.it, alejandro.kievsky@pi.infn.it

Abstract. Using two-nucleon and three-nucleon interactions derived in the framework of chiral perturbation theory (ChPT) with and without the explicit Δ isobar contributions, we calculate the energy per particle of symmetric nuclear matter and pure neutron matter employing the microscopic Brueckner–Hartree–Fock approach. Specifically, we present nuclear matter calculations using the new fully local in coordinate-space two-nucleon interaction at the next-to-next-to-next-to-leading-order (N3LO) of ChPT with Δ isobar intermediate states (N3LOΔ) recently developed by Piarulli *et al.* [1] supplemented with a local N2LO three-nucleon interaction with explicit Δ isobar degrees of freedom. We show that for this combination of two- and three-nucleon interactions it is possible to obtain a good saturation point of symmetric nuclear matter. We also calculate the nuclear symmetry energy and compare our results with the available empirical constraints on this quantity.

1. Introduction

The use of effective field theory (EFT) to describe the nuclear interaction [2] is nowdays one of most powerful tool to describe nucleonic systems (for comprehensive and thorough reviews on this subject see Refs. [3, 4]). The considerable advantage of using such method lies in the fact that two-body, three-body as well as many-body nuclear interactions can be calculated perturbatively, i.e. order by order, according to a well defined scheme based on a low-energy effective QCD Lagrangian which retains the symmetries of QCD, and in particular the approximate chiral symmetry. Within this chiral perturbation theory (ChPT) the details of the QCD dynamics are contained in parameters, the so called low-energy constants (LECs), which are fixed by low-energy experimental data.

Recently Piarulli et al. [1] have developed a fully local in coordinate-space two-nucleon chiral potential which includes the Δ isobar intermediate state. This new potential represents the fully local version of the minimally non-local chiral interaction reported in Ref. [5]. It has been shown by various authors [6, 7] that a Δ-full ChPT has a better convergence with respect to the Δ-less ChPT. In addition, the Δ-full ChPT naturally leads to three-nucleon forces (TNFs) induced by two-pion exchange with excitation of an intermediate Δ (the celebrated Fujita–Miyazawa three-nucleon force [8]).

In this paper, we present calculations of the equation of state (EOS) of symmetric nuclear matter and pure neutron matter using the local chiral potential of Ref. [1] and employing the microscopic Brueckner–Bethe–Goldstone (BBG) [9, 10, 11] many-body theory within the Brueckner–Hartree–Fock (BHF) approximation. The present work contains some of the results published in Ref. [12] and represents a development with respect to our previous works

[13, 14] where ChPT nuclear interactions have been used in BHF calculations of nuclear matter properties.

2. Chiral nuclear interactions

Let us now focus on the specific interactions we have employed in the present work. Among the large variety of nucleon-nucleon (NN) interactions derived in the framework of ChPT, as the two-body nuclear interaction, we have used the fully local chiral potential at N3LO including Δ isobar excitations in intermediate state (hereafter N3LOΔ) recently proposed in Ref. [1]. Originally this potential was presented in Ref. [5] in a minimal non-local form. Notice that Ref. [1] reports different parametrizations of the local potential obtained by fitting the low energy NN experimental data using different long- and short-range cutoffs. In the calculations presented in this work, we use the *model b* described in Ref. [1] (see their Tab. II) which fits the Granada database [15] of proton-proton (*pp*) and neutron-proton (*np*) scattering data up to an energy of 125 MeV in the laboratory reference frame and has a χ^2/datum$\sim$ 1.07.

To compare with other NN interactions, we have also employed the N3LO chiral NN potential by Entem and Machleidt (EM) [16], considering two different values of the cutoff, $\Lambda = 500$ MeV and $\Lambda = 450$ MeV, employed to regularize the high momentum components of the interaction. Notice that for consistency reasons, the same value of the cutoff has been employed in each calculation, both in the two- and three-nucleon interactions. However the considered shape of the cutoff in the two-body and in the three-body interaction is in general different (see Ref. [12] for more details).

Concerning the TNF, we have used the N2LO potential by Epelbaum et al. [17] in its local version given by Navratil [18]. We note that the non locality of the N2LO three-nucleon interaction depends only on the cutoff used to regularize the potential. The N2LO TNF depends on factors c_1, c_3, c_4, c_D and c_E which are the so called low energy constants. The N2LO interaction keeps the same operatorial structure both including or not the Δ degrees of freedom [7]. We note that the constants c_1, c_3 and c_4 are already fixed at the two-body level by the N3LO interaction. However when including the Δ isobar in the three-body potential, the parameters c_3 and c_4 take additional contribution from the Fujita–Miyazawa diagram. Such a diagram appears at the NLO and is clearly not present in the theory without the Δ (see discussion in Ref. [12] about this issue). The values of the constants c_i for the TNFs that we have considered in the present work are reported in Tab. 1.

The remaining parameters c_D and c_E are not determined by the two-body data and have to be fixed constraining some specific observables in few-nucleon systems or to reproduce the empirical saturation point of symmetric nuclear matter. In particular, for the interaction model N3LO+N2LO(450) we have set $c_D = -0.24$ and $c_E = -0.11$ in order to reproduce the binding energies of ^{3}H and ^{3}He and the Gammow-Teller matrix element for the ^{3}H β-decay considering contributions to the axial nuclear current up to order N3LO [19]. For the interaction model N3LO+N2LO(500), we have adopted a recent constraint on c_D and c_E employing the same strategy of Ref. [19] but considering contributions to the axial nuclear current up to order N4LO [20]. We note that this parametrization has also the valuable property to reproduce the neutron–deuteron doublet scattering length.

Finally, for the very recent [1] interaction model N3LOΔ+N2LOΔ no calculation for few-body nuclear systems has been done so far. Thus we have fitted the LECs c_D and c_E to get a good saturation point for symmetric nuclear matter.

3. The Brueckner–Hartree–Fock approach with averaged three-body forces

The BHF approach is the lowest order of the BBG many-body theory [9, 10, 11]. In this theory, the ground state energy of nuclear matter is evaluated in terms of the so-called hole-line expansion, where the various terms can be represented by Goldstone diagrams grouped according

Table 1. Values of the low energy constants of the TNFs models used in the present calculations. In the first row, we report the parametrizations of the N2LO three-body force with the Δ isobar excitations [1]. Notice that the values c_1, c_3 and c_4 have been kept fixed. In the third and in the forth rows we report the N2LO TNF parametrizations obtained in conjunction with the EM [16] N3LO two-nucleon potential with $\Lambda = 500$ MeV (third row) and with $\Lambda = 450$ MeV (forth row). The LECs c_1, c_3 and c_4 are expressed in GeV^{-1}, whereas c_D and c_E are dimensionless.

TNF	c_D	c_E	c_1	c_3	c_4
N2LOΔ	-0.10	1.30	-0.057	-3.63	3.14
N2LO500	-1.88	-0.48	-0.810	-3.20	5.40
N2LO450	-0.11	-0.24	-0.810	-3.40	3.40

to the number of independent hole-lines (*i.e.* lines representing empty single particle states in the Fermi sea). The expansion is derived by means of the in-medium two-body scattering Brueckner G-matrix which describes the effective interaction between two nucleons in presence of the surrounding nuclear medium.

In the case of asymmetric nuclear matter with neutron density ρ_n, proton density ρ_p, total nucleon density $\rho = \rho_n + \rho_p$ and isospin asymmetry $\beta = (\rho_n - \rho_p)/\rho$ (asymmetry parameter), one has different G-matrices describing the nn, pp and np in medium effective interactions. They are obtained by solving the well known Bethe–Goldstone equation [9]. In the present work we consider spin unpolarized nuclear matter. Spin polarized nuclear matter within the BHF approach has been considered, for example, in Ref. [21, 22].

We make use of the so-called continuous choice [23] for the single-particle potential $U_\tau(k)$ when solving the Bethe–Goldstone equation. As shown in Refs. [24, 25], the contribution of the three-hole-line diagrams to the energy per particle E/A is minimized in this prescription and a faster convergence of the hole-line expansion for E/A is achieved [24, 25, 26] with respect to the so-called gap choice for $U_\tau(k)$.

As it is well known, within the most advanced non-relativistic quantum many-body approaches [27], it is not possible to reproduce the empirical saturation point of symmetric nuclear matter, $\rho_0 = 0.16 \pm 0.01$ fm^{-3}, $E/A|_{\rho_0} = -16.0 \pm 1.0$ MeV, when using two-body nuclear interactions only. In addition, TNFs are crucial in the case of dense β-stable nuclear matter to obtain a stiff EOS [28, 29] compatible with the measured masses, $M = 1.97 \pm 0.04\, M_\odot$ [30] and $M = 2.01 \pm 0.04\, M_\odot$ [31] of the neutron stars in PSR J1614-2230 and PSR J0348+0432 respectively.

Within the BHF approach TNFs cannot be used directly in their original form. This is because it would be necessary to solve the three-body Faddeev equations in the nuclear medium (Bethe–Faddeev equations) [32, 33] and currently this is a task still far to be achieved. To circumvent this problem an effective density dependent two-body force is built starting from the original three-body one by averaging over one of the three nucleons [34, 35].

In the present work, we consider the in medium effective NN force derived in Ref. [36] (see Ref. [12] for more details on the average).

4. Results and discussion

In this section we present and discuss the results of our calculations for the EOS, i.e. the energy per particle E/A as a function of the density ρ, for symmetric nuclear matter (SNM) and pure neutron matter (PNM) using the chiral nuclear interaction models and the BHF approach described in the previous two sections. In all the calculations performed in this work, we have considered partial wave contributions up to a total two-body angular momentum $J_{max} = 8$.

IOP Conf. Series: Journal of Physics: Conf. Series **981** (2018) 012009 doi:10.1088/1742-6596/981/1/012009

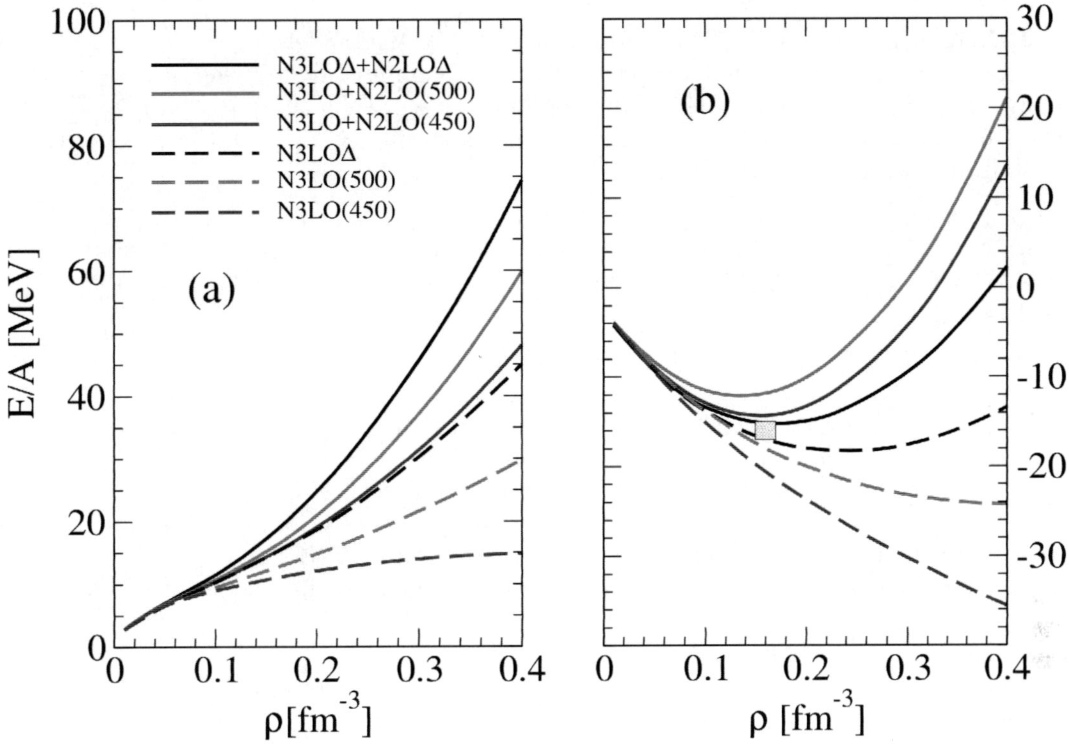

Figure 1. (Color online) Energy per particle of pure neutron matter [panel (a)] and symmetric nuclear matter [panel (b)] as a function of the nucleonic density for the models described in the text. Continuous lines have been obtained using two- plus three-body interactions, while the dashed lines have been obtained considering only the two-body interaction. The empirical saturation point of nuclear matter $\rho_0 = 0.16 \pm 0.01$ fm^{-3}, $E/A|_{\rho_0} = -16.0 \pm 1.0$ MeV is denoted by the grey box in the panel (b).

In Fig. 1 we show the energy per particle of PNM [panel (a)] and SNM [panel (b)] for the considered interaction models. The dashed lines, in both panels, refer to the calculations performed employing the two-body potential without any TNF, whereas the continuous lines refer to the calculations where the contribution of the TNFs to the energy per nucleon has been included.

Focusing first on the case of PNM (Fig. 1(a)), we note sizable differences between the three energy per nucleon curves produced by the different NN interactions. The model N3LOΔ (black upper dashed line) gives indeed a much stiffer EOS than the N3LO ones for both cutoff values, $\Lambda = 500$ MeV (red middle dashed line) and $\Lambda = 450$ MeV (blue lower dashed line). This behaviour is both due to the local form of the potential and to the inclusion of Δ isobar.

In addition, looking at Tab. 1, we see that the values of c_1 and c_3 are very similar for the considered models. Thus we expect a comparable effect of TNFs on the EOS for PNM as confirmed by our results.

The EOS for symmetric nuclear matter is shown in Fig. 1(b). When only two-body interactions are included, the models based on the EM N3LO potential [16] give unsatisfactory nuclear matter saturation properties. More specifically the model N3LO(500) (red middle dashed line) gives a saturation point ($\rho_0 = 0.41$ fm^{-3}, $E/A|_0 = -24.25$ MeV), whereas the EOS curve for the model N3LO(450) (blue lower dashed line) shows no saturation point up to density of

Table 2. Nuclear matter properties at saturation point for the models described in the text. In the first column we report the model name; in the other columns we give the saturation density (ρ_0) of symmetric nuclear matter, the corresponding value of the energy per particle E/A, the symmetry energy, its slope parameter L and the incompressibility K_∞. All these values refer to the calculated saturation density.

Model	ρ_0(fm^{-3})	E/A (MeV)	E_{sym} (MeV)	L (MeV)	K_∞ (MeV)
N3LOΔ+N2LOΔ	0.171	-15.23	35.39	76.0	190
N3LO+N2LO(500)	0.135	-12.12	25.89	38.3	153
N3LO+N2LO(450)	0.156	-14.32	29.20	39.8	205

~ 0.5 fm^{-3}. The EOS for the N3LOΔ NN interaction [1] (black upper dashed line) has instead a very different trend. In this case the saturation point turns out to be (0.24 fm^{-3}, -18.27 MeV).

The overall repulsive effect introduced by the inclusion of TNFs produces a significant improvement of the calculated SNM saturation point (see the continuous lines in Fig. 1(b)) with respect to the results described above for the case with no TNFs.

These results clearly show that in the case of Δ-full chiral nuclear interactions the contribution to the energy per particle generated by the TNFs is strongly reduced in comparison to the case where the EOS is obtained from Δ-less chiral interactions.

In Tab. 2 we report the calculated values of the saturation points of SNM for the interaction models considered in the present work. All the models, with the exception of the N3LO+N2LO(500) one, provide satisfacory saturation points.

The energy per nucleon of asymmetric nuclear matter can be accurately reproduced [37, 38, 39] using the so called parabolic (in the asymmetry parameter β) approximation

$$\frac{E}{A}(\rho,\beta) = \frac{E}{A}(\rho,0) + E_{sym}(\rho)\beta^2 \tag{1}$$

where $E_{sym}(\rho)$ is the nuclear symmetry energy [40]. Using Eq. (1), the symmetry energy can be calculated as the difference between the energy per particle of pure neutron matter ($\beta = 1$) and symmetric nuclear matter ($\beta = 0$).

The symmetry energy, calculated within this prescription, is plotted as function of the nucleon density ρ in Fig. 2. The symmetry energy calculated with the new local chiral potential of Ref. [1] (continuous line in Fig. 2) is systematically above the other curves and has a larger slope with respect to the one calculated with the N3LO+N2LO interaction model. In the same figure, we show E_{sym} (triangles) as obtained from recent calculations [41] of asymmetric neutron-rich matter with two- and three-body interactions determined respectively at N3LO and N2LO of the chiral perturbation theory. The results of Ref. [41] confirmes the validity of the quadratic approximation (Eq.(1)) for describing the EOS highly asymmetric matter. The two bands in Fig. 2 represent the constraints on the symmetry energy obtained by Danielewicz and Lee [42] using the excitation energies to isobaric analog states (IAS) in nuclei (black-dashed band labeled IAS) and with the additional constraints from neutron skin thickness Δr_{np} of heavy nuclei [43, 44] (red-dashed band labeled IAS+Δr_{np}).

The symmetry energy obtained in the present work for the N3LOΔ+N2LOΔ interaction model is in a very good agreement with the experimental constraints [42] reported in Fig. 2, whereas E_{sym} as calculated with the N3LO+N2LO model (with both $\Lambda = 500$ MeV and 450 MeV) lies slightly below the IAS+Δr_{np} red-dashed region.

In Tab. 2 we report the symmetry energy and the so called slope parameter

$$L = 3\rho_0 \frac{\partial E_{sym}(\rho)}{\partial \rho}\bigg|_{\rho_0} \tag{2}$$

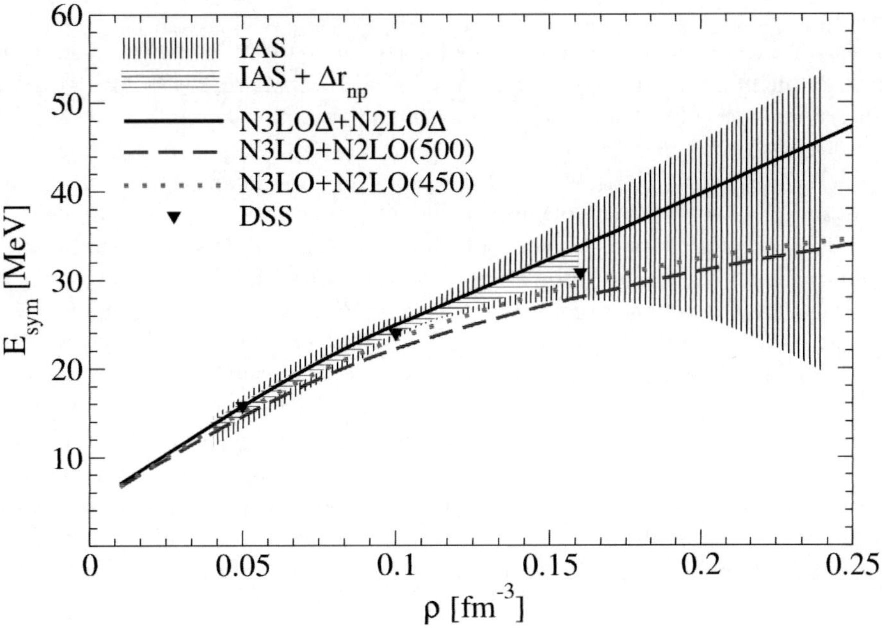

Figure 2. (Color on line) Nuclear symmetry energy as a function of the nucleonic density for the three interaction models used in the present work. The triangles labeled DSS represent the results of Ref. [41]. The black-dashed band, labeled IAS, represents the constraints on the symmetry energy obtained in Ref. [42] using the excitation energies of isobaric analog states (IAS) in nuclei. The additional constraints from neutron skin thickness Δr_{np} of heavy nuclei give the more limited region covered by the red-dashed band labeled IAS+Δr_{np} [41].

at the calculated saturation density ρ_0 (second column in Tab. 2) for the interaction models considered in the present work. As we can see our calculated $E_{sym}(\rho_0)$ and L are in a satisfactory agreement with the values obtained by other BHF calculations with two- and three-body interactions (see e.g. [45, 46]) and with the values extracted from various experimental data, $E_{sym}(\rho_0) = 29.0 - 32.7$ MeV, and $L = 40.5 - 61.9$ MeV, as summarized in Ref. [47].

The incompressibility K_∞ of symmetric nuclear matter at saturation density is given by:

$$K_\infty = 9\rho_0^2 \frac{\partial^2 E/A}{\partial \rho^2}\Big|_{\rho_0} . \tag{3}$$

The incompressibility K_∞ is usually extracted from experimental data of giant monopole resonance (GMR) energies in medium-mass and heavy nuclei. This analysis gives $K_\infty = 210 \pm 30$ MeV [48] or more recently $K_\infty = 240 \pm 20$ MeV [49]. Recently the authors of Ref. [50] performed a re-analysis of GMR data finding 250 MeV$< K_\infty <$ 315 MeV. The incompressibility K_∞, at the calculated saturation point for the various interaction models used in the present work, is reported in the last column of Tab. 2. These calculated values for K_∞ are rather low when compared with the empirical values extracted from GMR in nuclei. This is a common feature with many other BHF nuclear matter calculations with two- and three-body interactions (see e.g. [46].

5. Summary

The EOS of nuclear matter is a basic ingredient for the description of a large variety of physical and astrophysical phenomena [11, 51, 52, 53, 54, 55]. In this work we have investigated the EOS of symmetric nuclear matter and pure neutron matter using three interaction models fully derived in chiral effective field theory and using the microscopic Brueckner–Hartree–Fock many-body approach. In particular, we have tested the new fully local chiral potential at order N3LO which includes the Δ isobar contributions in the intermediate states of the NN interaction [1]. We have also considered two versions of the N3LO chiral NN potential by Entem and Machleidt [16], which differ in the value of the cutoff employed in the calculations. All the two-nucleon interactions have been supplemented with TNFs required to satisfactorily reproduce the empirical saturation point of symmetric nuclear matter. Our results for various nuclear matter properties at saturation density are in good agreement with the available experimental data except for the incompressibility K_∞ which is underestimated with respect to the highly uncertain empirical value [48, 49, 50]. A remarkable agreement has been found between our calculated symmetry energy and the recent experimental constraints for the low density behavior of this quantity obtained [42] using the excitation energies to isobaric analog states in nuclei with the additional constraints from neutron skin thickness of heavy nuclei [43, 44]. In addition, we have found that the inclusion of the Δ isobar in the NN potential diminishes the strength of the TNF needed to get a good saturation point of symmetric nuclear matter. In conclusion, the chiral nuclear interaction models considered in this work provide solid basis both for the physics light nuclei and low density nuclear matter.

Acknowledgments

This work has been partially supported by "NewCompstar", COST Action MP1304.

References

[1] Piarulli M, Girlanda L, Schiavilla R, Kievsky A, Lovato A, Marcucci L E, Pieper S C, Viviani M and Wiringa R B 2016 *Phys. Rev. C* **94** 054007
[2] Weinberg S 1979 *Physica A* **96** 327; 1990 *Phys. Lett. B* **251** 288; 1991 *Nucl. Phys. B* **363** 3; 1992 *Phys. Lett. B* **259** 114
[3] Epelbaum E, Hammer H-W and Meißner U G 2009 *Rev. Mod. Phys.* **81** 1773
[4] Machleidt R and Entem D R 2011 *Phys. Rep.* **503** 1
[5] Piarulli M, Girlanda L, Schiavilla R, Navarro Perez R, Amaro J E and Ruiz Arriola E 2015 *Phys. Rev. C* **91** 024003
[6] Kaiser N, Gerstendörfer S and Weise W 1998 *Nucl. Phys. A* **637** 395
[7] Krebs H, Epelbaum E and Meißner U G 2007 *Eur. Phys. J. A* **32** 127
[8] Fujita J and Miyazawa H 1957 *Prog. Theor, Phys.* **17** 360
[9] Day B D 1967 *Rev. Mod. Phys.* **39** 719
[10] Baldo M, Bombaci I, Giansiracusa G, Lombardo U, Mahaux C and Sartor R 1990 *Phys. Rev. C* **41** 1748
[11] Baldo M and Burgio G F 2012 *Rep. Progr. Phys.* **75** 026301
[12] Logoteta D, Bombaci I and Kievsky A 2016 *Phys. Rev. C* **94** 064001
[13] Logoteta D, Vidaña I, Bombaci I and Kievsky A 2015 *Phys. Rev. C* **91** 064001
[14] Logoteta D, Bombaci I and Kievsky A 2016 *Phys. Lett. B* **758** 449
[15] Navarro Pérez R, Amaro J E and Ruiz Arriola E 2013 *Phys. Rev. C* **88** 064002; Erratum 2015 *Phys. Rev. C* **91** 029901
[16] Entem D R and Machleidt R 2003 *Phys. Rev. C* **68** 041001(R)
[17] Epelbaum E, Nogga A, Glöckle W, Kamada H and Meißner U G and Witała H 2002 *Phys. Rev. C* **66** 064001
[18] Navratil P 2007 *Few-Body Syst.* **41** 117
[19] Marcucci L E, Kievsky A, Rosati S, Schiavilla R and Viviani M 2012 *Phys. Rev. Lett.* **108** 052502
[20] Baroni A, Girlanda L, Kievsky A, Marcucci L E and Viviani M 2016 *Phys. Rev. C* **94** 024003.
[21] Vidaña I and Bombaci I 2002 *Phys. Rev. C* **66** 045801
[22] Bombaci I, Polls A, Ramos A, Rios A and Vidaña I 2006 *Phys. Lett. B* **632** 638
[23] Jeukenne J P, Lejeunne A and Mahaux C 1976 *Phys. Rep.* **25** 83
[24] Song H Q, Baldo M, Giansiracusa G and Lombardo U 1998 *Phys. Rev. Lett.* **81** 1584

[25] Baldo M, Giansiracusa G, Lombardo U and Song H Q 2000 *Phys. Lett. B* **473** 1
[26] Baldo M, Bombaci I, Giansiracusa G and Lombardo U 1990 *J. Phys. G: Nucl. Part. Phys.* **16** L263
[27] Bombaci I, Fabrocini A, Polls A and Vidaña I 2005 *Phys. Lett. B* **609** 232
[28] Baldo M, Bombaci I and Burgio G F 1997 *Astron. and Astrophys.* **328** 274
[29] Akmal A, Pandharipande V R and Ravenhall D G 1998 *Phys. Rev. C* **58** 1804
[30] Demorest P, Pennucci T, Ransom S, Roberts M and Hessels J 2010 *Nature* **467** 1081
[31] Antoniadis J et al. 2013 *Science* **340** 1233232
[32] Bethe H A 1965 *Phys. Rev.* **138** 804B
[33] Rajaraman R and Bethe H A 1967 *Rev. Mod. Phys.* **39** 745
[34] Loiseau B A, Nogami Y and Ross C K 1971 *Nucl. Phys. A* **401** 601
[35] Grangé P, Lejeunne A, Martzolff B and Mathiot J-F 1989 *Phys. Rev. C* **40** 1040
[36] Holt J W, Kaiser N and Weise W 2010 *Phys. Rev. C* **81** 024002
[37] Bombaci I and Lombardo U 1991 *Phys. Rev. C* **44** 1892
[38] Zuo W, Bombaci I and Lombardo U 1999 *Phys. Rev. C* **60** 024605
[39] Zuo W, Bombaci I and Lombardo U 2014 *Eur. Phys. J. A* **50** 12
[40] *Eur. Phys. J. A* **50** (2) 2014, Topical issue on Nuclear Symmetry Energy, edited by Li B A, Ramos A, Verde G and Vidaña I
[41] Drischler C, Somà A and Schwenk A 2014 *Phys. Rev. C* **89** 025806
[42] Danielewicz P and Lee J 2014 *Nucl. Phys. A* **922** 1
[43] Roca-Maza X *et al.* 2013 *Phys. Rev. C* **87** 034301
[44] Zhang Z and Chen L-W 2013 *Phys. Lett. B* **726** 234
[45] Li Z H, Lombardo U, Schulze H-J, Zuo W, Chen L W and Ma H R 2006 *Phys. Rev. C* **74** 047304
[46] Li Z H and Schulze H-J 2008 *Phys. Rev. C* **78** 028801
[47] Lattimer J M 2014 *Gen. Rel. Grav.* **46** 1713
[48] Blaizot J P, Gogny D and Grammaticos B 1976 *Nucl. Phys. A* **265** 315
[49] Sholmo S, Kolomietz V K and Colò G 2006 *Eur. Phys. J. A* **30** 23
[50] Stone J R, Stone N J and Moszkowski S A 2014 *Phys. Rev. C* **89** 044316
[51] Oertel M, Hempel M, Klähan T and Typel S 2017, *Rev. Mod. Phys.* **89** 015007
[52] Logoteta D, Providencia C, Vidaña I and Bombaci I 2012 *Phys. Rev. C* **85** 055807
[53] Logoteta D, Bombaci I, Providencia C and Vidaña I 2012 *Phys. Rev. D* **85** 023003
[54] Logoteta D and Bombaci I 2013 *Phys. Rev. D* **88** 063001
[55] Bombaci I and Logoteta D 2013 *Monthly Notices of the Royal Astronomical Society* **433** L79

IOP Conf. Series: Journal of Physics: Conf. Series **981** (2018) 012010 doi:10.1088/1742-6596/981/1/012010

Meson properties in asymmetric matter

Andrea Mammarella, Massimo Mannarelli
Laboratori Nazionali del Gran Sasso, via G. Acitelli 22, Assergi (AQ), Italy
E-mail: andrea.mammarella@lngs.infn.it

Abstract. In this work we study dynamic and thermodynamic (at $T = 0$) properties of mesons in asymmetric matter in the framework of Chiral Perturbation Theory. We consider a system at vanishing temperature with nonzero isospin chemical potential and strangeness chemical potential; meson masses and mixing in the normal phase, the pion condensation phase and the kaon condensation phase are described. We find differences with previous works, but the results presented here are supported by both theory group analysis and by direct calculations. Some pion decay channels in the normal and the pion condensation phases are studied, finding a nonmonotonic behavior of the decay width as a function of μ_I. Furthermore, pressure, density and equation of state of the system at $T = 0$ are studied, finding remarkable agreement with analogue studies performed by lattice calculations.

1. Introduction

The properties of strongly interacting matter in an isospin and/or strangeness rich medium are relevant in a wide range of phenomena including the astrophysics of compact stars and heavy-ion collisions. It is known that depending on the value of the isospin chemical potential, μ_I, and on the value of the strangeness chemical potential, μ_S, three different phases can be realized: the normal phase, the pion condensed (πc) phase and the kaon condensed (Kc) phase [1, 2, 3]. The realization of a mesonic condensate can drastically change the low energy properties of matter, including the mass spectrum and the lifetime of mesons.

Previous analysis of the meson condensed phases by QCD-like theories were developed in [4, 5]. Pion condensation in two-flavor quark matter was studied in [2, 6] and in three-flavor quark matter in [3]. In particular, the phase diagram as a function of μ_I and μ_S was presented in [3]. Finite temperature effects in $SU(2)_L \times SU(2)_R$ chiral perturbation theory (χPT) have been studied in [7, 8, 9]. One remarkable property of quark matter with nonvanishing isospin chemical potential is that it is characterized by a real measure, thus the lattice realization can be performed with standard numerical algorithms [10, 11]. The πc phase and the Kc phase have been studied by NJL models in [12, 13, 14] and by random matrix models in [15]. All these models find results in qualitative and quantitative agreement, and in particular, the phase diagram of matter has been firmly established. However, regarding the low energy mass spectrum in three-flavor quark matter, we found that it was only studied in [3]. Our results are in disagreement with those of [3], the most relevant difference is in the mixing between mesonic states. Regarding the pion decay, previous works focused on density and temperature effects in standard decay channels [16, 17, 18], but not all the decay channels have been considered.
The thermodynamic properties of the meson condensed phases have been studied by LQCD

simulations in [19, 20]. Previously, various results on the πc phase were derived in [9] by an NJL model. In particular, the equation of state (EoS) of the NJL model was presented. Recently in [21, 22, 23] a perturbative analysis of QCD at large isospin density has been presented. Those pQCD results are consistent with LQCD for $\mu_I \gtrsim 3m_\pi$ [23], where m_π is the pion mass. At smaller values of μ_I it seems that pQCD underestimates the energy density and is not able to capture the condensation mechanism. However, for small values of μ_I (and μ_S), χPT can be used. Although the χPT isospin density of the system has been determined several years ago [2], to the best of our knowledge a careful study of the EoS of imbalanced matter has never been derived within this framework.

In this article we briefly review how to include chemical potentials in χPT [24, 25, 26, 27, 28], then we describe some phenomenological properties related to this inclusion, like the existence of the different phases already listed, the meson masses and mixing. We will also show how the inclusion of chemical potentials affects pion decay channels [29]. Finally we will study the effect of nonzero isospin and strangeness chemical potentials on the pressure, density and state equation of the system [30].

The paper is organized as follows. In Sec. 2 I describe the model and how it predicts different phases. In Sec. 3 we show how group theory tools can be used to calculate the mass eigenstates and the meson mixing in the condensed phases and we list and discuss the results obtained. In Sec.4 we discuss the impact of chemical potentials on charged pion decays. In Sec. 5 we study the termodynamic properties of the system. Finally, in Sec. 6, we summarize the results.

2. Model

2.1. Lagrangian and definitions

In this section we briefly review the model that we are going to use in the following. It is the one described in [3]. The general $\mathcal{O}(p^2)$ Lorentz invariant Lagrangian density describing the pseudoscalar mesons can be written as

$$\mathcal{L} = \frac{F_0^2}{4}\mathrm{Tr}(D_\nu \Sigma D^\nu \Sigma^\dagger) + \frac{F_0^2}{4}\mathrm{Tr}(X\Sigma^\dagger + \Sigma X^\dagger)\,, \tag{1}$$

where Σ corresponds to the meson fields, $X = 2B_0(s + ip)$ describes scalar and pseudoscalar external fields and the covariant derivative is defined as

$$D_\mu \Sigma = \partial_\mu \Sigma - \frac{i}{2}[v_\mu, \Sigma] - \frac{i}{2}\{a_\mu, \Sigma\}\,, \tag{2}$$

with v_μ and a_μ the external vectorial and axial currents, respectively. The Lagrangian has two free parameters F_0 and B_0, related to the pion decay and to the quark-antiquark condensate, respectively, see for example [24, 25, 26, 27, 28].

The Lagrangian density is invariant under $SU(N_f)_L \times SU(N_f)_R$ provided the meson field transforms as

$$\Sigma \to R\Sigma L^\dagger\,, \tag{3}$$

and the chiral symmetry breaking corresponds to the spontaneous global symmetry breaking $SU(N_f)_L \times SU(N_f)_R \to SU(N_f)_{L+R}$. In standard χPT, the mass eigenstates are charge eigenstates as well. Thus mesons are particles with a well defined mass and charge. The presence of a medium can change this picture. In particular, if the vacuum carries an electric charge, then the mass eigenstates will not typically be charge eigenstates. The effects of a medium can be taken into account by considering appropriate external currents in Eq. (1).

At vanishing temperature the ground state is determined by maximizing the Lagrangian density with respect to the external currents. The pseudoscalar mesons are then described as oscillations around the vacuum. We use the same nonlinear representation of [3] corresponding to

$$\Sigma = u\bar{\Sigma}u \qquad \text{with} \qquad u = e^{iT\cdot\phi/2}\,, \tag{4}$$

where T_a are the $SU(N_f)$ generators and $\bar{\Sigma}$ is a generic $SU(N_f)$ matrix to be determined by maximizing the static Lagrangian. The reasoning behind the above expression is that under $SU(N_f)_L \times SU(N_f)_R$ mesons can be identified as the fluctuations of the vacuum as in Eq. (3) with $\theta_a^R = -\theta_a^L = \phi_a$.

In the following we will assume that $a_\mu = 0$, $p = 0$, $X = 2GM$, where M is the $N_f \times N_f$ diagonal quark mass matrix and G is a constant, that with these conventions is equal to B_0. Moreover, we will assume that

$$v^\nu = 2\mu\delta^{\nu 0} \tag{5}$$

meaning that the vectorial current consists of the quark chemical potential, with μ a $SU(N_f)$ matrix in flavor space. Its explicit expression is:

$$\mu = \text{diag}\,(\mu_u, \mu_d, \mu_s) = \text{diag}\left(\frac{1}{3}\mu_B + \frac{1}{2}\mu_I, \frac{1}{3}\mu_B - \frac{1}{2}\mu_I, \frac{1}{3}\mu_B - \mu_S\right) = \frac{\mu_B - \mu_S}{3}I + \frac{\mu_I}{2}\lambda_3 + \frac{\mu_S}{\sqrt{3}}\lambda_8\,, \tag{6}$$

It is important to remark that this model only holds for $|\mu_B| \lesssim 940$ MeV, $|\mu_I| \lesssim 770$ MeV and $|\mu_S| \lesssim 550$ MeV [3].

2.2. Ground state and different phases

To find the ground state we have to substitute (4) in the Lagrangian (1). It is not necessary to use a complete SU(3) parametrization for $\bar{\Sigma}$, but is sufficient:

$$\bar{\Sigma} = \begin{pmatrix} 1 & 0 & 0 \\ 0 & \cos\beta & -\sin\beta \\ 0 & \sin\beta & \cos\beta \end{pmatrix} \begin{pmatrix} \cos\alpha & \sin\alpha & 0 \\ -\sin\alpha & \cos\alpha & 0 \\ 0 & 0 & 1 \end{pmatrix} \begin{pmatrix} 1 & 0 & 0 \\ 0 & \cos\beta & \sin\beta \\ 0 & -\sin\beta & \cos\beta \end{pmatrix}, \tag{7}$$

because $\bar{\Sigma}$ has to be orthogonal to the chemical potential in the SU(3) generator space [29]. Substituting (7) in (1) and maximizing, we find three different vacua, that implies that there are three ground states and therefore three different phases [3]:

- Normal phase:

$$\mu_I < m_\pi \qquad\qquad \mu_S < m_K - \frac{1}{2}\mu_I\,, \tag{8}$$

  characterized by

$$\alpha_N = 0, \quad \beta_N \in (0, \pi), \quad \bar{\Sigma}_N = \text{diag}(1, 1, 1)\,. \tag{9}$$

- Pion condensation phase:

$$\mu_I > m_\pi \qquad\qquad \mu_S < \frac{-m_\pi^2 + \sqrt{(m_\pi^2 - \mu_I^2)^2 + 4m_K^2\mu_I^2}}{2\mu_I}\,, \tag{10}$$

  characterized by $\cos\alpha_\pi = \left(\frac{m_\pi}{\mu_I}\right)^2$, $\beta_\pi = 0$,

$$\bar{\Sigma}_\pi = \begin{pmatrix} \cos\alpha_\pi & \sin\alpha_\pi & 0 \\ -\sin\alpha_\pi & \cos\alpha_\pi & 0 \\ 0 & 0 & 1 \end{pmatrix} = \frac{1 + 2\cos\alpha_\pi}{3}I + i\lambda_2\sin\alpha_\pi + \frac{\cos\alpha_\pi - 1}{\sqrt{3}}\lambda_8\,.$$

- Kaon condensation phase:

$$\mu_S > m_K - \frac{1}{2}\mu_I \qquad \mu_S > \frac{-m_\pi^2 + \sqrt{(m_\pi^2 - \mu_I^2)^2 + 4m_K^2\mu_I^2}}{2\mu_I}, \tag{11}$$

characterized by $\cos\alpha_K = \left(\frac{m_K}{\frac{1}{2}\mu_I + \mu_S}\right)^2$, $\beta_K = \pi/2$,

$$\bar{\Sigma}_K = \begin{pmatrix} \cos\alpha & 0 & \sin\alpha \\ 0 & 1 & 0 \\ -\sin\alpha & 0 & \cos\alpha \end{pmatrix} = \frac{1 + 2\cos\alpha_K}{3} I + \frac{\cos\alpha_K - 1}{2\sqrt{3}}\left(\sqrt{3}\lambda_3 - \lambda_8\right) + i\lambda_5 \sin\alpha_K .$$

Note that the kaon condensation can only happen for

$$\mu_S > \bar{\mu}_S = m_K - \frac{m_\pi}{2} . \tag{12}$$

3. Meson Masses and Mixing

3.1. *Mixing*

As we have seen in Sec. 2.2 the ground state in the condensed phases is not diagonal and thus has SU(3) charges that cause symmetry breaking. It is useful to study the breaking pattern to learn something about the possible meson mixing in the condensed phases.

The starting Lagrangian has an $SU(3)_L \times SU(3)_R$ simmetry, broken to $SU(3)_V$ by the quark masses. The introduction of chemical potentials via the external current (5) further reduces this symmetry to $U(1)_{L+R} \times U(1)_{L+R}$, meaning that the λ_3 and λ_8 terms in (5) break isospin and hypercharge conservation. When the system enters one of the condensed phases the vacuum acquires a charge and thus there is no symmetry left.

In all of these cases we can use the quantum numbers of the $SU(2)$ subgroups of $SU(3)$ to label the states and to find out the ones that can mix. We need only two of them, because they are not independent. The states that can mix and the related quantum numbers are shown in table 1

Mixing states	(T, U)
π_+, π_-	$(1, 1/2)$
K_+, K_-	$(1/2, 1/2)$
$K_0, \bar{K}_0$	$(1/2, 1)$

Table 1. Mixing mesons with the corresponding T-spin and U-spin quantum numbers. These quantum numbers label the $SU(3)$ subspace spanned by the corresponding mesonic states. The π_0 and the η do not appear because they are not U-spin eigenstates.

Unfortunately, this is not sufficient to determine if π_0 and η can mix, because they do not have a well defined U and V spin, so we have to study deeply how the ground state affects them. Let us first consider the normal phase. In the normal phase there is no operator that can induce the mixing of the mesonic states, thus the mesonic states remain unchanged but the Q_3 and Q_8 charges will induce Zeeman-like mass splittings.

In any of the condensed phases, there is an additional charge that is spontaneously induced, and the corresponding operator will lead to mixing.

Let us first focus on isospin (or T-spin). We have to consider two cases. Suppose that the vacuum has a charge that commutes with T^2, as in the πc phase, say the charge corresponding to $T_2 = i(T_- - T_+)$, see Eq. (4). The $T_\pm$ operators can induce mixing among the charged pions and among the kaons. On the other hand, T-spin conservation does not allow the $|\pi_0\rangle = |T = 1, T_3 = 0\rangle$ to mix with the $|\eta\rangle = |T = 0, T_3 = 0\rangle$.

Now suppose instead that the vacuum has a charge that does not commute with T^2 as in the Kc phase. Any operator that does not commute with isospin will commute with U-spin or with V-spin. In the Kc phase $Q_5|0\rangle \neq 0$, then the vacuum is not invariant under this charge. However, since $[T_5, U] = 0$ it follows that U-spin is conserved. The lowering and raising operators inducing the mixing will be $U_\pm$. Regarding the π_0 and the η, in this case we have that $|U = 1, U_3 = 0\rangle$ and $|U = 0, U_3 = 0\rangle$ do not mix. Since $|U = 1, U_3 = 0\rangle = \frac{|\pi_0\rangle + \sqrt{3}|\eta\rangle}{2}$ and $|U = 0, U_3 = 0\rangle = \frac{\sqrt{3}|\pi_0\rangle - |\eta\rangle}{2}$, these will be the mass eigenstates.

3.2. Masses

The mass eigenstates are found diagonalizing the Lagrangian in the different phases. They present the mixing predicted by the group theory analysis of the previous subsection. They have been calculated in [29]. Here we only show plots of their μ_I dependence for different values of μ_S in figure 1. Note that these results differs from the ones of [3]. Because our results agree with the group theory reasoning explained above, we assert that the results in [3] are affected by some calculation error.

4. Pion decays

It has been shown how the introduction of chemical potentials can change meson masses and mixing. Here I will describe how it affects the charged pions decay. The processes are:

$$\pi^+ \to \ell^+ \nu_\ell\,, \tag{13}$$
$$\pi^- \to \ell^- \bar{\nu}_\ell\,, \tag{14}$$

that in the Standard Model (SM) have the decay width:

$$\Gamma^0_{\pi\to\ell\nu_\ell} = \frac{G_F^2 F_0^2 V_{ud}^2 m_\ell^2 m_\pi}{4\pi}\left(1 - \frac{m_\ell^2}{m_\pi^2}\right)^2\,, \tag{15}$$

where G_F is the Fermi constant, V_{ud} is the ud CKM matrix element, m_ℓ and m_π are the lepton and pion masses.

For the following calculation I will assume $\mu_S = 0$ for simplicity.

In the normal phase chemical potentials do not mix states, so the only change is that we have to replace m_π with $m_{\pi\pm} = m_\pi \mp \mu_I$ [29].

In the pion condensation phase the $\tilde{\pi}_+$ is massless, so it does not decay, while the $\tilde{\pi}_-$ is a mixture of $\pi_\pm$ so it can decay in both $\ell^+\nu_\ell$ and $\ell^-\bar{\nu}_\ell$. The related decay widths are:

$$\frac{\Gamma_{\tilde{\pi}_-\to\ell^+\nu_\ell}}{\Gamma^0_{\pi\to\ell\nu_\ell}} = \frac{|U_{21}^*\cos\alpha + iU_{22}^*|^2}{2}\frac{m_{\tilde{\pi}^-}}{m_\pi}\left(\frac{1 - m_\ell^2/m_{\tilde{\pi}^-}^2}{1 - m_\ell^2/m_\pi^2}\right)^2\,, \tag{16}$$

$$\frac{\Gamma_{\tilde{\pi}_-\to\ell^-\bar{\nu}_\ell}}{\Gamma^0_{\pi\to\ell\nu_\ell}} = \frac{|U_{21}^*\cos\alpha - iU_{22}^*|^2}{2}\frac{m_{\tilde{\pi}^-}}{m_\pi}\left(\frac{1 - m_\ell^2/m_{\tilde{\pi}^-}^2}{1 - m_\ell^2/m_\pi^2}\right)^2\,, \tag{17}$$

where:

$$\begin{pmatrix} \tilde{\pi}_+ \\ \tilde{\pi}_- \end{pmatrix} = \begin{pmatrix} U_{11} & U_{12} \\ U_{21} & U_{22} \end{pmatrix}\begin{pmatrix} \pi_+ \\ \pi_- \end{pmatrix}. \tag{18}$$

The Us are functions of μ_I and of the $\tilde{\pi}$ energy k_0. Their expressions are quite involved and thus we are not going to show them here. They can be found in [29].

The plot of these decay widths are shown in figure 2.

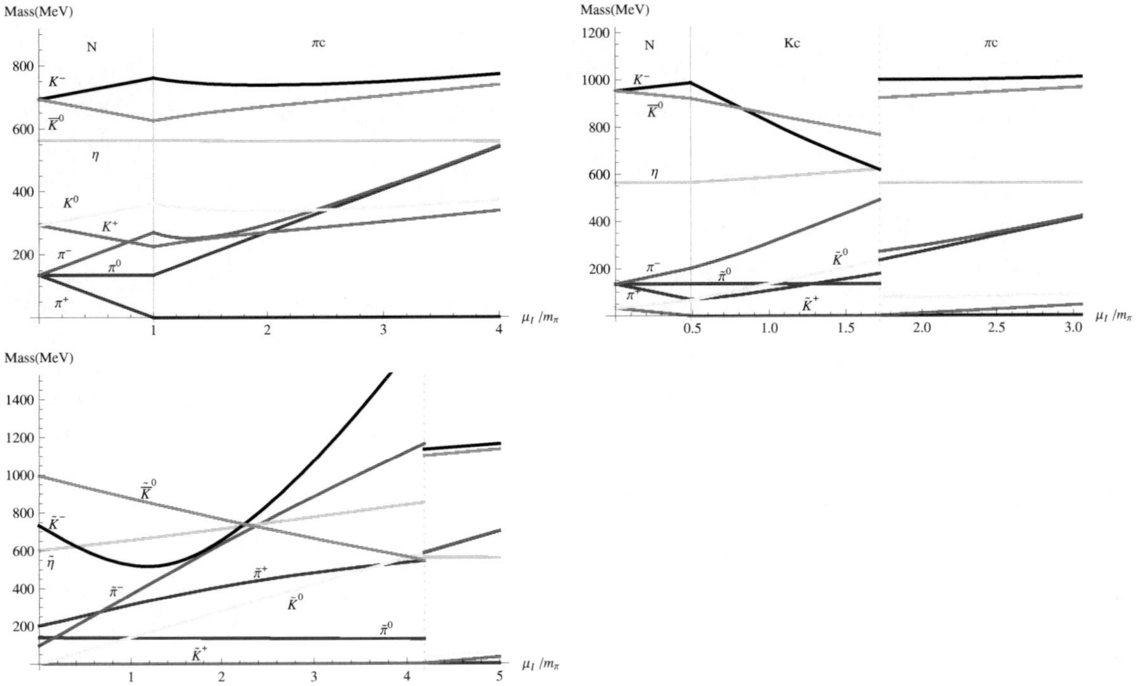

Figure 1. (color online) Mass spectrum of the pseudoscalar mesonic octet. Top panel, results obtained for $\mu_S = 200$ MeV. The vertical solid line represents the second order phase transition between the normal phase and the pion condensation phase. In this case the strange quark chemical potential is below the threshold value for kaon condensation, 425 MeV, thus the kaon condensed phase does not take place for any value of μ_I. Middle panel, results obtained for $\mu_S = 460$ MeV. The vertical solid line represents the second order phase transition from the normal phase to the kaon condensation phase. The dashed line corresponds to the first order phase transition between the kaon condensed phase to the pion condensed phase. Bottom panel, results obtained for $\mu_S = 550$ MeV, corresponding to the largest value of μ_S.

5. Thermodynamic

Because we are studying the zero temperature termodynamics all the fluctuations can be neglected, so the pressure of the system can be found substituting $\Sigma \to \bar{\Sigma}$ in (1). In the different phases we get:

$$p_{\rm LO}^{N} = 0 \qquad p_{\rm LO}^{\pi c} = \frac{F_0^2 \mu_I^2}{2}\left(1 - \frac{m_\pi^2}{\mu_I^2}\right)^2 \qquad p_{\rm LO}^{Kc} = \frac{F_0^2 \mu_K^2}{2}\left(1 - \frac{m_K^2}{\mu_K^2}\right)^2$$

where $\mu_K = \mu_I/2 + \mu_S$. Number density and equation of state can be derived using:

$$n_{i,{\rm LO}} = \frac{\partial p_{\rm LO}}{\partial \mu_i} \qquad \text{and} \qquad \epsilon = \mu_I n_I + \mu_S n_S - p\,, \tag{19}$$

where the index i can represent I=isospin or K=kaon. In the πc phase we obtain:

$$n_{I,{\rm LO}}^{\pi c} = F_0^2 \mu_I \left(1 - \frac{m_\pi^4}{\mu_I^4}\right) \qquad \text{and} \qquad \epsilon_{\rm LO}^{\pi c} = \frac{F_0^2 \mu_I^2}{2}\left(1 + 2\frac{m_\pi^2}{\mu_I^2} - 3\frac{m_\pi^4}{\mu_I^4}\right) \tag{20}$$

IOP Conf. Series: Journal of Physics: Conf. Series **981** (2018) 012010 doi:10.1088/1742-6596/981/1/012010

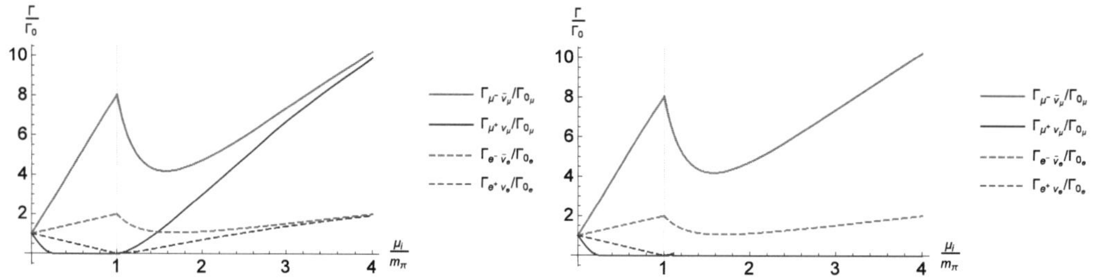

Figure 2. (color online) Leptonic decay rates of charged pions in normal phase and in the condensed phase normalized to the value in vacuum. The phase transition between the normal phase and the pion condesed phase corresponds to the solid vertical line. Top, results obtained for vanishing leptonic chemical potential. Bottom, results obtained assuming weak equilibriub. In this case the decay in positively charged leptons is Pauli blocked. In both plots the thick solid line represents $\Gamma_{\mu^-\bar{\nu}_\mu}/\Gamma_{0_\mu}$, the thin solid line represents $\Gamma_{\mu^+\nu_\mu}/\Gamma_{0_\mu}$, the thick dashed line represents $\Gamma_{e^-\bar{\nu}_e}/\Gamma_{0_e}$ and the thin dashed line represents $\Gamma_{e^+\nu_e}/\Gamma_{0_e}$

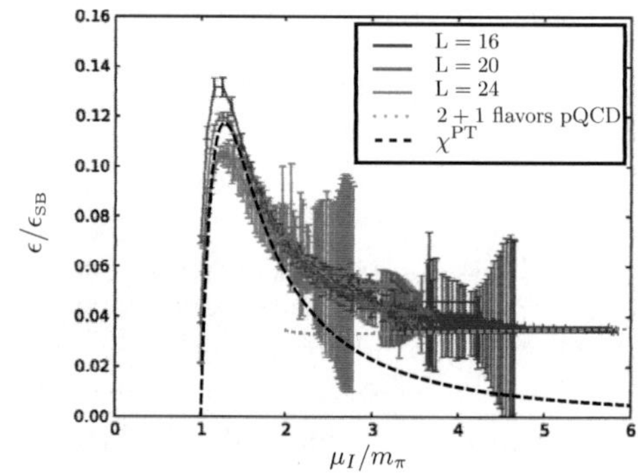

Figure 3. (color online) Energy density over the Stefan-Boltzmann limit. The lattice points have been obtained at T = 20 MeV; the colors correspond to the different lattice volumes considered in [19]. The orange dotted line corresponds to the pQCD results of [23]. The dashed black line corresponds to our results.

while the analogous results in the kaon condensation phase can be obtained substituting $m_\pi \to m_K$ and $\mu_I \to \mu_K$, see [20, 30]. For comparing the χPT energy density with the results obtained by LQCD simulations in [19] and by pQCD in [23], we divide it by the Stefan-Boltzmann limit, which has been defined in [19] as $\epsilon_{SB} = 9\mu_I^4/(4\pi^2)$. In Fig. 3 we report our ratio $\epsilon_{\rm LO}^{\pi c}/\epsilon_{\rm SB}$ and the results of [19] and [23]. We immediately notice that the χPT curve perfectly captures the peak structure at low μ_I, while it begins to depart from the LQCD results after $\mu_I \sim 2m_\pi$, indicating the breakdown of the LO approximation. An interesting result is that within our framework we can obtain an analytic expression for the position of the peak, which for the πc phase is given by $\mu_I^{\rm peak} = \left(\sqrt{13}-2\right)^{1/2} m_\pi \simeq 1.276\, m_\pi$ and is independent of f_π. This result is very close to the LQCD results obtained in [19], where the values $\mu_I^{\rm peak} = \{1.20, 1.25, 1.275\} m_\pi$ have been obtained considering different spatial volumes L^3 with side $L = \{16, 20, 24\}$, respectively. The continuum-linearly-extrapolated value for the peak position is $\mu_I^{\rm peak} = 1.30\ m_\pi$.

IOP Conf. Series: Journal of Physics: Conf. Series **981** (2018) 012010 doi:10.1088/1742-6596/981/1/012010

6. Conclusion

We have shown how meson physics in presence of chemical potentials can be described using Chiral Perturbation Theory. There are three different phases: a normal phase, a pion condensation phase and a kaon condensation phase. The condensed phases ground states have a charge and thus can generate mixing among mesons.
In this context we have illustrated how the mixing is influenced by model symmetries and that groups theory constraints the mixing possibilities. These results, obtained by group theory alone, is expected to hold in any theory describing meson states. Then we have described the masses of mesons in the three phases, calculated using the Lagrangian (1). These masses are in perfect agreement with the group theory reasoning.
We have also described how the charged pion decay is influenced by the chemical potentials, showing that their inclusion in the model leads to a significant asymmetry between the decay in $\ell^+\nu_\ell$ and the decay in $\ell^-\bar{\nu}_\ell$.
Finally we have derived the pressure and equation of state of the system at $T = 0$, comparing them with the equivalent results obtained in lattice simulations and obtaining a very good agreement.
These results can be applied for example in the physics of compact stars, the study of the cosmic ray and the study of in medium nuclear decays.

References

[1] Migdal A B, Saperstein E, Troitsky M and Voskresensky D 1990 *Phys.Rept.*, **192** 179–437
[2] Son D T and Stephanov M A 2001 *Phys.Rev.Lett.*, **86** 592–595
[3] Kogut J and Toublan D 2001 *Phys.Rev.*, **D64**, 034007
[4] Kogut J, Stephanov M A, and Toublan D 1999 *Phys.Lett.*, **B464** 183–191
[5] Kogut J, Stephanov M A, Toublan D, Verbaarschot J and Zhitnitsky A 2000 *Nucl.Phys.*, **B582** 477–513
[6] Son D T and Stephanov M A 2001 *Phys.Atom.Nucl.*, **64** 834–842
[7] Loewe M and Villavicencio C 2003 *Phys.Rev.* **D67** 074034
[8] Loewe M and Villavicencio C 2004 *Phys.Rev.* **D70** 074005
[9] He L Y, Jin M and Zhuang P 2005 *Phys.Rev.* **D71** 116001
[10] Alford M G, Kapustin A and Wilczek F 1999 *Phys.Rev.* **D59** 054502
[11] Kogut J and Sinclair D 2002 *Phys.Rev.* **D66** 034505
[12] Toublan D and Kogut J 2003 *Phys.Lett.* **B564**, 212–216
[13] Barducci A, Casalbuoni R, Pettini G, and Ravagli L 2004 *Phys.Rev.* **D69** 096004
[14] Barducci A, Casalbuoni R, Pettini G, and Ravagli L 2005 *Phys.Rev.* **D71** 016011
[15] Klein B, Toublan D, and Verbaarschot J 2005 *Phys.Rev.* **D72** 015007
[16] Barducci A, Casalbuoni R, De Curtis S, Gatto R and Pettini G 1990 *Phys.Rev.* **D42** 1757–1763
[17] Dominguez C, Loewe M and Rojas J 1994 *Phys.Lett.* **B320** 377–380
[18] Loewe M and Villavicencio C 2011 Pion stability in a hot dense media Preprint hep-ph/1107.3859
[19] Detmold W, Orginos K and Shi Z 2012 *Phys. Rev.* **D86** 054507
[20] Detmold W, Orginos K, Savage M J and Walker-Loud A 2008 *Phys. Rev.* **D78** 054514
[21] Andersen J O, Haque N, Mustafa M G and Strickland M 2015 *Phys. Rev.* **D93** 054045
[22] Cohen T D and Sen S 2015 *Nucl. Phys.* **A942** 39
[23] Graf T, Schaffner-Bielich J and Fraga E S 2015 *Phys. Rev.* **D93** 085030
[24] Gasser J and Leutwyler H 1984 *Annals Phys.* **158**, 142
[25] Leutwyler H 1994 *Annals Phys.* **235** 165–203
[26] Ecker G 1995 *Prog.Part.Nucl.Phys.* **35** 1–80
[27] Scherer S 2003 *Adv.Nucl.Phys.* **27** 277
[28] Scherer S and Schindler M R 2005 A Chiral Perturbation Theory Primer Preprint hep-ph/0505265
[29] Mammarella A and Mannarelli M, 2015 Phys. Rev. D **92** 8 085025
[30] Carignano S, Mammarella M and Mannarelli M, 2016 Phys. Rev. D **93** 5 051503

IOP Conf. Series: Journal of Physics: Conf. Series **981** (2018) 012011 doi:10.1088/1742-6596/981/1/012011

Spatially modulated chiral condensates

Stefano Carignano

INFN, Laboratori Nazionali del Gran Sasso, Via G. Acitelli, 22, I-67100 Assergi (AQ), Italy

E-mail: carignano@lngs.infn.it

Abstract. I discuss some properties of spatially dependent chiral condensates, focusing on one-dimensional modulations. After briefly introducing a generic formalism for studying inhomogeneous solutions in dense quark matter, I consider a plane-wave and a sinusoidal ansatz, and discuss the relationship between their order parameters and free energies, as well as a comparison of their dispersion relations. I then introduce a real kink crystal, and discuss its relationship with the sinusoidal ansatz by performing a Fourier analysis of this kind of solution.

1. Introduction

While the properties of quantum chromodynamics (QCD) at finite density are still poorly known, its phase structure is expected to be extremely rich. Matter at low temperatures and asymptotically large baryonic chemical potentials should behave as a color-superconductor, but other pairing mechanisms might become competitive as the density of the system decreases. In particular, in the past few years some growing consensus has been building around the idea that at densities around a few times nuclear matter saturation density and low temperatures, spatially inhomogeneous chiral condensates might form, giving rise to crystalline phases. Explicit model calculations indeed corroborate this hypothesis by suggesting that an inhomogeneous "island" appears in this region, where the favored structure for the chiral condensate is a spatially modulated one (for a recent review, see [1]). A typical phase diagram obtained within this kind of calculations is shown in Fig. 1.

The phenomenon of inhomogeneous chiral symmetry breaking at finite densities stems from pairing of quarks and holes with equal momenta on patches of the Fermi surface. The formed pairs thus carry a nonvanishing total momentum $|\mathbf{P}_{pair}| \sim 2\mu$, ($\mu$ being the quark chemical potential) and the corresponding condensate is spatially non-uniform. Superposition of paired quarks on different patches of the Fermi surface can then in principle give rise to more involved crystalline structures [2].

Recent calculations performed within QCD-inspired effective models have shown that the favored shape for the chiral condensate in the inhomogeneous island is that of a modulation which varies in only one spatial direction [3, 4]. In this contribution, I will discuss a few properties of some of these one-dimensional solutions within an effective model of QCD.

2. NJL model studies of inhomogeneous chiral condensates

The theoretical framework employed in this work for the characterization of inhomogeneous chiral condensates in dense quark matter is the Nambu–Jona-Lasinio (NJL) model, a popular tool often employed in the study of low-energy properties of strong-interaction matter as well

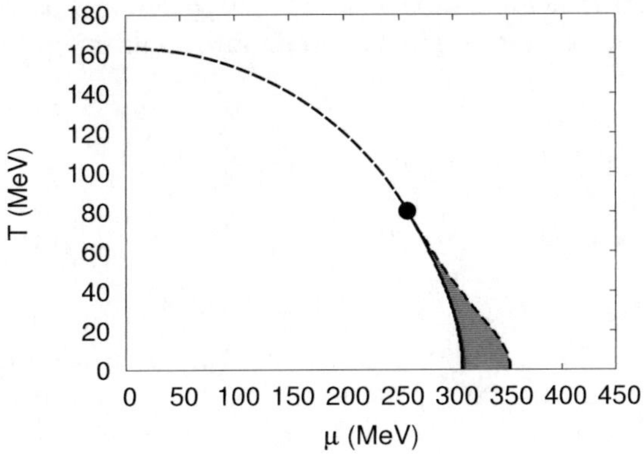

Figure 1. Phase diagram obtained within NJL model calculations when allowing for inhomogeneous chiral condensates. The shaded area denotes a region where spatially modulated chiral condensates are favored over the homogeneous chirally broken and restored solutions. The actual size of the inhomogeneous window depends on the ansatz considered for the spatial dependence of the chiral condensate, as well as on model parameters.

as its phase diagram at finite temperature and density. In its simplest form, the NJL-model Lagrangian reads [5]

$$\mathcal{L}_{\rm NJL} = \bar{\psi}(i\gamma^\mu\partial_\mu - m)\psi + G_S\left((\bar{\psi}\psi)^2 + (\bar{\psi}i\gamma^5\tau^a\psi)^2\right) , \tag{1}$$

where ψ represents a quark field with $N_f = 2$ flavor and $N_c = 3$ color degrees of freedom, γ^μ and γ^5 are Dirac matrices, and τ^a $(a = 1,2,3)$ are isospin Pauli matrices. The $SU(2)_L \times SU(2)_R$ chiral symmetry of $\mathcal{L}_{\rm NJL}$ is explicitly broken by the bare quark mass m, as well as spontaneously by non-vanishing condensates of the form $\phi_\sigma \equiv \langle\bar{\psi}\psi\rangle$ or $\phi_\pi^a \equiv \langle\bar{\psi}i\gamma^5\tau^a\psi\rangle$.

In the following I will consider the chiral limit, $m = 0$, and work within the standard mean-field approximation where fluctuations of the order parameters are neglected. This allows to obtain the model thermodynamic potential per volume V for a given temperature T and chemical potential μ, as a function of the scalar and pseudoscalar mean-fields:

$$\begin{aligned}\Omega(T,\mu) &= -\frac{T}{V}\mathrm{Log}\int\mathcal{D}\bar{\psi}\mathcal{D}\psi\exp\left(\int_{V_4} d^4x_E\left(\mathcal{L}_{\rm NJL} + \mu\bar{\psi}\gamma^0\psi\right)\right)\\ &\xrightarrow{\rm MFA} -\frac{T}{V}\mathbf{Tr}\,\mathrm{Log}\left(\frac{S^{-1}}{T}\right) + \frac{T}{V}\int_{V_4} d^4x_E\, G_S(\phi_\sigma^2 + \phi_{\pi,a}^2)\end{aligned} \tag{2}$$

where the integral is performed in Euclidean space-time, $x_E = (\tau, \mathbf{x})$ with the imaginary time $\tau = it$, and extends over the four-volume $V_4 = [0, \frac{1}{T}] \times V$. The functional trace of the inverse quark propagator S^{-1} runs over V_4 and internal (color, flavor, and Dirac) degrees of freedom.

In order to investigate inhomogeneous chiral-symmetry breaking, within this approach the mean-fields are allowed to be spatially dependent, but are still assumed to be static so that the temporal part of the functional trace can be evaluated as a standard sum over fermionic Matsubara frequencies.

From a technical point of view, the inversion of the quark propagator requires the diagonalization of the effective Hamiltonian

$$\mathcal{H} = \gamma^0\left[-i\gamma^i\partial_i - 2G_S(\phi_\sigma(\mathbf{x}) + i\gamma^5\tau_a\phi_\pi^a(\mathbf{x}))\right] . \tag{3}$$

IOP Conf. Series: Journal of Physics: Conf. Series **981** (2018) 012011 doi:10.1088/1742-6596/981/1/012011

If charged pion condensation is neglected ($\phi_\pi^1 = \phi_\pi^2 = 0\,, \phi_\pi^3 \equiv \phi_\pi$), after expliciting the Dirac structure it is possible to factorize the Hamiltonian into a direct product of two isospectral blocks in flavor space [6],

$$\mathcal{H} = H_+(M) \otimes H_+(M^*)\,, \quad \text{with} \qquad H_+(M) = \begin{pmatrix} i\sigma^i\partial_i & M(\mathbf{x}) \\ M^*(\mathbf{x}) & -i\sigma^i\partial_i \end{pmatrix}\,, \tag{4}$$

where σ^i are Pauli matrices and one customarily defines a constituent mass function

$$M(\mathbf{x}) = -2G_S(\phi_\sigma(\mathbf{x}) + i\phi_\pi(\mathbf{x}))\,. \tag{5}$$

In presence of spatially modulated condensates, the diagonalization of $\mathcal{H}$ becomes extremely involved, as quark momenta are no longer conserved and the Hamiltonian is not diagonal in momentum space. In practice, it is then useful to restrict the analysis to periodic modulations of the chiral condensate, an assumption which allows to perform a Fourier decomposition of the mass function:

$$M(\mathbf{x}) = \sum_{\mathbf{q}_k} \Delta_k e^{i\mathbf{q}_k\cdot\mathbf{x}}\,, \tag{6}$$

with momenta $\mathbf{q}_k$ forming a reciprocal lattice (RL) in momentum space. Due to the crystal symmetries, only quark momenta differing by elements of the RL are paired in $\mathcal{H}$, so that one can effectively diagonalize a Hamiltonian with a discrete momentum structure. The components of one of its momentum blocks are given by

$$[H_+]_{\mathbf{p}_m,\mathbf{p}_n} = \begin{pmatrix} -\vec{\sigma}\cdot\mathbf{p}_m\,\delta_{\mathbf{p}_m,\mathbf{p}_n} & \sum\limits_{\mathbf{q}_k}\Delta_k\delta_{\mathbf{p}_m,\mathbf{p}_n+\mathbf{q}_k} \\ \sum\limits_{\mathbf{q}_k}\Delta_k^*\delta_{\mathbf{p}_m,\mathbf{p}_n-\mathbf{q}_k} & \vec{\sigma}\cdot\mathbf{p}_m\,\delta_{\mathbf{p}_m,\mathbf{p}_n} \end{pmatrix}\,, \tag{7}$$

where it can be clearly seen that the inhomogeneous chiral condensate couples different quark momenta.

Any periodic modulation of arbitrary spatial dimension can in principle be implemented within this approach by fixing an underlying lattice structure via the definition of the $\mathbf{q}_k$ and considering an appropriate number of Fourier coefficients. The favored solutions can then be obtained by minimizing the model thermodynamic potential with respect to the variational parameters $\{\Delta_k, \mathbf{q}_k\}$. In practice, however, the numerical diagonalization of the quark Hamiltonian in momentum space becomes extremely demanding from a numerical point of view as the mass ansatz becomes more involved. A dramatic simplification of the problem comes if one considers solutions for which the chiral condensate is modulated only along one spatial direction (which one can take without loss of generality along the z axis), while remaining constant along the others. In particular, it was shown in [6] that it is usually possible to solve the eigenvalue problem by diagonalizing a dimensionally-reduced Hamiltonian with $p_x = p_y = 0$ and subsequently boost the obtained energies along the transverse directions.

A systematic NJL model analysis on the favored type of modulation including higher-dimensional ansätze has been performed in [3] at zero temperature, and in [4] in proximity of the chiral critical point within a Ginzburg-Landau (GL) expansion study. Both agree on finding that the thermodynamically favored solution is a one-dimensional one, in particular a so-called real kink crystal (RKC) which can be expressed in terms of Jacobi elliptic functions. This is in contrast with the situation for the 1+1-dimensional NJL model (NJL$_2$), where the ground state is given by a plane wave modulation of the chiral condensate often referred to as "chiral spiral" [7]. On the other hand, RKC solutions are found to be favored in the Gross-Neveu model [8], which has only a discrete chiral symmetry (see also [9] for a comparison of the pairing mechanisms in the two models). From a technical point of view, it is the particular form of the Hamiltonian in the NJL model in 3+1 dimensions (Eq. (4)) containing two blocks differing only by the substitution $M \leftrightarrow M^*$ which makes these real solutions the favored ones [6].

IOP Conf. Series: Journal of Physics: Conf. Series **981** (2018) 012011 doi:10.1088/1742-6596/981/1/012011

3. Comparison of one-dimensional modulations

As previously mentioned, in order to perform a model analysis of inhomogeneous condensation, a given ansatz for the spatial dependence of the mean-fields must be chosen.

Among all possible one-dimensional solutions, the simplest is a so-called "(dual-)chiral density wave" (CDW) [10], a plane wave ansatz which can be seen as analogous of the Fulde-Ferrell solutions introduced in the context of superconductivity and color superconductivity:

$$M_{CDW}(\mathbf{x}) = \Delta e^{iqz}\,, \tag{8}$$

with two variational parameters Δ and q, characterizing the amplitude and the wave number of the modulation. This can be seen as a one-dimensional truncated version of the general ansatz of Eq. (6) where only the first Fourier component Δ_1 is allowed to be nonzero.

A very important feature of the CDW is that for this ansatz the explicit spatial dependence of the condensates in the quark Hamiltonian (Eq. (3)) can be removed by means of a local chiral rotation of the quark fields [11, 12] (see also [1]), thus allowing to evaluate the eigenvalue spectrum analytically. The resulting eigenvalues of the dimensionally reduced Hamiltonian are the positive and negative square roots of

$$E^2_{\pm}(p_z) = p_z^2 + \Delta^2 + \frac{q^2}{4} \pm \sqrt{\Delta^2 q^2 + (qp_z)^2} = \left(\sqrt{p_z^2+\Delta^2} \pm \frac{q}{2}\right)^2\,. \tag{9}$$

An alternative ansatz often considered in literature is a simple real sinusoidal modulation of the mass function, for which the pseudoscalar part ϕ_π is set to zero:

$$M_{COS}(\mathbf{x}) = \Delta\cos(qz)\,. \tag{10}$$

From a technical point of view, this ansatz can be seen as a superposition of two plane waves propagating in opposite directions, and as such its spatial dependence cannot be rotated away using the technique employed for a CDW. In fact, no analytical expression for the eigenvalue spectrum in presence of this modulation is known, so that a numerical diagonalization of the model Hamiltonian in momentum space must be performed, using, for example, the techniques developed in [13, 3].

After minimization of the thermodynamic potential, one can observe a very close relationship between the order parameters and the free energies of the CDW and COS modulations. Indeed, as can be seen in Fig. 2, throughout almost the entire inhomogeneous window one finds that $\Delta_{COS} = 2\Delta_{CDW}$, while q_{COS} and q_{CDW} are similar but not equal and start overlapping only close to the chiral restoration phase transition, a behavior already predicted within a GL analysis in [6]. Another interesting result is that the COS solution is always favored by a factor of 2 in free energy compared to the CDW one, as shown in Fig. 3.

In order to understand why the real COS solution is favored over the complex CDW one, it might be also of interest to compare the dispersion relations for the two modulations. The eigenvalues of the dimensionally-reduced Hamiltonians for the two are plotted in Fig. 4. There it can be seen how the CDW spectrum still exhibits branches which are analogous to those of chirally restored quark matter, whereas for the COS modulation these also acquire a gap. This additional pairing present for the COS modulation allows for a larger energy gain than in the CDW case, since twice as many branches become deformed and allow for the system to further lower its energy.

The final part of my discussion is devoted to the thermodynamically favored modulation in the NJL model, which is known to be given by a real-kink crystal. This particular kind of ansatz can be expressed as [8] (see also [1])

$$M(z) = \Delta\sqrt{\nu}\,\mathrm{sn}(\Delta z|\nu)\,, \tag{11}$$

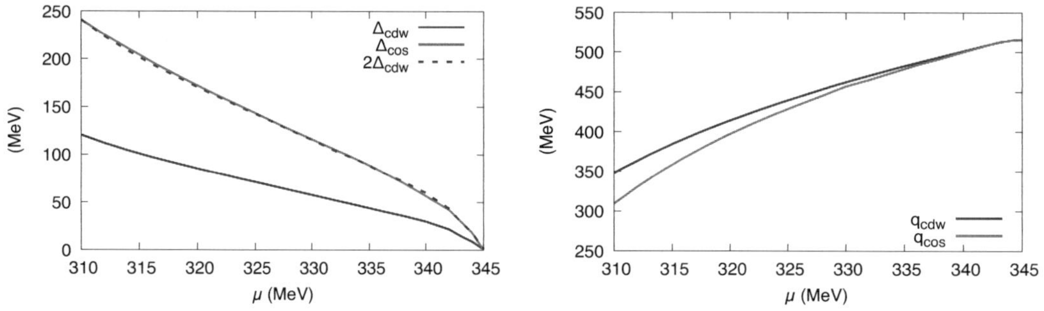

Figure 2. Left: comparison of amplitudes at $T = 0$ for CDW and COS modulation. The solid blue line denotes the thermodynamically favored Δ for the CDW modulation, the dashed blue line is twice that value, the solid red line is the favored Δ for the COS modulation. Right: comparison of wave numbers. The blue line is the favored q for the CDW, the red one is the q for the COS.

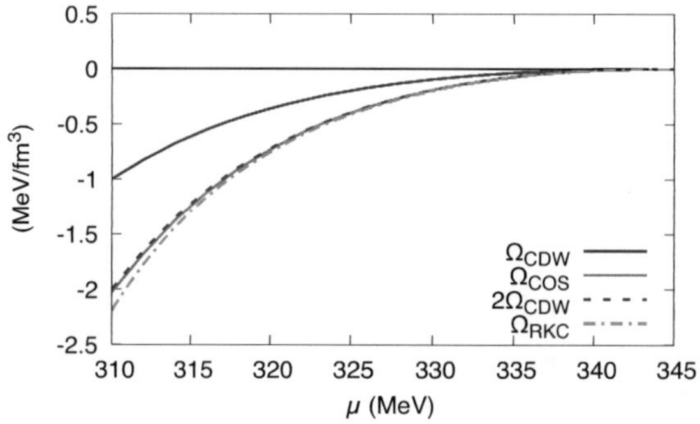

Figure 3. Free energy associated with different modulations of the chiral condensate in the region where inhomogeneous phases are thermodynamically favored, as a function of quark chemical potential ($T = 0$). Solid blue line: CDW, Red line: COS. The dashed blue line, corresponding to twice the CDW free energy, is basically overlapping the COS one throughout the whole inhomogeneous window. The purple dash-dotted line denotes the RKC free energy, which is slightly lower than the COS one for lower μ but rapidly becomes degenerate with it.

where sn is a Jacobi elliptic function and the parameter ν, called elliptic modulus and ranging from 0 to 1, describes the shape and periodicity of the modulation.

By inspecting the behavior of the mass function as chemical potential increases, one can see that at the onset of the inhomogeneous phase this solution has the limit $M(z) \to \Delta \tanh(\Delta z)$, which can be seen as a single soliton interpolating between the two homogeneous chirally broken solutions. As μ increases, solitons start stacking up against each other and the order parameter progressively assumes a sinusoidal shape [14, 15]. This kind of solution and its behavior can be encompassed within the framework introduced in the previous section by evaluating the Fourier coefficients associated with it, as function of chemical potential. The resulting values are presented in Table 1. There one can see, as expected, that as the chemical potential increases the higher Fourier coefficients rapidly drop to zero, and the RKC solution assumes the form of a simple cosine.

IOP Conf. Series: Journal of Physics: Conf. Series **981** (2018) 012011 doi:10.1088/1742-6596/981/1/012011

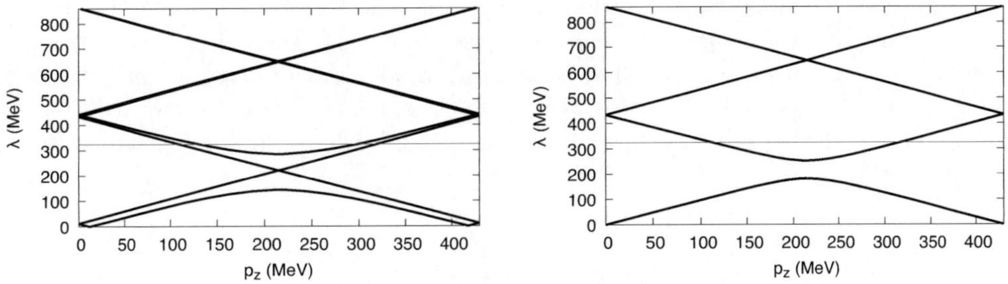

Figure 4. Dispersion relations for the CDW (left) and COS (right) modulations. Both have been evaluated for values of the order parameters corresponding to the respective minima at $T = 0, \mu = 325$ MeV (recall at the minimum $q_{COS} \simeq q_{CDW}$ and $\Delta_{COS} \simeq 2\Delta_{CDW}$). The chemical potential is plotted as reference as an horizontal blue line.

Table 1. Values of the higher order nonzero harmonics for a RKC modulation at different values of chemical potential close to the onset of the inhomogeneous phase ($T = 0$). Results are normalized to the value of the first harmonic ($n = 1$) in order to show their relative weights. Being a real modulation, one also finds $\Delta_{-n} = \Delta_n$. Even coefficients are found to be zero.

μ	Δ_3/Δ_1	Δ_5/Δ_1	Δ_7/Δ_1	Δ_9/Δ_1
308.0	-0.12936	0.01959	-0.002978	0.000452
310.0	-0.04672	0.00229	-0.000112	0.000005
315.0	-0.02057	0.00043	-0.000009	0.000000
320.0	-0.01174	0.00013	-0.000001	0.000000

4. Conclusions

In this contribution I discussed some properties of one-dimensional modulations of the chiral condensate in dense quark matter within NJL model calculations. In particular, I focused on two possible one-dimensional solutions, namely a chiral density wave and a single cosine, which, in the framework of a Fourier expansion of the order parameter, can be seen as prototypes for complex and real ansätze, respectively. I showed that the thermodynamically favored values for the wave numbers of the chiral density wave and that of the cosine are very similar throughout the whole inhomogeneous window, while the amplitude of the cosine is twice that of a plane wave. Furthermore, by inspecting the quasiparticle dispersion relations I argued why a real sinusoidal modulation is favored over a complex plane wave one. In particular, the fact that for the COS modulation twice as many branches as for the CDW become gapped could explain the fact that the energy gain for the former is twice the one of the latter.

Finally, I considered the solution which is found to be the most favored throughout the whole inhomogeneous window, a real kink crystal, and investigated its relationship with a simpler real sinusoidal modulation by performing a Fourier decomposition of this type of ansatz. The RKC effectively reduces to a simple cosine rapidly after the onset of the inhomogeneous phase. This in turn implies that for most intents and purposes the RKC solution can be reasonably approximated by considering a single cosine, with the exception of the immediate vicinity of the phase transition associated with the onset of inhomogeneous condensation. In particular, it is important to remember that the nature of such phase transition depends strongly on the type of modulation considered. Among those discussed here, only the RKC can smoothly reduce in the thermodynamic limit to an homogeneous chirally broken solution. Indeed, in order to reproduce an homogeneous solution which has an infinite period, an arbitrarily large set of Fourier modes needs to be considered. As such, only the RKC solution can realize a second-order transition to the homogeneous chirally broken phase, whereas for the other simpler ansätze a first-order transition must occur.

5. Acknowledgments

I would like to thank M. Buballa for helpful discussions.

References

[1] Buballa M and Carignano S 2015 *Prog. Part. Nucl. Phys.* **81** 39–96 (*Preprint* 1406.1367)
[2] Kojo T, Hidaka Y, Fukushima K, McLerran L D and Pisarski R D 2012 *Nucl. Phys. A* **875** 94–138 (*Preprint* 1107.2124)
[3] Carignano S and Buballa M 2012 *Phys. Rev. D* **86** 074018 (*Preprint* 1203.5343)
[4] Abuki H, Ishibashi D and Suzuki K 2012 *Phys.Rev. D* **85** 074002 (*Preprint* 1109.1615)
[5] Nambu Y and Jona-Lasinio G 1961 *Phys. Rev.* **124** 246
[6] Nickel D 2009 *Phys. Rev. D* **80** 074025 (*Preprint* 0906.5295)
[7] Schön V and Thies M 2000 Contribution to the Festschrift in honor of Boris Ioffe, edited by M. Shifman. (*Preprint* hep-th/0008175)
[8] Schnetz O, Thies M and Urlichs K 2004 *Annals Phys.* **314** 425–447 (*Preprint* hep-th/0402014)
[9] Kojo T 2014 *Phys. Rev.* **D90** 065030 (*Preprint* 1406.4630)
[10] Nakano E and Tatsumi T 2005 *Phys. Rev. D* **71** 114006
[11] Dautry F and Nyman E 1979 *Nucl. Phys. A* **319** 323
[12] Kutschera M, Broniowski W and Kotlorz A 1990 *Phys.Lett.* **B237** 159
[13] Nickel D and Buballa M 2009 *Phys. Rev. D* **79** 054009 (*Preprint* 0811.2400)
[14] Carignano S, Nickel D and Buballa M 2010 *Phys. Rev. D* **82** 054009 (*Preprint* 1007.1397)
[15] Buballa M and Carignano S 2013 *Phys. Rev. D* **87** 054004 (*Preprint* 1210.7155)

IOP Conf. Series: Journal of Physics: Conf. Series **981** (2018) 012012 doi:10.1088/1742-6596/981/1/012012

A Unified Equation of State on a Microscopic Basis : Implications for Neutron Stars Structure and Cooling

G. F. Burgio

INFN Sezione di Catania, Via S. Sofia 64, I-95123 Catania, Italy

E-mail: fiorella.burgio@ct.infn.it

Abstract. We discuss the structure of Neutron Stars by modelling the homogeneous nuclear matter of the core by a suitable microscopic Equation of State, based on the Brueckner-Hartree-Fock many-body theory, and the crust, including the pasta phase, by the BCPM energy density functional which is based on the same Equation of State. This allows for a unified description of the Neutron Star matter over a wide density range. A comparison with other unified approaches is discussed. With the same Equation of State, which features strong direct Urca processes and using consistent nuclear pairing gaps as well as effective masses, we model neutron star cooling, in particular the current rapid cooldown of the neutron star Cas A. We find that several scenarios are possible to explain the features of Cas A, but only large and extended proton ${}^{1}S_0$ gaps and small neutron ${}^{3}PF_2$ gaps can accommodate also the major part of the complete current cooling data.

1. Introduction

In the past few years, the interplay between the observations and the theoretical predictions about neutron stars (NS) has stimulated an impressive progress in the field, and it is expected to help answering many fundamental questions on the properties of very dense matter and the corresponding elementary processes that can occur. Therefore it is necessary to construct models based on a sound theoretical background, in order to reduce the uncertainty on the possible conclusions one can draw from these studies.

The equation of state (EoS) of neutron-rich matter is a basic input needed to compute most properties of NS. A large amount of experimental data on nuclei, heavy ion collisions, and astrophysical observations has been used to constrain the nuclear EoS and, at the same time, to understand the structure and properties of NS [1]. Unfortunately, a direct link of measurements and observations with the underlying EoS is very difficult, and a big uncertainty does exist about a proper interpretation of the data. To reduce the uncertainty on these types of analyses, it is helpful to develop a unified theory of matter, which is able to describe on a microscopic level the complete structure of NS from the outer crust to the core. This is not a simple task, since the methods developed for homogeneous nuclear matter cannot be easily extended to nuclei and to the non-homogeneous matter present in NS crust. Recently [2, 3, 4], the BCPM (Barcelona-Catania-Paris-Madrid) nuclear energy density functional has been developed in order to describe finite nuclei. This functional has been obtained from the *ab initio* Brueckner-Hartree-Fock (BHF) calculations in nuclear matter within an approximation inspired by the Kohn-Sham formulation of density functional theory [5]. We employ the BHF EoS and the corresponding BCPM functional to describe the whole NS structure, and compare with the results obtained with the few other methods that describe the whole NS structure within a unified theoretical scheme. In this paper we limit the treatment to only nucleon degrees of freedom, neglecting the possibility of the appearance of exotic components like hyperons and quarks, for which the uncertainty is too large to perform a fruitful comparison with other approaches.

The interpretation of the signals coming from the astrophysical observations on the processes and phenomena that occur in NS does not require just a reliable Equation of State, but also additional theoretical tools. One of the most important is the temperature-vs.-age cooling diagram, in which currently a few (~ 20) observed NS are located. NS cooling is over a vast domain of time ($10^{-10} - 10^5$ yrs) dominated by neutrino emission due to several microscopic processes [6]. The theoretical analysis of these reactions requires the knowledge of the relevant beta-stable nuclear EoS, along with the superfluid properties of the stellar matter, i.e., the gaps and critical temperatures in the different pairing channels. The observation of very rapid cooling of the supernova remnant Cas A, for which different analyses deduce a temperature decline of about 2 to 5 percent during the last ten years [7, 8, 9], has stimulated an intense theoretical activity. Several theoretical scenarios have been proposed to explain this observation, e.g. a fine-tuned small neutron 3PF2 (n3P2) gap [10, 11, 12], that generates strong cooling at the right moment due to the superfluid neutron pair breaking and formation (PBF) mechanism [13]; a strongly reduced thermal conductivity of the stellar matter that delays the heat propagation from the core to the crust to a time compatible with the age of Cas A [14]; a strong dissipation of the small-scale magnetic field in the crust which produces Joule heating [15]; a transition from a fully gapped two-flavor color-superconducting phase to a gapless/ crystalline phase [16]. A common feature of all these cooling scenarios is that EoS and pairing gaps are treated in disjoint and inconsistent manner, i.e., a given EoS is combined with pairing gaps obtained within a different theoretical approach and using different input interactions. Moreover all those approaches exclude from the beginning the possibility of very fast direct Urca (DU) cooling, although many microscopic nuclear EoS do reach easily the required proton fractions for this process [17, 18, 19]. In a recent paper [20] EoS and nuclear pairing gaps have been obtained with exactly the same nuclear (in-medium) interaction, as well as nucleon effective masses [21], which affect the neutrino emissivities. Moreover the DU cooling process as predicted by the BHF microscopic nuclear EOS have been consistently taken into account.

The present article is organized as follows. In Sect. II we present the models of the EOS considered in this work respectively for the NS crust and core, with a discussion of the results obtained for the NS structure. In Sect. III we analyze several possible cooling scenarios derived within our consistent approach, and discuss the results. Finally, our concluding remarks are drawn in Sect. IV.

2. EOS of nuclear matter

In this section we remind briefly the Brueckner-Hartree-Fock (BHF) method for the nuclear matter EOS [22, 1]. The main ingredient is the calculation of the in-medium G-matrix nucleon-nucleon (NN) interaction from the bare NN potential V,

$$G[\rho;\omega] = V + \sum_{k_a k_b} V \frac{|k_a k_b\rangle Q \langle k_a k_b|}{\omega - e(k_a) - e(k_b)} G[\rho;\omega] \,, \tag{1}$$

where $\rho = \sum_{k<k_F}$ is the nucleon number density, and ω the starting energy. The single-particle (s.p.) energy (assuming $\hbar = 1$)

$$e(k) = e(k;\rho) = \frac{k^2}{2m} + U(k;\rho) \tag{2}$$

and the Pauli operator Q determine the propagation of intermediate baryon pairs. The BHF approximation for the s.p. potential $U(k;\rho)$ using the *continuous choice* prescription is

$$U(k;\rho) = \mathrm{Re} \sum_{k'<k_F} \langle kk'|G[\rho;e(k)+e(k')]|kk'\rangle_a \,, \tag{3}$$

where the subscript a indicates antisymmetrization of the matrix element. Due to the occurrence of $U(k)$ in Eq. (2), these equations constitute a coupled system that has to be solved in a self-consistent manner

IOP Conf. Series: Journal of Physics: Conf. Series **981** (2018) 012012 doi:10.1088/1742-6596/981/1/012012

for several momenta of the particles involved, at the considered densities. In the BHF approximation the energy per nucleon is given by

$$\frac{E}{A} = \frac{3}{5}\frac{k_F^2}{2m} + \frac{1}{2\rho} \sum_{k,k' \leq k_F} \langle kk'|G[\rho; e(k) + e(k')]|kk'\rangle_a. \qquad (4)$$

In this scheme, the only input quantity needed is the bare NN interaction V in the Bethe-Goldstone equation (1). In this work we use the Argonne V_{18} NN ipotential [23] as the two-nucleon interaction. The resulting nuclear EoS can be calculated with good accuracy in the Brueckner two hole-line approximation with the continuous choice for the single-particle potential, being the results in this scheme quite close to the calculations which include also the three hole-line contribution [24].

Since non-relativistic calculations, based on purely two-body interactions, fail to reproduce the correct saturation point of symmetric nuclear matter, the bare NN interaction has to be supplemented by a suitable three-nucleon force (TBF). For that, TBF are reduced to a density dependent two-body force by averaging over the generalized coordinates (position, spin and isospin) of the third particle, assuming that the probability of having two particles at a given distance is reduced according to the two-body correlation function [25, 26, 27]. In this work we will illustrate results obtained using a phenomenological approach to the TBF, which is based on the so-called Urbana model [28, 29]. Within the BHF approach, those TBF produce a shift of about +1 MeV in energy and -0.1 fm^{-3} in density, and is obtained by tuning the two parameters contained in the TBF in order to get an optimal saturation point [25, 26].

Besides a purely phenomenological model, microscopic TBF have also been derived and a tentative approach proposed using the same meson-exchange parameters as the underlying NN potential [30, 31]. However, at present the theoretical status of microscopically derived TBF is still quite rudimentary, Alternatively, latest nuclear matter calculations [32, 33] using chiral inspired TBF [34, 35], showed that the considered TBF models are not able to reproduce simultaneously the correct saturation point and the properties of three- and four-nucleon systems. Therefore, the role of TBF is still an important open question, but it may be greatly reduced if the NN potential is based on a realistic constituent quark model [36, 37] which can explain at the same time few-nucleon systems and nuclear matter, including the observational data on Neutron Stars and the experimental data on heavy-ions collisions.

In the past years, the BHF approach has been extended in order to include the hyperon degrees of freedom [38, 39], which are expected to appear in neutron star matter already at relatively low densities, thus producing a softening of the EoS with a strong decrease of the maximum mass. Accurate theoretical calculations of hypernuclear matter, starting from the available information on both nucleon and hyperon interactions, were performed in the past years. In the Brueckner scheme, we used the phenomenological nucleon-hyperon potentials [40, 41, 42] as fundamental input, while the hyperon-hyperon potentials were neglected due to the lack of appropriate experimental data. We found very low maximum masses of hyperon stars, below 1.4 $M_\odot$ ($M_\odot = 2 \times 10^{33}$g). Several proposals of this so-called *hyperon puzzle* have been put forward [43, 44, 45], but many inconsistencies still remain, and the solution to this problem is still far from being understood.

2.1. The crust region

A NS interior consists of three main regions, namely, an outer crust, an inner crust, and a core. The outer crust region [46] consists of matter composed by neutron-rich nuclei permeated by a degenerate electron gas, at densities approximately between 10^4 g/cm^3, where atoms become completely ionized, and 4×10^{11} g/cm^3, where neutrons start to drip from the nuclei. The nuclei arrange themselves in a solid body-centred cubic (bcc) lattice in order to minimize the Coulomb repulsion, and are stabilized against β decay by the surrounding electron gas. At the low densities of the beginning of the outer crust, the Coulomb lattice is populated by ^{56}Fe nuclei. As the density of matter increases with increasing depth in the crust, it becomes energetically favourable for the system to lower the proton fraction through electron captures with the energy excess carried away by neutrinos. The system progressively evolves towards

IOP Conf. Series: Journal of Physics: Conf. Series **981** (2018) 012012 doi:10.1088/1742-6596/981/1/012012

a lattice of more and more neutron-rich nuclei as it approaches the bottom of the outer crust, until the neutron drip density is reached and the inner crust of the NS begins.

The structure of the inner crust consists of a Coulomb lattice of nuclear clusters permeated by the gases of free electrons and neutrons, such that the whole system is charge neutral. The neutron fraction increases with growing matter density. At the bottom layers of the inner crust, the nuclear clusters may adopt non-spherical shapes in order to minimize their energy [47, 48, 49, 50], and the equilibrium nuclear shape may change from sphere, to cylinder, slab, tube (cylindrical hole), and bubble (spherical hole) before going into uniform matter. These shapes are generically known as "nuclear pasta".

The EoS derived in the BHF scheme was used to construct the BCPM (Barcelona-Catania-Paris-Madrid) nuclear energy density functional for describing finite nuclei, as explained in ref. [2]. The BCPM functional is built up with a bulk part obtained directly from the BHF results in symmetric and neutron matter via the local density approximation. Additional ingredients of the functional are a phenomenological surface energy term, necessary to describe correctly the nuclear surface, together with the Coulomb, spin-orbit, and a pairing contribution to describe open-shell nuclei. The resulting functional has only four open parameters, which have been determined by minimizing the root-mean-square (rms) deviation σ_E between the theoretical and experimental binding energies of a set of well-deformed nuclei in the rare-earth, actinide, and super-heavy mass regions of the nuclear chart [3]. The predictive power of the functional is then tested by computing the binding energies of 467 known spherical nuclei. A fairly reasonable rms deviation $\sigma_E = 1.3$ MeV between theory and experiment is found. This gives confidence to the use of this functional for the study of the NS crust, where very asymmetric nuclei appear. In particular, the EoS of the outer crust requires the masses of neutron-rich nuclei, and those are obtained through Hartree-Fock-Bogoliubov calculations with the BCPM functional when they are unknown experimentally. To compute the inner crust, Thomas-Fermi (TF) calculations in Wigner-Seitz cells are performed with the same BCPM functional. Different shapes of the elementary cells of the lattice are tested in order to determine the most favourable structure. Existence of nuclear pasta is predicted in a range of average baryon densities between $\simeq 0.067$ fm^{-3} and $\simeq 0.0825$ fm^{-3}, where the transition to the core takes place. More details on the BCPM functional can be found in ref. [2].

2.2. *The high-density EOS*

The core is the internal region at densities larger than 1.5×10^{14} g/cm^3, where matter forms a homogeneous liquid composed of neutrons plus a certain fraction of protons, electrons, and muons that maintain the system in β equilibrium. Deep in the core, at still higher densities, strange baryons and even deconfined quarks may appear [51, 52]. In the center of a NS the baryon density can reach values that are several times larger than the nuclear saturation density. The core includes most of the NS mass and therefore the high-density EoS is crucial. We adopt a conventional description of the core as being composed by homogeneous asymmetric nuclear matter and a gas of electrons and muons. The EoS for the liquid core is derived in the framework of the BHF theory just described.

In order to study the structure of the NS core, we have to calculate the composition and the EoS of cold, neutrino-free, catalyzed matter in the following standard way. The Brueckner calculation yields the energy density of lepton/baryon matter as a function of the different partial densities,

$$\begin{aligned}\varepsilon(n_n,n_p,n_e,n_\mu) &= (n_n m_n + n_p m_p) + (n_n + n_p)\frac{E}{A}(n_n,n_p) \\ &+ \varepsilon(n_\mu) + \varepsilon(n_e)\,,\end{aligned} \tag{5}$$

where we have used ultrarelativistic and relativistic approximations for the energy densities of electrons and muons, respectively. In practice, it is sufficient to compute only the binding energy of symmetric nuclear matter and pure neutron matter, since within the BHF approach it has been verified [53, 54, 55] that a parabolic approximation for the binding energy of nuclear matter with arbitrary proton fraction

$x_p = n_p/n_b$, $n_b = n_n + n_p$, is well fulfilled,

$$\frac{E}{A}(n_b, x_p) \approx \frac{E}{A}(n_b, x_p = 0.5) + (1 - 2x_p)^2 E_{\text{sym}}(n_b)\,, \tag{6}$$

where the symmetry energy E_{sym} can be expressed in terms of the difference of the energy per particle between pure neutron ($x_p = 0$) and symmetric ($x_p = 0.5$) matter:

$$E_{\text{sym}}(n_b) = -\frac{1}{4}\frac{\partial(E/A)}{\partial x_p}(n_b, 0) \approx \frac{E}{A}(n_b, 0) - \frac{E}{A}(n_b, 0.5)\,. \tag{7}$$

Knowing the energy density Eq. (5), the various chemical potentials (of the species $i = n, p, e, \mu$) can be computed straightforwardly,

$$\mu_i = \frac{\partial \varepsilon}{\partial n_i}\,, \tag{8}$$

and the equations for beta-equilibrium,

$$\mu_i = b_i \mu_n - q_i \mu_e\,, \tag{9}$$

(b_i and q_i denoting baryon number and charge of species i) and charge neutrality,

$$\sum_i n_i q_i = 0\,, \tag{10}$$

allow one to determine the equilibrium composition n_i at given baryon density n_b and finally the EoS,

$$P(n_b) = n_b^2 \frac{d}{dn_b}\left(\frac{\varepsilon(n_i(n_b))}{n_b}\right) = n_b \frac{d\varepsilon}{dn_b} - \varepsilon = n_b \mu_n - \varepsilon\,. \tag{11}$$

2.3. Results and Discussion

In this Section, we will report on the main results obtained for the EoS, and compare with the few nuclear EoS that have been devised and used to cover the whole NS structure.

A partly phenomenological approach, based on the compressible liquid drop model, has been developed by Lattimer and Swesty (LS) [56]. It can cover the whole range of density, including the crust and the pasta phase, and it gives a complete description of the NS matter structure. This EoS is derived from a macroscopic functional, and it is compatible with an accurate mass formula throughout the nuclear table. There are different versions of this EoS, and here we present results obtained with the SKa version, which is characterized by an incompressibility K = 264 MeV.

Another approach with similar characteristics has been developed by Shen et al. (SH) [57, 58], and it is based on a nuclear relativistic mean field (RMF) model. The crust was described in the Thomas-Fermi scheme using the variational method with trial profiles for the nucleon densities. For the SH EoS the incompressibility is K = 281 MeV. The LS and SH EoS are widely used in astrophysical calculations for both neutron stars and supernova simulations due to their numerical simplicity and the large range of tabulated densities and temperatures.

Other approaches employ Skyrme forces to calculate the whole structure of NS. In ref. [59] the force SLy4 was used, and since this force was adjusted to reproduce the pure neutron matter EOS of Friedman and Pandharipande [60], this approach contains some microscopic input. In this case the incompressibility is about K = 230 MeV. This EoS is also known as the Douchin-Haensel EoS. More recently a similar approach was followed in refs. [61, 62], where the Skyrme forces BSk19, BSk20 and BSk21 were used, each one adjusted to a different microscopic neutron matter EOS with different stiffness at high density. Analytical fits of these neutron star EoS have been constructed in order to facilitate their inclusion in

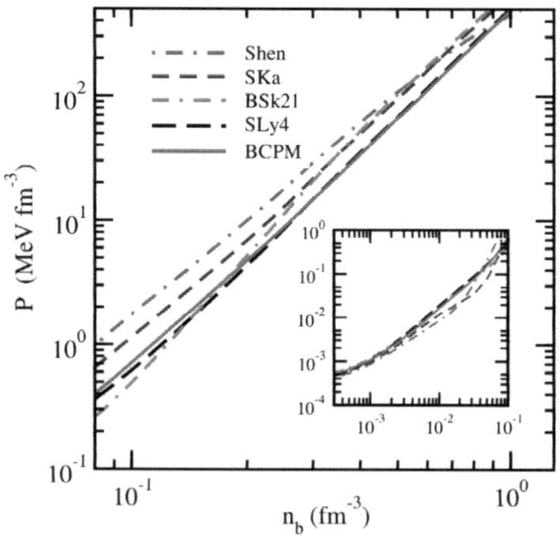

Figure 1. (Color online) The pressure P is displayed vs. the baryon density for several EoS. In the inset the inner crust EoS is plotted. See text for details.

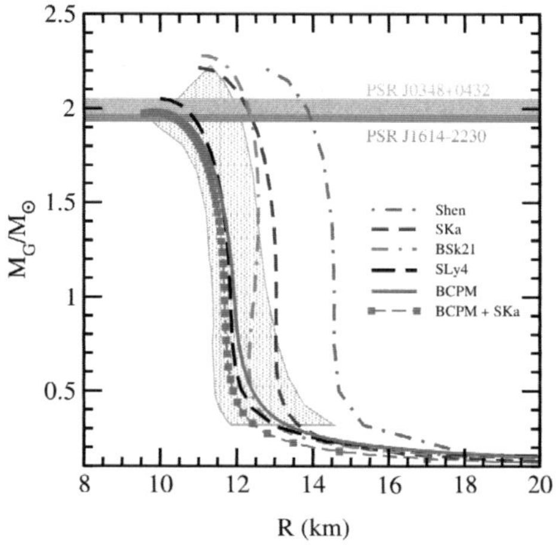

Figure 2. (Color online) The gravitational mass M_G is plotted vs. the radius R for the cases displayed in Fig.1.

astrophysical simulations [63]. Here we consider the BSk21 version of that EoS as a representative example, being characterized by an incompressibility $K = 246$ MeV.

In Fig. 1 we compare the EoS derived from the BHF theory and the BCPM functional (solid red) with the just mentioned EoS. The density range extends from the crust-core density, located around $0.08\,\text{fm}^{-3}$, up to typical NS core densities. We notice a remarkable similarity with the SLy4 EoS (long-dashed, black), whereas a strong difference with the SKa EoS (short-dashed, blue), the SH EoS (dot-dashed, green), and the BSk21 EoS (dot-dashed-dashed, magenta) is observed. In the inset we report the EoS of the inner crust starting from neutron drip $n_{drip} = 3\times10^{-4}\,\text{fm}^{-3}$. We observe that the pressures from the Lattimer-Swesty EoS and the Shen EoS differ significantly from BCPM and SLy4 EoS. However, the BCPM and SLy4 pressures in the inner crust show an impressive concordance with BSk21 that remains within the core region up to about $0.2\,\text{fm}^{-3}$, but is not maintained in the extrapolation to higher densities, where the BSk21 and Lattimer-Swesty models predict the stiffest EoS. We recall that, despite the crust contains a small fraction of the NS mass, its EoS and its matching with the core EoS can be relevant for the determination of the NS radius. Once the EoS of nuclear matter is known, one can use the well-known Tolman-Oppenheimer-Volkoff equations for spherically symmetric NS:

$$\begin{aligned}\frac{dP}{dr} &= -G\frac{\varepsilon m}{r^2}\left(1+\frac{P}{\varepsilon}\right)\left(1+\frac{4\pi P r^3}{m}\right)\left(1-\frac{2Gm}{r}\right)^{-1}\\ \frac{dm}{dr} &= 4\pi r^2\varepsilon,\end{aligned} \tag{12}$$

where G is the gravitational constant, P the pressure, ε the energy density, m the mass enclosed within a radius r, and r the (relativistic) radius coordinate. To close the equations we need the relation between pressure and density, $P = P(\varepsilon)$, i.e. just the EoS. Integrating these equations one gets the mass and radius of the star for a given central density. As shown in Fig.2, the considered EoS are compatible with the largest mass observed up to now, which is close to $2.01\pm0.04M_\odot$ in PSR J0348+0432 [64], and displayed in Fig.2, along with the previously observed mass of PSR J1614-2230 [65] having $M_G = 1.97\pm0.04M_\odot$. For completeness, we also display in the orange hatched band the probability distribution for M_G and R deduced from five PRE (Photospheric Radius Expansion) burst sources and five QLMXB sources, after a Bayesian analysis [66]. We see that, except the Shen EoS, all EoS are compatible with the observational

IOP Conf. Series: Journal of Physics: Conf. Series **981** (2018) 012012 doi:10.1088/1742-6596/981/1/012012

data. Moreover we find that the mass-radius relationship calculated with SLy4 EoS is surprisingly close to the one obtained with BHF EoS and the consistent BCPM functional for the crust, which can be explained by the close value of the respective bulk incompressibility. We also notice that the maximum mass calculated with the BCPM and the SLy4 EoS is characterized by a radius of about 10 km, which is somewhat smaller than the radius calculated with the other considered EoS, in line with recent analyses of observations on quiescent low-mass X-ray binaries (QLMXB) [67] and X-ray bursters [68], for which more studies could however be needed [69]. In order to study the role of the crust for the values of the NS radius, we have performed calculations using the BHF EoS for the core and the LS-SKa EoS for the crust. The results are reported in the same Fig. 2 (red squares). One can see that the use of LS-SKa crust EOS makes only a relatively small effect on the mass-radius relationship, being most relevant for masses smaller than $1\,M_\odot$, where the crust has a prominent relevance. This shows the importance of using the EoS derived from the same theoretical scheme both for the crust and the core. In fact, matching the core EoS with a crust EoS from a different model can introduce uncertainties for the NS structure.

3. Neutron star cooling simulations

A unified description of the EoS turns out to be very important also in the NS cooling simulations. We recall that the main cooling regulators are the neutrino emissivities and the nuclear pairing gaps. In most of the cooling simulations, the EoS and pairing gaps are treated in a disjoint and inconsistent way, i.e. they are obtained using different NN in-medium interaction. Recently a tentative step towards a consistent description has been performed [20].

We remind that in a non-superfluid scenario three main mechanisms of neutrino emission are usually taken into account, which are the direct Urca (DU), the modified Urca (MU), and the NN bremsstrahlung (BNN) processes. By far the most efficient mechanism of NS cooling is the DU process, for which the derivation of the neutrino emissivity Q under the condition of β equilibrium gives

$$Q^{(DU)} \approx 4.0 \times 10^{27} \left(\frac{\rho_p}{\rho_0}\right)^{1/3} \left(\frac{m_n^*}{m_n}\right) \left(\frac{m_p^*}{m_p}\right) T_9^6\, \Theta(k_{F_p} + k_{F_e} - k_{F_n})\,, \quad (13)$$

where T_9 is the temperature in units of 10^9K, and the emissivity $Q^{(DU)}$ is given in units of erg cm^{-3} s^{-1}. If muons are present, then the corresponding DU process may also become possible, in which case the neutrino emissivity is increased by a factor of 2. In Eq.(13) ρ_p is the proton population, which depends on the behaviour of the symmetry energy, while m_n^*, m_p^* are the neutron and proton effective masses. In the BHF approach they are easily calculated as

$$\frac{m^*(k)}{m} = \frac{k}{m}\left[\frac{de(k)}{dk}\right]^{-1}, \quad (14)$$

i.e. they are derived consistently from the BHF single-particle energy $e(k)$, Eq. (2), [21]. Although the effect is not large compared to other uncertainties, such a consistent treatment is hard to find in previous works. We recall that the BHF EoS predicts a density onset value for DU process close to $\rho = 0.44\,\mathrm{fm}^{-3}$, and therefore with our EoS fast cooling is active in nearly all stars with $M/M_\odot > 1.10$. This is at variance with the APR EoS [70], widely used in cooling simulations, where the DU process is delayed to $\rho = 0.82\,\mathrm{fm}^{-3}$, and therefore it is active only in the heaviest stars, $M/M_\odot > 2.03$. This has strong consequences for the cooling behaviour.

Of vital importance for any cooling simulation is the knowledge of the 1S0 and 3PF2 pairing gaps for neutrons and protons in beta-stable matter, which on one hand block the DU and MU reactions, and on the other hand open new cooling channels by the pair breaking and formation (PBF) mechanism for stellar matter in the vicinity of the critical temperature [6]. In this work we focus on the most important proton 1S0 (p1S0) and neutron 3PF2 (n3P2) pairing channels. As already stressed, the gaps should be computed in a framework that is consistent with the determination of the EOS, i.e., be based on the same

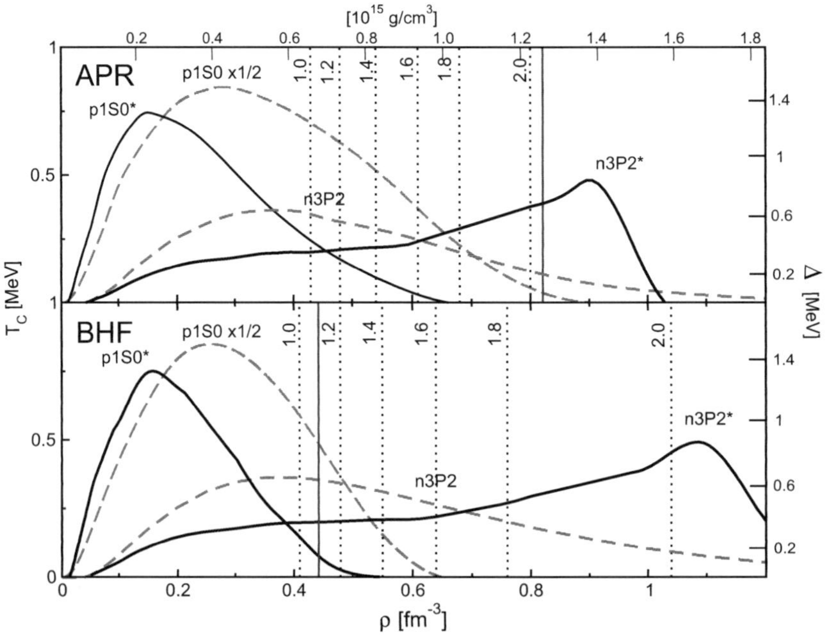

Figure 3. (Color online) Pairing gaps in NS matter in the p1S0 and n3P2 channels, including (*) or not effective mass effects. In the upper (lower) panel results are displayed for the APR(BHF) EoS. Vertical dotted lines indicate the central density of NS with different masses $M/M_{\odot} = 1.0, \ldots, 2.0$. DU onset occurs at the vertical solid (blue) lines.

NN interaction and using the same medium effects (TBF and effective masses). Indeed in ref. [71] the same Argonne V_{18} + UIX Urbana nuclear interaction and BHF s.p. spectra were employed for the calculation of the gaps by solving the (angle-averaged) gap equation on the BCS level [72, 73, 74, 75, 76].

Fig. 3 displays the p1S0 and n3P2 pairing gaps as a function of baryonic density of beta-stable matter for the APR (upper panel) and the fully consistent BHF (lower panel) EoS models. Pairing gaps calculated with only the two-body force V_{18} together with the kinetic s.p. energies are represented by dashed-red curves, whereas curves denoted by p1S0* and n3P2* are obtained using the BHF s.p. spectra. Since polarization corrections are uncertain [77, 78, 79], we use scaling factors for the proton and neutron gaps, s_p and s_n, respectively. Qualitatively one observes in Fig. 3 the natural scaling effect of the different proton fractions for the BHF and APR EoS, such that the p1S0 gaps extend to larger (central) densities for the APR model, due to the lower proton fraction in that case. Therefore the blocking effect on the cooling extends up to higher densities and NS masses for the APR model. In both cases, the n3P2$^{(*)}$ gaps extend up to very large density and can thus provide an efficient means to block this cooling process, in particular for the n3P2* model comprising medium effects. The price to pay is an enhanced PBF cooling rate close to the critical temperature in that case.

A further important ingredient of the cooling simulations is the (leptonic and baryonic) thermal conductivity [6]. Recently it has been conjectured [14, 80, 81, 82] that the conductivities could be substantially suppressed by in-medium effects, and this could explain the rapid Cas A cooling. Hence we introduce a further scale factor s_κ for the total thermal conductivity. In the next subsection we present some results for certain choices of the three global parameters, i.e. s_p, s_n, s_κ.

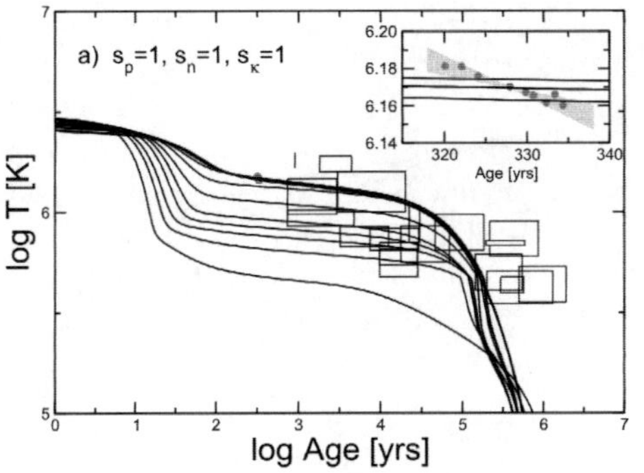

Figure 4. (Color online) Cooling curves without scaling factors for different NS masses $M/M_\odot = 1.0, 1.1, \ldots, 2.0$. BCS gaps with free s.p. spectra are employed. Red dots show the Cas A cooling data (enlarged in the inset).

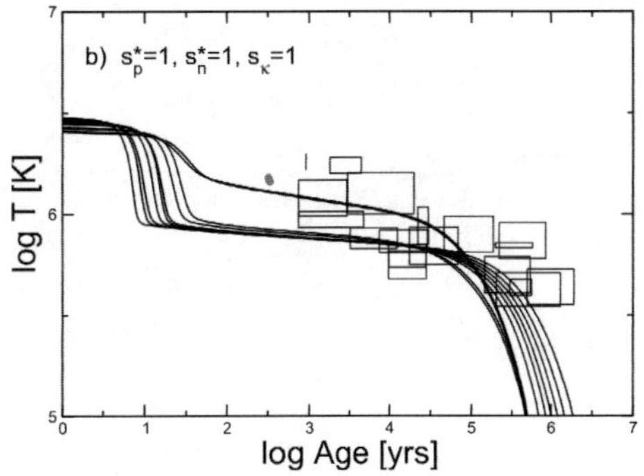

Figure 5. (Color online) Same as Fig.4, but BCS gaps are calculated with the BHF s.p. spectra.

3.1. Neutron star cooling and Cas A

We carried out the NS cooling simulations employing the widely used code `NSCool` [83], and including in our set of observational cooling data the (age, temperature) information of the 19 isolated NS sources listed in [84].

We begin by showing the results obtained with the original pairing gaps shown in Fig. 3, and without any modification of the conductivities, i.e., putting $s_p = s_n = s_k = 1$. Only the BHF EoS was used. Fig. 4 shows the temperature vs age results (11 curves for NS with masses $1.0, 1.1, \ldots, 2.0$) obtained with the BCS gaps without any medium modification (dashed red curves in the lower panel of Fig. 3), while Fig. 5 employs the BCS gaps with BHF effective masses (black curves in Fig. 3), and is indicated by the notation $s_p^* = s_n^* = 1$ here and in the following. One observes results in line with the features of the pairing gaps, namely in Fig. 4 light ($M < 1.4M_\odot$) NS cool slower and heavy ($M > 1.7M_\odot$) NS cool faster than in Fig. 5. This is due to the larger overall values of the corresponding BCS gaps in the low-density ($n < 0.6\,\text{fm}^{-3}$) region and the smaller n3P2 value in the high-density ($n > 0.7\,\text{fm}^{-3}$) domain, see Fig. 3, which cause, respectively, a stronger or weaker blocking of the dominant DU process in light or heavy stars. In both scenarios, it is possible to explain cold stars due to the DU mechanism in the BHF model, but they cannot reproduce the particular cooling properties of Cas A: While the first one can fit its current age and temperature as a $M = 1.2M_\odot$ NS, neither reproduces the apparent very fast current cooldown [7, 8, 9] shown in the inset of Fig. 4.

Alternative scenarios have been proposed, e.g. one based on an appropriately chosen n3P2 gap, which causes strong cooling due to the opening of the neutron PBF process at the current age/temperature of the star. The BHF EOS including strong DU reactions also allows this interpretation by choosing the scaling factors $s_p = 2.0$, $s_n = 0.132$, $s_\kappa = 1$ for a $1.4M_\odot$ star, corresponding to maximum values of the gaps $\Delta_p \approx 6\,\text{MeV}$ and $\Delta_n \approx 0.1\,\text{MeV}$, i.e., Δ_p is larger than usually chosen while Δ_n is in line with the equivalent results of [85, 10, 11, 12]. The results are shown in Fig. 6, which displays the sequence of cooling curves for NS masses $M/M_\odot = 1.0, 1.1, \ldots, 2.0$, the 1.4 case corresponding to Cas A by construction. In this case the neutron $^1S_0$ gap has been rescaled by a factor 0.04 for fine tuning.

As in similar investigations [13, 85], one notes that the rapid cooldown caused by the n3P2 PBF renders even more difficult the reproduction of very old and warm stars, which are not covered by any cooling curve. As already mentioned, a possible alternative scenario predicts a strong suppression of the lepton and baryon thermal conductivities due to in-medium effects [14, 82]. The reduced conductivity serves to

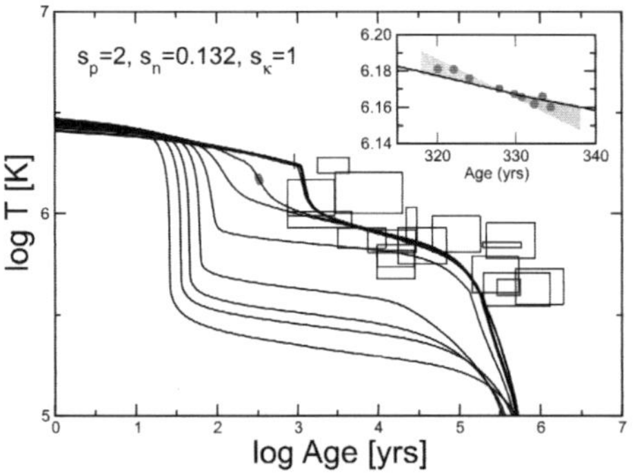

Figure 6. (Color online) Same as Fig. 4, but for the PBF cooling scenario. The M = $1.4 M_\odot$ cooling curve passes through Cas A by construction.

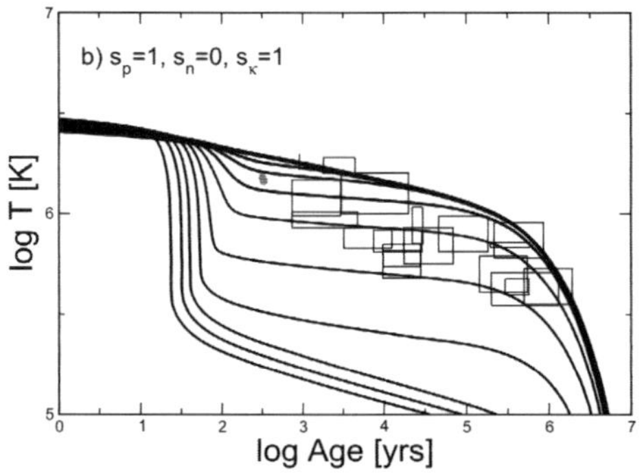

Figure 7. Same as Fig. 4, using $s_p = 1$, $s_n = 0$ and $s_k = 1$.

delay the temperature decline up to the current age of Cas A without need to introduce fine-tuned nPBF cooling. Our EoS including DU cooling also accommodates this possibility, and a rather satisfactory fit of all cooling data including Cas A, is obtained employing the parameter set $s_p^* = 1$, $s_n^* = 0$, $s_K = 0.135$, where the size of s_K is comparable to the values of about 0.2 deduced in [14, 82]. In this scenario a rather small value of the n3P2 gap seems to be required, as otherwise old hot (and also young cold) NS cannot be obtained, even if Cas A is reproduced. More details are reported in ref.[20].

The previous results have demonstrated that it is difficult to satisfy simultaneously the slow cooling of old NS and the rapid cooling of Cas A, in which case some doubts have been expressed about the validity of the data analysis [86], and a future revision towards much slower or no cooling at all is not excluded.

We therefore study finally a scenario without the Cas A constraint (apart from reproducing its current age and temperature with a reasonable mass) in our strong DU model, trying to cover the full range of current cooling data. Considering that the use of the unscaled BCS gaps in Fig. 4 yields already a reasonable reproduction of most young NS, and considering the fact that a finite n3P2 gap produces too strong PBF cooling, simply switching off this channel yields a nearly perfect coverage of all current cooling data, as shown in Fig. 7. In this scenario Cas A turns out to be a $1.31 M_\odot$ NS. Thus the BCS p1S0 gap alone is able to suppress sufficiently the DU cooling, provided that it extends over a large enough density range. Though in our case the p1S0 gap is perhaps somewhat large, we have investigated the effect of its rescaling, finding excellent cooling simulations, with the predicted mass of Cas A being between $1.18 M_\odot$ and $1.46 M_\odot$. Therefore a precise information on the masses of the NS in the cooling diagram is necessary, without which no theoretical cooling model can be verified.

4. Conclusions

We have presented a study of NS structure by focusing on microscopic models which are able to describe on the same theoretical framework both the core and the crust regions. To this purpose we have performed calculations of NS structure based on the EoS derived within the Brueckner-Hartree-Fock scheme. This EoS is the basis of the BCPM, a recent and accurate nuclear energy density functional [2] devised to reproduce nuclear binding energies throughout the nuclear mass table and, as such, used also to calculate the structure of NS crust. The resulting EoS has been used to calculate the overall NS structure, from the crust to the core, and the results compared with other more phenomenological models that also are able to describe the whole NS structure. In particular, we have chosen the Lattimer-Swesty [56], Shen [57, 58], BSk21 [61, 62], and the Douchin-Haensel EoS [59]. The role of the NS crust EoS has been emphasized as far as the radius for for the lightest NS masses is concerning. This indicates the relevance of using models that are based on the same physical framework both for the crust and the core.

Following the same point of view, we have studied NS cooling using the microscopic BHF EoS, which features strong DU reactions already for $M/M_{\odot} > 1.10$, and using compatible p1S0 and n3P2 pairing gaps as well as nucleon effective masses. The current substantial theoretical uncertainty regarding gaps and thermal conductivity was modelled in a rather simple way by introducing three global scale factors. We have found that it is possible to reproduce the apparent fast cooling of Cas A by either fine tuning the n3P2 pairing gap or reducing the thermal conductivity. In general it is difficult to fit simultaneously the rapid cooling of Cas A and old and warm NS, but once the Cas A constraint is relaxed, all current cooling data can be fit by just assuming the p1S0 BCS gap and a vanishing n3P2 gap.

In summary, it is very difficult to draw quantitative conclusions from the current NS cooling data on the properties of matter at large densities if the masses of the cooling objects are unknown, due to the large variety of required microphysics input.

Acknowledgments
We acknowledge useful discussions with M. Baldo, H.-J. Schulze, X. Viñas and W. Ho. Partial support comes from "NewCompStar," COST Action MP1304.

5. References

[1] Baldo M and Burgio G F 2012 *Rep. Prog. Phys.* **75** 026301
[2] Baldo M, Robledo L M, Schuck P and Viñas X 2013 *Phys. Rev. C* **87** 064305
[3] Baldo M, Schuck P and Viñas X 2008 *Physics Letters B* **663** 390–394
[4] Baldo M, Robledo L, Schuck P and Viñas X 2010 *Journal of Physics G Nuclear Physics* **37** 064015
[5] Kohn W and Sham L J 1965 *Physical Review* **140** 1133–1138
[6] Yakovlev D G, Kaminker A D, Gnedin O Y and Haensel P 2001 *Phys. Rep.* **354** 1–155
[7] Ho W C G and Heinke C O 2009 *Nature* **462** 71–73
[8] Heinke C O and Ho W C G 2010 *Astrophys. J. Lett.* **719** L167–L171
[9] Elshamouty K G, Heinke C O, Sivakoff G R, Ho W C G, Shternin P S, Yakovlev D G, Patnaude D J and David L 2013 *Astrophys. J.* **777** 22
[10] Page D, Prakash M, Lattimer J M and Steiner A W 2011 *Phys. Rev. Lett.* **106** 081101
[11] Yakovlev D G, Ho W C G, Shternin P S, Heinke C O and Potekhin A Y 2011 *Mon. Not. R. Astron. Soc.* **411** 1977–1988
[12] Shternin P S, Yakovlev D G, Heinke C O, Ho W C G and Patnaude D J 2011 *Mon. Not. R. Astron. Soc.* **412** L108–L112
[13] Page D, Lattimer J M, Prakash M and Steiner A W 2009 *Astrophys. J.* **707** 1131–1140
[14] Blaschke D, Grigorian H, Voskresensky D N and Weber F 2012 *Phys. Rev. C* **85** 022802
[15] Bonanno A, Baldo M, Burgio G F and Urpin V 2014 *Astron. Astrophys.* **561** L5
[16] Sedrakian A 2013 *Astron. Astrophys.* **555** L10
[17] Li Z H and Schulze H J 2008 *Phys. Rev. C* **78** 028801
[18] Burgio G F and Schulze H J 2010 *Astron. Astrophys.* **518** A17
[19] Taranto G, Baldo M and Burgio G 2013 *Phys. Rev. C* **87** 045803
[20] Taranto G, Burgio G F and Schulze H J 2016 *Mon. Not. R. Astron. Soc.* **456** 1451–1458
[21] Baldo M, Burgio G F, Schulze H J and Taranto G 2014 *Phys. Rev. C* **89** 048801
[22] Baldo M (ed) 1999 *Nuclear Methods and The Nuclear Equation of State* (Singapore: World Scientific)
[23] Wiringa R B, Stoks V and Schiavilla R 1995 *Phys. Rev. C* **51** 38–51
[24] Song H, Baldo M, Giansiracusa G and Lombardo U 1998 *Phys. Rev. Lett.* **81** 1584–1587
[25] Baldo M, Bombaci I and Burgio G 1997 *Astron.Astrophys.* **328** 274–282
[26] Zhou X R, Burgio G F, Lombardo U, Schulze H J and Zuo W 2004 *Phys. Rev. C* **69** 018801
[27] Lovato A, Benhar O, Fantoni S, Illarionov A Y and Schmidt K E 2011 *Phys. Rev.* **C83** 054003
[28] Carlson J, Pandharipande V and Wiringa R B 1983 *Nucl. Phys. A* **401** 59–85
[29] Schiavilla R, Pandharipande V and Wiringa R B 1986 *Nucl. Phys. A* **449** 219–242
[30] Li Z, Lombardo U, Schulze H J and Zuo W 2008 *Phys. Rev. C* **77** 034316
[31] Li Z and Schulze H 2012 *Phys. Rev. C* **85** 064002
[32] Lovato A, Benhar O, Fantoni S and Schmidt K E 2012 *Phys. Rev. C* **85** 024003
[33] Logoteta D, Vidaña I, Bombaci I and Kievsky A 2015 *Phys. Rev. C* **91** 064001
[34] Coon S A and Han H K 2001 *Few-Body Systems* **30** 131–141
[35] Navrátil P 2007 *Few-Body Systems* **41** 117–140
[36] Baldo M and Fukukawa K 2014 *Phys. Rev. Lett.* **113** 242501
[37] Fukukawa K, Baldo M, Burgio G F, Lo Monaco L and Schulze H J 2015 *Phys. Rev. C* **92** 065802

[38] Baldo M, Burgio G F and Schulze H 1998 *Phys. Rev. C* **58** 3688–3695
[39] Baldo M, Burgio G and Schulze H 2000 *Phys. Rev. C* **61** 055801
[40] Maessen P M M, Rijken T A and de Swart J J 1989 *Phys. Rev. C* **40** 2226–2245
[41] Stoks V G J and Rijken T A 1999 *Phys. Rev. C* **59** 3009–3020
[42] Schulze H J and Rijken T 2011 *Phys. Rev.* **C84** 035801
[43] Lonardoni D, Pederiva F and Gandolfi S 2014 *Phys. Rev. C* **89** 014314
[44] Vidaña I, Logoteta D, Providencia C, Polls A and Bombaci I 2011 *Europhys. Lett.* **94** 11002
[45] Yamamoto Y, Furumoto T, Yasutake N and Rijken T A 2014 *Phys. Rev. C* **90** 045805
[46] Baym G, Pethick C and Sutherland P 1971 *Astrophys. J.* **170** 299
[47] Baym G, Bethe H A and Pethick C J 1971 *Nuclear Physics A* **175** 225–271
[48] Ravenhall D G, Pethick C J and Wilson J R 1983 *Phys. Rev. Lett.* **50**(26) 2066–2069
[49] Lorenz C P, Ravenhall D G and Pethick C J 1993 *Phys. Rev. Lett.* **70**(4) 379–382
[50] Oyamatsu K 1993 *Nuclear Physics A* **561** 431–452
[51] Shapiro S L and Teukolsky S A 1983 *Black holes, white dwarfs, and neutron stars: The physics of compact objects* (New York: Wiley-Interscience)
[52] Haensel P, Potekhin A Y and Yakovlev D G 2006 *Neutron Stars 1: Equation of State and Structure* (New York: Springer)
[53] Bombaci I and Lombardo U 1991 *Phys. Rev. C* **44** 1892–1900
[54] Baldo M, Bombaci I and Burgio G F 1997 *Astron. Astrophys.* **328** 274–282
[55] Baldo M, Burgio G F and Schulze H J 2000 *Phys. Rev. C* **61** 055801
[56] Lattimer J M and Swesty D F 1991 *Nuclear Physics A* **535** 331–376
[57] Shen H, Toki H, Oyamatsu K and Sumiyoshi K 1998 *Progress of Theoretical Physics* **100** 1013–1031
[58] Shen H, Toki H, Oyamatsu K and Sumiyoshi K 1998 *Nucl. Phys. A* **637** 435–450
[59] Douchin F and Haensel P 2001 *Astron. Astrophys.* **380** 151–167
[60] Friedman B and Pandharipande V R 1981 *Nucl. Phys. A* **361** 502–520
[61] Pearson J M, Chamel N, Goriely S and Ducoin C 2012 *Phys. Rev. C* **85** 065803
[62] Fantina A F, Chamel N, Pearson J M and Goriely S 2013 *Astron. Astrophys.* **559** A128
[63] Potekhin A Y, Fantina A F, Chamel N, Pearson J M and Goriely S 2013 *Astron. Astrophys.* **560** A48
[64] Antoniadis J, Freire P, Wex N, Tauris T, Lynch R and et al 2013 *Science* **340** 1233232
[65] Demorest P B, Pennucci T, Ransom S M, Roberts M S E and Hessels J W T 2010 *Nature* **467** 1081–1083
[66] Lattimer J M and Steiner A W 2014 *European Physical Journal A* **50** 40
[67] Guillot S and Rutledge R E 2014 *Astrophys. J. Lett.* **796** L3
[68] Güver T and Özel F 2013 *Astrophys. J. Lett.* **765** L1
[69] Lattimer J M and Steiner A W 2014 *Astrophys. J.* **784** 123
[70] Akmal A, Pandharipande V R and Ravenhall D G 1998 *Phys. Rev. C* **58** 1804–1828
[71] Zhou X R, Schulze H J, Zhao E G, Pan F and Draayer J P 2004 *Phys. Rev. C* **70** 048802
[72] Amundsen L and Østgaard E 1985 *Nucl. Phys. A* **442** 163–188
[73] Baldo M, Cugnon J, Lejeune A and Lombardo U 1992 *Nucl. Phys. A* **536** 349–365
[74] Elgarøy Ø, Engvik L, Hjorth-Jensen M and Osnes E 1996 *Nucl. Phys. A* **607** 425–441
[75] Khodel V A, Khodel V V and Clark J W 1998 *Phys. Rev. Lett.* **81** 3828–3831
[76] Baldo M, Elgarøy Ø, Engvik L, Hjorth-Jensen M and Schulze H J 1998 *Phys. Rev. C* **58** 1921–1928
[77] Lombardo U and Schulze H J 2001 Superfluidity in Neutron Star Matter *Physics of Neutron Star Interiors* (*Lecture Notes in Physics, Berlin Springer Verlag* vol 578) ed Blaschke D, Glendenning N K and Sedrakian A p 30
[78] Sedrakian A and Clark J W 2006 *Nuclear Superconductivity in Compact Stars: BCS Theory and Beyond* (World Scientific Publishing Co) p 135
[79] Baldo M and Schulze H J 2007 *Phys. Rev. C* **75** 025802
[80] Shternin P S and Yakovlev D G 2007 *Phys. Rev. D* **75** 103004
[81] Shternin P S and Yakovlev D G 2008 *Astron. Lett.* **34** 675–685
[82] Blaschke D, Grigorian H and Voskresensky D N 2013 *Phys. Rev. C* **88** 065805
[83] Page D 2010 Nscool code available at http://www.astroscu.unam.mx/neutrones/nscool
[84] Beznogov M V and Yakovlev D G 2015 *Mon. Not. R. Astron. Soc.* **447** 1598–1609
[85] Ho W C G, Elshamouty K G, Heinke C O and Potekhin A Y 2015 *Phys. Rev. C* **91** 015806
[86] Posselt B, Pavlov G G, Suleimanov V and Kargaltsev O 2013 *Astrophys. J.* **779** 186

IOP Conf. Series: Journal of Physics: Conf. Series **981** (2018) 012013 doi:10.1088/1742-6596/981/1/012013

Three–Dimensional parton structure of light nuclei

Sergio Scopetta[1], Alessio Del Dotto[2,3], Leonid Kaptari[4], Emanuele Pace[5], Matteo Rinaldi[6] and Giovanni Salmè[2]

[1] Dipartimento di Fisica e Geologia, Università di Perugia and INFN, Sezione di Perugia, Italy
[2] INFN, Sezione di Roma, Italy
[3] University of South Carolina, Columbia, SC 29208, USA
[4] Bogoliubov Lab. Theor. Phys., 141980, JINR, Dubna, Russia
[5] Dipartimento di Fisica, Università di Roma Tor Vergata and INFN, Sezione di Roma Tor Vergata, Italy
[6] Departamento de Fìsica Teòrica, Universidad de Valencia and IFIC, CSIC, Valencia, Spain

Abstract. Two promising directions beyond inclusive deep inelastic scattering experiments, aimed at unveiling the three dimensional structure of the bound nucleon, are reviewed, considering in particular the ^{3}He nuclear target. The 3D structure in coordinate space can be accessed through deep exclusive processes, whose non-perturbative part is encoded in generalized parton distributions. In this way, the distribution of partons in the transverse plane can be obtained. As an example of a deep exclusive process, coherent deeply virtual Compton scattering off ^{3}He nuclei, important to access the neutron generalized parton distributions (GPDs), will be discussed. In Impulse Approximation (IA), the sum of the two leading twist, quark helicity conserving GPDs of ^{3}He, H and E, at low momentum transfer, turns out to be dominated by the neutron contribution. Besides, a technique, able to take into account the nuclear effects included in the Impulse Approximation analysis, has been developed. The spin dependent GPD $\tilde{H}$ of ^{3}He is also found to be largely dominated, at low momentum transfer, by the neutron contribution. The knowledge of the GPDs H, E and $\tilde{H}$ of ^{3}He is relevant for the planning of coherent DVCS off ^{3}He measurements. Semi-inclusive deep inelastic scattering processes access the momentum space 3D structure parameterized through transverse momentum dependent parton distributions. A distorted spin-dependent spectral function has been recently introduced for ^{3}He, in a non-relativistic framework, to take care of the final state interaction between the observed pion and the remnant in semi-inclusive deep inelastic electron scattering off transversely polarized ^{3}He. The calculation of the Sivers and Collins single spin asymmetries for ^{3}He, and a straightforward procedure to effectively take into account nuclear dynamics and final state interactions, will be reviewed. The Light-front dynamics generalization of the analysis is also addressed.

1. Introduction

The nucleus is a unique laboratory for studies of the QCD hadron structure. Indeed, nuclear targets are required, e.g., for the extraction of the neutron information from light nuclei and a precise flavor separation of parton distributions (PDFs), for the measurement of nuclear PDFs, relevant for the analysis of heavy ions collisions aimed at producing quark-gluon plasma, or to study in-medium fragmentation, mandatory to clarify the dynamics of hadronization. Unfortunately, inclusive Deep Inelastic Scattering (DIS) off nuclei has proven to be unable to answer fundamental questions, such as the quantitative microscopic explanation of the European Muon Collaboration (EMC) effect [1], i.e., the medium modification of PDFs, the quantitative

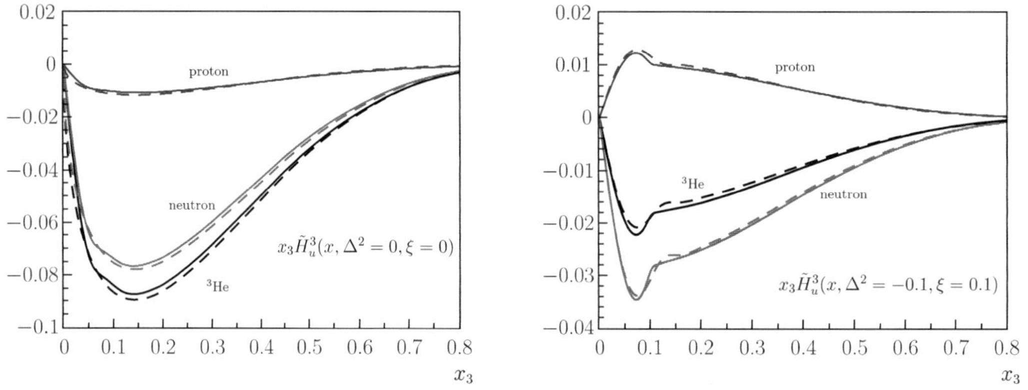

Figure 1. The $q = u$ flavor GPD $x_3\tilde{H}^{3,u}(x, \Delta^2, \xi)$, where $x_3 = (M_3/M)x$ and $\xi_3 = (M_3/M)\xi$, shown in the forward limit (left panel) and at $\Delta^2 = -0.1$ GeV^2 and $\xi_3 = 0.1$ (right panel), together with the neutron and the proton contribution. Solid lines correspond to the full IA result, Eq. (2), while the dashed ones represent the approximation Eq. (6).

understanding of the parton structure of the neutron, the study in-medium of the distribution of parton transverse momentum, relevant for studies of hadronization as well as of chiral-odd quantities, such as the transversity PDFs or the Sivers and Collins functions and asymmetries. Thanks to novel coincidence measurements, possible at high luminosity facilities such as Jefferson Laboratory (JLab), a new era in the knowledge of the parton structure of nuclei has started [2]. In particular, two promising directions beyond inclusive measurements, aimed at accessing the three dimensional (3D) structure of the bound nucleon, are deep exclusive processes off nuclei, and semi-inclusive deep inelastic scattering (SIDIS) involving nuclear targets. In deep exclusive processes, such as deeply virtual Compton Scattering (DVCS), one accesses the 3D structure in coordinate space in terms of generalized parton distributions (GPDs) [3]; in SIDIS, the momentum space 3D structure is obtained by studying transverse momentum dependent parton distributions (TMDs) [3]. In the following, we summarize relevant results obtained by our group, for the ^{3}He target, in the last few years, mainly with the aim of extracting the neutron information. The next section is devoted to the analysis of ^{3}He GPDs, entering the DVCS cross sections of experiments which could be planned at the high-luminosity facilities; in the third section, an analysis of SIDIS off ^{3}He, taking into account final state interactions between the produced meson and the remnants, will be reviewed. In the last section, perspectives are addressed and conclusions are drawn.

2. Deeply virtual Compton Scattering off ^{3}He

Initially introduced in Ref. [4], Generalized Parton Distributions (GPDs) are a crucial source of unique information. They represent in particular a tool to shed light on the so called "Spin Crisis" problem. GPDs measurements will allow indeed to access the parton total angular momentum [5]. Then, by subtracting from the latter the helicity quark contribution, measured in other independent processes, the parton orbital angular momentum (OAM) could be then estimated.

Deeply Virtual Compton Scattering (DVCS), i.e. the process $eH \longmapsto e'H'\gamma$ when $Q^2 \gg M^2$ ($Q^2 = -q \cdot q$) is the transferred momentum between the leptons e and e', Δ^2 the one between hadrons H and H' with momenta P and P', and M is the nucleon mass), is one of the cleanest processes to access GPDs. The so called skewedness, $\xi = -\Delta^+/(P^+ + P'^+)$ [1] is another relevant kinematical variable.

Despite serious difficulties to extract GPDs from experiments, data for proton and nuclear targets are being analyzed, see, i.e., Refs. [6, 7]. The measurement of GPDs for nuclei could be very useful to distinguish between different mechanisms of nuclear medium modifications of the nucleon structure, an impossible task analyzing DIS experiments only (this discussion started in Ref. [8]). As usual, the neutron measurement, which requires nuclear targets, is important. It permits in facts, together with the proton measurement, a flavor decomposition of GPDs. In studying observables related to the neutron polarization, ^{3}He plays a special role, due to its spin structure (see, e.g., Refs. [9, 10]). This is valid also for GPDs studies. Indeed, among the light nuclei ^{3}He is the only one for which the quantity $\tilde{G}_M^{3,q}(x,\Delta^2,\xi) = H_q^3(x,\Delta^2,\xi) + E_q^3(x,\Delta^2,\xi)$, i.e., the sum of its quark-helicity independent GPDs H_q and E_q, could be dominated by the neutron [11]: in facts, the isoscalar targets ^{2}H and ^{4}He are not suitable to this aim. In Refs. [11], it has been also shown to what extent this fact can be used to extract the neutron information.

The analysis of ^{3}He GPDs in Impulse Approximation (IA) can be found in Refs. [12], where, for the GPD H of ^{3}He, H_q^3, convolution-like formulas in terms of the corresponding nucleon quantities has been derived. Later, the treatment has been extended to $\tilde{G}_M^{3,q}$ (see Ref. [11] for details), and to the quark helicity flip GPD $\tilde{H}_q^3$ [13], yielding

$$\tilde{G}_M^{3,q}(x,\Delta^2,\xi) = \sum_N \int dE \int d\vec{p}\; [P^N_{+-,+-} - P^N_{+-,-+}]\,(\vec{p},\vec{p}\,',E)\frac{\xi'}{\xi}\tilde{G}_M^{N,q}(x',\Delta^2,\xi'), \tag{1}$$

and

$$\tilde{H}_q^3(x,\Delta^2,\xi) = \sum_N \int dE \int d\vec{p}\,[P^N_{++,++} - P^N_{++,--}]\,(\vec{p},\vec{p}\,',E)\frac{\xi'}{\xi}\tilde{H}_q^N(x',\Delta^2,\xi')\;, \tag{2}$$

respectively.

In the last two equations, x' and ξ' are the variables for the GPDs of the bound nucleons, $p\,(p' = p + \Delta)$ is its 4-momentum in the initial (final) state and, eventually, proper components appear of the one body, off diagonal, spin dependent spectral function:

$$P^N_{SS',ss'}(\vec{p},\vec{p}\,',E) = \frac{1}{(2\pi)^6}\frac{M\sqrt{ME}}{2}\int d\Omega_t \sum_{s_t}\langle \vec{P}'S'|\vec{p}\,'s',\vec{t}s_t\rangle_N \langle \vec{p}s,\vec{t}s_t|\vec{P}S\rangle_N\;, \tag{3}$$

where $S, S'(s,s')$ are respectively the nuclear (nucleon) spin projections in the initial (final) state, and $E = E_{min} + E_R^*$, being E_R^* the excitation energy of the interacting two-body recoiling system. The main ingredient in the definition Eq. (3) is the intrinsic overlap integral

$$\langle \vec{p}\;s,\vec{t}\;s_t|\vec{P}S\rangle_N = \int d\vec{y}\,e^{i\vec{p}\cdot\vec{y}}\langle \chi_N^s, \Psi_t^{s_t}(\vec{x})|\Psi_3^S(\vec{x},\vec{y})\rangle \tag{4}$$

between the ^{3}He wave function, Ψ_3^S, and the final state, described by two different wave functions: *i)* the eigenfunction $\Psi_t^{s_t}$, with eigenvalue $E = E_{min} + E_R^*$, of the state s_t of the intrinsic Hamiltonian of the system of two *interacting* nucleons with relative momentum $\vec{t}$, which can be either a deuteron nucleus or a scattering state, and *ii)* the plane wave describing the nucleon N in IA. For a numerical evaluation of Eqs. (1) and (2), we used the overlaps Eq. (4) appearing in Eq. (3) and corresponding to the analysis presented in Ref. [14] in terms of wave functions [15] evaluated using the AV18 interaction [16].

[1] In this paper, $a^\pm = (a^0 \pm a^3)/\sqrt{2}$.

For the nucleonic $\tilde{G}_M^{N,q}$ and $\tilde{H}_q^N$, use has been made of a simple model [17], extended to evaluate also spin dependent GPDs (see Ref. [11, 13] for details). Since ^{3}He data are not yet available, one can verify only a few general GPDs properties, i.e., the forward limit and the first moments. In particular the calculation of $H_q^3(x,\Delta^2,\xi)$ fulfills these constraints [12]. Since there is no observable forward limit for $E_q^3(x,\Delta^2,\xi)$, for the $\tilde{G}_M^{3,q}(x,\Delta^2,\xi)$ calculation the only possible check is the first moment: $\sum_q \int dx\, \tilde{G}_M^{3,q}(x,\Delta^2,\xi) = G_M^3(\Delta^2)$; where $G_M^3(\Delta^2)$ is the magnetic form factor (ff) of ^{3}He. The obtained result was found to be in agreement with previous calculations (in particular, with the one-body part of the AV18 calculation presented in Ref. [18]) and, for the low values of Δ^2 which are relevant for the coherent process described here, i.e., $-\Delta^2 \le 0.15$ GeV^2, our results compare well also with the data.

As an example, let us discuss now, in a little more detail, our calculation of $\tilde{H}_q^3$ [13]. We checked first of all that the forward limit of our expression, Eq. (2), reproduces formally and quantitatively the results of Ref. [9] for polarized DIS off ^{3}He. On the other hand, the first moment of $\tilde{H}_q^3$ is related to the axial form factor of ^{3}He, a poorly known observable which does not permit therefore a direct check. Having fulfilled at least the forward limit constraint, we proceed now to analyze the proton and neutron contributions to the ^{3}He observable. The GPD $\tilde{H}_q^3$ is measured using a polarized target and, as a consequence, it should be dominated by the neutron contribution. Let us show now to what extent this property is obtained and how, thanks to this, the neutron information could be extracted.

Results of the numerical evaluation of Eq. (2) are shown in Fig. 1. In the forward limit, the neutron contribution strongly dominates the ^{3}He quantity but, increasing Δ^2, the proton contribution grows up, in particular for the u flavor (see solid lines in Fig. 1). A procedure to safely extract the neutron information from ^{3}He data is therefore necessary. This can be obtained by observing that Eq. (2) can be cast in the form

$$\tilde{H}_q^3(x,\Delta^2,\xi) = \sum_N \int_{x_3}^{\frac{M_A}{M}} \frac{dz}{z} h_N^3(z,\Delta^2,\xi) \tilde{H}_q^N\left(\frac{x}{z},\Delta^2,\frac{\xi}{z}\right), \tag{5}$$

where $h_N^3(z,\Delta^2,\xi)$ is a "light-cone spin dependent off-forward momentum distribution", which turns out to be strongly peaked around $z=1$, close to the forward limit. Therefore, in this region, for $x_3 = (M_A/M)x \le 0.7$ one has:

$$\begin{aligned} \tilde{H}_q^3(x,\Delta^2,\xi) &\simeq \sum_N \tilde{H}_q^N(x,\Delta^2,\xi) \int_0^{\frac{M_A}{M}} dz h_N^3(z,\Delta^2,\xi) \\ &= G_A^{3,p,point}(\Delta^2)\tilde{H}_q^p(x,\Delta^2,\xi) + G_A^{3,n,point}(\Delta^2)\tilde{H}_q^n(x,\Delta^2,\xi) \,. \end{aligned} \tag{6}$$

In the equation above, we have introduced the axial point like ff, $G_A^{3,N,point}(\Delta^2) = \int_0^{\frac{M_A}{M}} dz\, h_N^3(z,\Delta^2,\xi)$, which would give the nuclear axial ff if the proton and the neutron were point-like particles. These objects, at small values of Δ^2, depend slowly on the potential used in the calculation, so that theoretical errors in their computation are small. This can be realized observing that, in the forward limit, they reproduce the "effective polarizations" of the protons (p_p) and the neutron (p_n) in ^{3}He, whose values are rather similar if evaluated within different nucleon nucleon potentials (see Refs. [9, 14] for a comprehensive discussion). In particular, within the AV18 potential used here, the values $p_n = 0.878$ and $p_p = -0.024$ are obtained. Then, Eq. (6) can be used to extract the neutron contribution from possible sets of data for the proton and for ^{3}He:

$$\tilde{H}^{n,extr}(x,\Delta^2,\xi) \simeq \frac{1}{G_A^{3,n,point}(\Delta^2)} \left\{ \tilde{H}^3(x,\Delta^2,\xi) - G_A^{3,p,point}(\Delta^2)\tilde{H}^p(x,\Delta^2,\xi) \right\}. \tag{7}$$

This comparison between the free neutron GPDs, used as input in the calculation, and the ones extracted using our calculation for $\tilde{H}^3$ and the proton model for $\tilde{H}^p$, shows that the procedure works nicely even beyond the forward limit. The only theoretical input are axial point like ffs, which, as discussed above, are under good theoretical control. The procedure works for $x \leq 0.7$, the region where possible data could be gathered. Analogous extraction procedures have been demonstrated to hold for $\tilde{H}$ in ref. [11].

In closing this section, we have shown that coherent DVCS off ^{3}He at low momentum transfer Δ^2 as an ideal process to access the neutron GPDs; if data were taken at higher Δ^2, a relativistic treatment [19] and/or the inclusion of many body currents, beyond the present IA scheme, should be implemented. The next step of this investigation will be the evaluation of cross section asymmetries relevant to DVCS experiments at JLab kinematics, using the obtained GPDs H, E and $\tilde{H}$ of ^{3}He as a theoretical input. At the beginning, the leading twist analysis of DVCS for a spin 1/2 target, presented in Ref. [20], will be performed.

3. Extraction of neutron asymmetries from SIDIS experiments off ^3He

As discussed in the Introduction, information on the three-dimensional structure of the nucleon in momentum space can be obtained studying TMDs and transverse momentum dependent fragmentation functions (TMFFs) in SIDIS processes. The Sivers TMD [21] and the Collins TMFF [22], describing correlations between the spin and the momentum of a parton, are due to leading-twist final state interactions at the parton level and have therefore a very interesting dynamical content [23]. The Sivers and Collins functions are extracted from single spin asymmetries (SSAs) built from differential cross sections of SIDIS of unpolarized electrons off transversely polarized targets. The present data for the processes $\vec{p}(e, e'\pi)x$ [24] and $\vec{D}(e, e'\pi)x$ [25] show a strong flavor dependence and measurements with a $^3\vec{\text{He}}$ target have been proposed and performed (see Ref. [26] for the first data at JLab with a 6 GeV electron beam). Further accurate experiments are planned at the 12-GeV upgrade of JLab [27]. As in the GPDs case discussed in the previous section, neutron data are important to achieve the flavor separation of the Sivers and Collins distributions [28], and polarized ^{3}He targets play a special role. To obtain a reliable information one has to take into account the nuclear structure of ^{3}He as carefully as possible. Initially, the process can be described using the Plane Wave Impulse Approximation (PWIA), leading to dynamical nuclear effects described by the spin-dependent spectral function of ^{3}He, $\mathrm{P}_{\sigma,\sigma'}(\vec{p}, E)$, (see, e.g. [29]) that yields the probability distribution to find a nucleon with given removal energy (as a matter of facts, the spectator pair is interacting), three-momentum and polarization inside the nucleus. By using this formalism, it has been shown long time ago that, in spin dependent DIS, one can safely extract the neutron longitudinal asymmetry, A_n, from the corresponding ^{3}He observable, A_3^{exp} [9]. One obtains

$$A_n \simeq \left(A_3^{exp} - 2p_p f_p A_p^{exp}\right) / (p_n f_n) \tag{8}$$

with $p_{n(p)}$ the neutron (proton) effective polarization inside the polarized ^{3}He, and $f_{n(p)}$, the dilution factor. Values of p_n and p_p obtained using the AV18 interaction [16] are $p_p = -0.024$, $p_n = 0.878$ (see, e.g., [30]). In Ref. [30], an analogous extraction was demonstrated to work also in SIDIS, if the Bjorken limit and PWIA are assumed, and applied to the SSA of a transversely polarized $^3\vec{\text{He}}$ target, obtained from the process $^3\vec{\text{He}}(e, e'\pi)X$, in order to obtain the SSA of a transversely polarized neutron. In that approach, the SSAs of $^3\vec{\text{He}}$ can be expressed as convolutions of $\mathrm{P}_{\sigma,\sigma'}(\vec{p}, E)$ and proper combinations of suitable TMDs and TMFFs. The same extraction procedure has been also applied in combination with a Monte Carlo simulating the kinematics of the experiment E12-09-018 [31]. Indeed Eq. (8) has been used by the JLab Hall A Collaboration to extract, for the first time, the Collins and Sivers moments from a transversely polarized ^{3}He target [26]. Since existing measurements are limited in statistical accuracy and

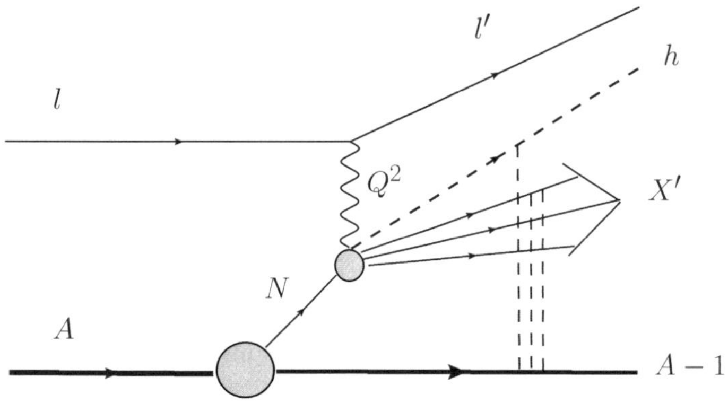

Figure 2. Final State Interaction between the fully interacting $(A-1)$ spectator system and the debris produced by the absorption of a virtual photon by a nucleon in the nucleus.

kinematics coverage, an extensive program of high precision measurements of SIDIS off ^{3}He will be part of the JLab program at 12 GeV [27]. The expected statistical accuracy is of the order of percent in a wide range of multi-dimensional kinematical binning; for this reason the PWIA could be no longer sufficient and Final State Interactions (FSI) between the detected meson and the remnants of the process, not considered in PWIA, may have a relevant role. Let us discuss now a proper way to introduce this mechanism, leading to a distorted spin-dependent spectral function of the ^{3}He. The JLab SIDIS experiments will exploit an electron beam at 8.8 and 11 GeV off ^{3}He polarized gaseous target; the relative energy between the $(A-1)$ system and the system of the detected pion and the remnant (depicted in Fig. 2) is a few GeV and therefore FSI can be treated within a generalized eikonal approximation (GEA). This framework was already successfully applied to unpolarized SIDIS [32], and in a recent paper the *distorted* spin-dependent spectral function has been calculated for the so-called *spectator* SIDIS, where a slow $(A-1)$ nucleon system, acting as a spectator of the photon-nucleon interaction, is detected, while the produced fast hadron is not [33]. In the following we discuss standard SIDIS, where all the possible state of the two-nucleon spectator system have to be considered.

The *distorted* spin-dependent spectral function for a polarized ^{3}He target can be written as

$$S^{N\,\mathbf{S}_3}_{\lambda\lambda'}(E,\mathbf{p}_{mis}) = \sum_{f_2} \sum\!\!\!\!\!\!\int_{\epsilon_2^*} \rho\,(\epsilon_2^*)\; \tilde{O}^{N\,\mathbf{S}_3\,f_2}_{\lambda\lambda'}(\,\epsilon_2^*,\mathbf{p}_{mis})\,\delta\,(E+M_3-m_N-M_2^*) \qquad (9)$$

with the product of distorted overlaps, given in terms of the intrinsic Jacobi coordinates $\mathbf{r}$ and $\boldsymbol{\rho}$, defined by

$$\begin{aligned} \tilde{O}^{N\,\mathbf{S}_3\,f_2}_{\lambda\lambda'}(\epsilon_2^*,\mathbf{p}_{mis}) &= \langle\lambda,\phi^{f_2}_{\epsilon_2^*}(\mathbf{r})e^{-i\mathbf{p}_{mis}\boldsymbol{\rho}}\mathcal{G}(\mathbf{r},\boldsymbol{\rho})|\Psi^{\mathbf{S}_3}_3(\mathbf{r},\boldsymbol{\rho})\rangle \\ &\times\; \langle\Psi^{\mathbf{S}_3}_3(\mathbf{r}',\boldsymbol{\rho}')|\lambda',\mathcal{G}(\mathbf{r}',\boldsymbol{\rho}')\phi^{f_2}_{\epsilon_2^*}(\mathbf{r}')e^{-i\mathbf{p}_{mis}\boldsymbol{\rho}'}\rangle. \end{aligned} \qquad (10)$$

where i) $\rho\,(\epsilon_2^*)$ is the density of the spectator pair with intrinsic energy ϵ_2^*, ii) $|\Psi^{\mathbf{S}_3}_3(\mathbf{r},\boldsymbol{\rho})\rangle$ is the ground state of the 3-nucleon system with polarization $\mathbf{S}_3$ iii) E is the removal energy $E=\epsilon_2^*+B_3$, $\mathbf{p}_{mis}$ is the three momentum of the spectator pair.

The Glauber operator in Cartesian coordinates is given by

$$\mathcal{G}(\mathbf{r}_1,\mathbf{r}_2,\mathbf{r}_3) = \mathcal{G}(\mathbf{r},\boldsymbol{\rho}) = \prod_{i=2,3}\left[1-\theta(\mathbf{r}_{i||}-\mathbf{r}_{1||})\Gamma\left(\mathbf{r}_{i\perp}-\mathbf{r}_{1\perp},\mathbf{r}_{i||}-\mathbf{r}_{1||}\right)\right]\;, \qquad (11)$$

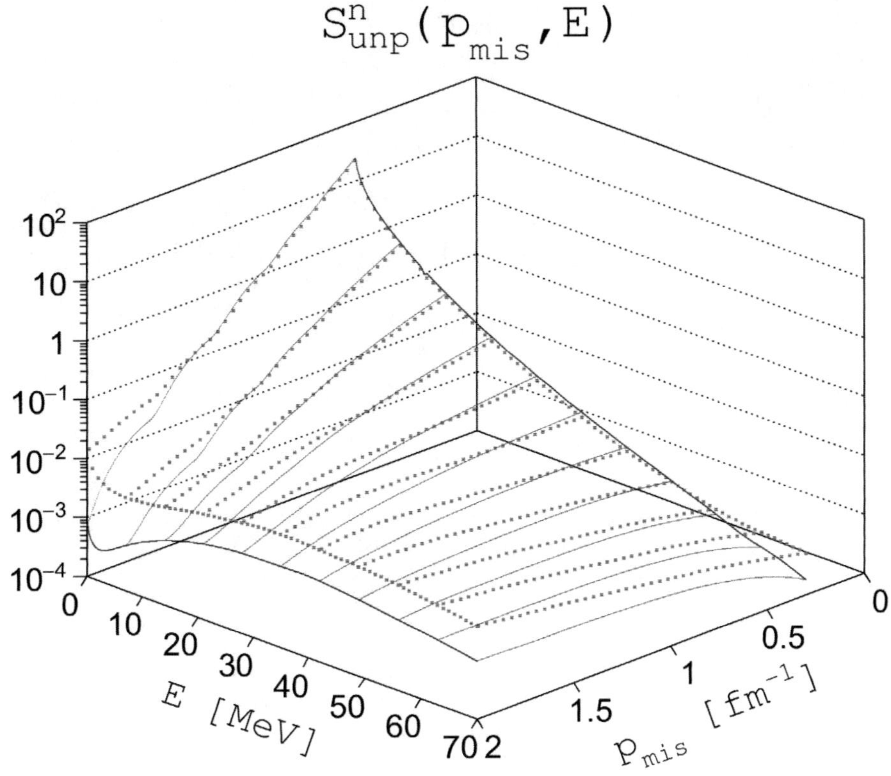

Figure 3. The ^{3}He spectral function, for the neutron, in the unpolarized case, as a function of $p_{mis} = |\mathbf{p}_{mis}|$ and of the removal energy E, in PWIA (full lines) and with FSI taken into account in the present GEA framework (dotted lines). The spectral functions are shown for the values of E and p_{mis} that contribute to the calculus of the SIDIS cross section when $A(p_N \cdot q)/(P_3 \cdot q) = 0.86$ at $\mathcal{E} = 11$ GeV and $Q^2 \simeq 7.6$ (GeV/c)^2 (see Ref. [34] for more details).

where $\hat{\mathbf{r}}_{i\perp}$ and $\hat{\mathbf{r}}_{i||}$ are the perpendicular and the parallel components of $\mathbf{r}_i$ with respect to the direction of the debris. The profile function

$$\Gamma(\mathbf{r}_{i\perp} - \mathbf{r}_{1\perp}, \mathbf{r}_{i||} - \mathbf{r}_{1||}) = \frac{(1 - i\,\eta)\;\sigma_{eff}(\mathbf{r}_{i||} - \mathbf{r}_{1||})}{4\,\pi\, b_0^2}\;\exp\left[-\frac{(\mathbf{r}_{i\perp} - \mathbf{r}_{1\perp})^2}{2\, b_0^2}\right] \quad , \tag{12}$$

unlike that occurring in the standard Glauber approach, depends not only on the impact parameter but also on the longitudinal separation through an effective cross section, σ_{eff}. For the latter we used the model proposed and exploited in previous works (for details on the model see Ref. [32, 33]). To recover the PWIA formulation, one has simply to put $\mathcal{G} \equiv 1$ in Eq. (11). In Fig. 3, a plot of the ^{3}He spectral function, evaluated within plane wave impulse approximation or taking FSI into account, is shown, for the neutron, in the unpolarized case.

The relevant part in the extraction of the transversely polarized neutron's SSAs is the transverse spectral function, given by a proper combination of non-diagonal elements of the spin-dependent spectral function defined in Eq. (9). In general, the transverse spin-dependent SF evaluated in PWIA, $\mathcal{S}^{\perp(PWIA)}$, and the corresponding distorted quantity, $\mathcal{S}^{\perp(FSI)}$, can be

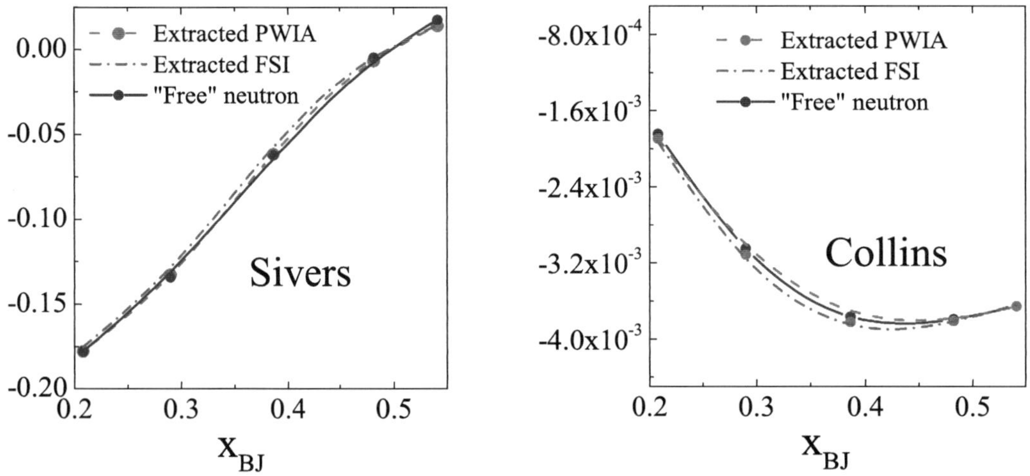

Figure 4. Neutron asymmetries extracted trough Eq. (8) from the Sivers (Left panel) and Collins (Right panel) ^{3}He asymmetries, with and without FSI taken into account in the actual kinematics of JLab at 12 GeV [27]. Preliminary results to appear in [34].

quite different and the corresponding effective polarizations can differ by 10-15%. Nevertheless, in Eq. (8) the effective polarizations occur in products with the dilution factors and to a large extent it turns out that, in the kinematical range explored at JLab, $p_p^{PWIA} f_p^{PWIA} \approx p_p^{FSI} f_p^{FSI}$, $p_n^{PWIA} f_n^{PWIA} \approx p_n^{FSI} f_n^{FSI}$ [43]. This allows one to safely adopt the usual extraction, as shown in Fig. 4, and therefore the goal of a sound flavor decomposition of TMDs seems definitely safe.

4. Conclusions and Perspectives

Let us summarize the status of our calculations of DVCS and SIDIS processes with ^{3}He targets. In section 2, we have shown realistic calculations of GPDs of ^{3}He, in plane wave impulse approximation, the essential ingredients for the evaluation of DVCS cross sections. Although DVCS off ^{3}He turns out to be promising for the extraction of neutron GPDs and relevant to understand the $A-$dependence of the modification of the parton structure of bound nucleons, for technical reasons, the present experimental program of JLab at 12 GeV will collect data for DVCS off deuteron and ^{4}He targets. Preliminary data of DVCS off ^{4}He, gathered with JLab at 6 GeV, are already available and should be published soon [35]. They deserve the attention of our community, which could perform precise calculations. Anyway, as explained in section 2, isoscalar targets such as deuteron or ^{4}He are not useful for a complete flavor separation of GPDs. Besides, a scalar nucleus as ^{4}He cannot be used for the measurement of helicity dependent GPDs, and the same quantity is hardly extracted from deuteron data. At the planned Electron Ion Collider [36], where slow nuclear recoiling systems will be easily detected and polarized nuclear beams will be naturally available, the proposal of polarized DVCS off ^{3}He and the measurement of the neutron helicity dependent GPD, $\tilde{H}$, using the technique described in Section 2, should become feasible.

As described in this talk for both DVCS and SIDIS, the standard theoretical description of few-nucleon systems, where nucleon and pion degrees of freedom are taken into account, has achieved a very high degree of sophistication. Nonetheless, one should try to fulfill, as much as possible, the relativistic constraints, dictated by covariance with respect to the Poincaré group, if processes involving nucleons with large 3-momentum are considered and a high precision is needed. At least, one should carefully deal with the boosts of the nuclear states, $|\Psi_{init}\rangle$ and

$|\Psi_{fin}\rangle$. In particular a relativistic treatment is important to precisely describe the JLab program @ 12 GeV for few-body systems (see, e.g., Refs. [37], [38]). To this aim, in the last few years, we have developed a relativistic description of ^{3}He using a Poincaré covariant spectral function built up within the light-front Hamiltonian dynamics (LFHD) [39].

Indeed, the Relativistic Hamiltonian Dynamics (RHD) of an interacting system with a fixed number of on-mass-shell constituents (see, e.g., [40]), associated to the Bakamijan-Thomas construction of the Poincaré generators [41], permits a fully Poincaré covariant description of DIS, SIDIS and DVCS off ^3He. The light-front (LF) form of RHD has been adopted [40], which has a subgroup structure of the LF boosts (with a separation of the intrinsic motion from the global one, which is very important for our aim) and a meaningful Fock expansion (see, e.g., Ref. [42]). Furthermore, within the LFHD, one can take advantage of the successful non-relativistic phenomenology that has been developed for the nuclear interaction.

The LF spin-dependent spectral function obtained from the LF wave functions for the two- and the three-nucleon systems has been defined in [39, 43]. With respect to previous attempts, the essential difference is the definition of the nucleon momentum: in our approach, it is the intrinsic momentum of the nucleon in the cluster formed by the nucleon and by the fully interacting (A-1) spectator system with given mass. This definition implements the macrocausality: if a system is separated into disjoint subsystems by a sufficiently large spacelike separation, then the subsystems behave as independent systems. The proposed formalism can find useful applications in deep inelastic scattering, since the LF momentum distribution fulfills both normalization and momentum sum rules, while the cluster separability introduces new binding effects with respect to previous approaches.

We are presently applying our relativistic spectral function to study the role of relativity in the EMC effect for ^{3}He. Preliminary results have been presented in Ref. [44]. We plan now to complete the LF analysis of the EMC effect for ^{3}He. As next steps, we will calculate a LF non-diagonal spectral function, to analyze DVCS cross sections, where relativity could play an essential role, if high transfer momentum is involved, as it happens for hadrons electromagnetic form factors (see, e.g., Ref. [45]). Eventually, FSI should be added to our LF relativistic description of ^{3}He, to study SIDIS processes and calculate single spin asymmetries.

Aknowledgmenys

S.S. thanks the organizers of the Conference "Theoretical Nuclear Physics in Italy, 2016" for the invitation to give this talk in a lively and timely event.

References

[1] Aubert JJ et al. (European Muon Collaboration) 1983 *Phys. Lett.* B **123** 275
[2] Dupré R and Scopetta S, 2016, *Eur. Phys. J.* A **52** no.6 159;
Dupré R and Scopetta S (organizers), Talks at the Workshop "New directions in nuclear Deep Inelastic Scattering", ECT*, Trento, Italy (2015)
[3] Diehl M 2016 *Eur. Phys. J.* A **52** no.6 149
[4] Mueller D *et al.* 1994 *Fortsch. Phys.* **42** 101
[5] Radyushkin A V 1996 *Phys. Lett.* B **380** 417; Ji X -D 1997 *Phys. Rev. Lett.* **78** 610
[6] Mazouz M *et al.* [Jefferson Lab Hall A Collaboration] 2007 *Phys. Rev. Lett.* **99** 242501
[7] Guidal M 2010 *Phys. Lett.* B **693** 17
[8] Berger E R, Cano F, Diehl M, and Pire, B 2001 *Phys. Rev. Lett.* **87** 142302
[9] Ciofi degli Atti C *et al.* 1993 *Phys. Rev.* C **48** R968-R972
[10] Ciofi degli Atti C and Scopetta S 1997 *Phys. Lett.* B **404** 223
[11] Rinaldi M and Scopetta S 2012 *Phys. Rev.* C 85 062201(R); Rinaldi M and Scopetta S 2013 *Phys. Rev.* C **87** 035208
[12] Scopetta S 2004 *Phys. Rev.* C **70** 015205; Scopetta S *Phys. Rev.* C **79** 025207
[13] Rinaldi M and Scopetta S 2014 *Few Body Syst* **55** 861-864
[14] Kievsky A, Pace E, Salmè G and Viviani M 1997 *Phys. Rev.* C **56** 64
[15] Kievsky A, Viviani M and Rosati S 1994 *Nucl. Phys.* A **577** 511

[16] Wiringa R B, Stoks V G J and Schiavilla R 1995 *Phys. Rev.* C **51** 38
[17] Musatov I V and Radyushkin A V 2000 *Phys. Rev.* D **61** 074027
[18] Marcucci L E, Riska D O and Schiavilla R. 1998 *Phys. Rev.* C **58** 3069
[19] Pace E, Salmè G, Scopetta S, Del Dotto A and Rinaldi M 2013 *Few Body Syst* **54** 1079
[20] Belitsky A V, Müller D, Niedermeier L and Schäfer A 2001 *Nucl. Phys.* B **593** 289
[21] Sivers D W 1990 *Phys. Rev.* D **41** 83 doi:10.1103/PhysRevD.41.83
[22] Collins J C 1993 *Nucl. Phys.* B **396** 161
[23] Brodsky S J, Hwang D S and Schmidt I 2002 *Phys. Lett.* B **530** 99-107
[24] Airapetian A. *et al.* [HERMES Collaboration] 2005 *Phys. Rev. Lett.* **94** 012002
[25] Alexakhin V Y *et al.* [COMPASS Collaboration] 2005 *Phys. Rev. Lett.* **94** 202002
[26] Qian X *at al.* [Jefferson Lab Hall A Collaboration] 2011 *Phys. Rev. Lett.* **107** 072003
[27] Cates G et al., E12-09-018 JLAB approved experiment, hallaweb.jlab.org/collab/PAC/PAC38//E12-09-018-SIDIS.pdf
[28] Signori A, Bacchetta A, Radici M and Schnell G 2013 *JHEP* **1311** 194
[29] Ciofi degli Atti C, Pace E and Salmè G 1992 *Phys. Rev.* **C 46** 1591
[30] Scopetta S 2007, *Phys. Rev.* **D 75** 054005
[31] Del Dotto A, Kaptari L P, Pace E, Salmè G, Scopetta S, Cisbani, E 2014 *EPJ Web of Conferences* **73** 02019
[32] Ciofi degli Atti C and Kaptari L P 2011 *Phys. Rev.* C **83**, 044602
[33] Kaptari L P, Del Dotto A, Pace E, Salmè G. and Scopetta S 2014 *Phys. Rev.* C **89** 035206-1,035206-18
[34] Del Dotto A, Kaptari L P, Pace E, Salmè G and Scopetta S to be published
[35] Hattawy M 2015 Thesis, Université Paris-Sud (unpublished)
[36] Accardi A *et al.* 2016 *Eur. Phys. J.* A **52** no.9, 268
[37] MARATHON Coll. E12-10-103 (Rating A): MeAsurement of the F_2^n/F_2^p, d/u Ratios and A=3 EMC Effect in Deep Inelastic Electron Scattering Off the Tritium and Helium MirrOr Nuclei, hallaweb.jlab.org/12GeV/
[38] Gao H et al., PR12-09-014 (Rating A), Target Single Spin Asymmetry in Semi-Inclusive Deep-Inelastic (e, e'$\pi^{\pm}$) Reaction on a Transversely Polarized ^3He Target, hallaweb.jlab.org/12GeV/; Chen JP et al., PR12-11-007 (Rating A): Asymmetries in Semi-Inclusive Deep-Inelastic (e, e'$\pi^{\pm}$) Reactions on a Longitudinally Polarized ^3He Target.
[39] Del Dotto A, Pace E, Salmè G and Scopetta S 2017 *Phys. Rev.* C **95** 014001
[40] Keister B D and Polyzou W N 1991 *Adv. Nucl. Phys.* **21** 225-550
[41] Bakamjian B and Thomas L K 1953 *Phys. Rev.* **92** 1300-1310.
[42] Brodsky S J, Pauli H C and Pinsky S S 1998 *Phys. Rept.* **301** 299-406
[43] Pace E, Del Dotto A, Rinaldi M, Salmè G and Scopetta S 2013 *Few Body Syst.* **54** 1079-1082.
[44] Pace E, Del Dotto A, Kaptari L P, Rinaldi M, Salmè G and Scopetta S, 2016 *Few Body Syst.* **57** no.7, 601-606
[45] Cardarelli F, Pace E, Salmè G and Simula S 1995 *Phys. Lett.* B **357** 267

IOP Conf. Series: Journal of Physics: Conf. Series **981** (2018) 012014 doi:10.1088/1742-6596/981/1/012014

Nucleon 3D structure from double parton scattering: a Light-Front quark model analysis

Matteo Rinaldi

Instituto de Fisica Corpuscular (CSIC-Universitat de Valencia), Parc Cientific UV, C/ Catedratico Jose Beltran 2, E-46980 Paterna (Valencia), Spain.

E-mail: `mrinaldi@ific.uv.es; matteo.rinaldi@pg.infn.it`

Abstract. Double parton distribution functions (dPDFs) represent a tool to explore the 3D partonic structure of the proton. They can be measured in high energy proton-proton and proton nucleus collisions and encode information on how partons inside a proton are correlated among each other. dPFDs are studied here in the valence region by means of a constituent quark model scenario within the relativistic Light Front approach, where two particle correlations are present without any additional prescription. Furthermore, a study of the QCD evolution at high energy scale, of the model results, has been completed in order to compare our predictions with future data analyses. In closing, results on the evaluation of the so called σ_{eff}, crucial ingredient for the description of double parton scattering, where dPDFs can be accessed, are presented and discussed.

1. Introduction

The observation of multiple parton interactions (MPI), occurring in high energy hadron-hadron collisions, will be an important goal of the experimental and theoretical analyses of the processes studied, e.g., at the LHC. In particular, in these kind of events, more than one parton of a hadron interact with partons of the other colliding hadron. Naturally, MPI contribution to the total cross section is suppressed with respect to the single parton interaction. However, the measurement and estimate of MPI cross sections is an important challenge being MPI events a background for the search of new Physics. Furthermore, from a hadronic point of view, we are interested on MPI due to the possibility of accessing new fundamental information on the partonic proton structure. We focus our studies on the the double parton scattering (DPS), the most simple case of MPI, which can be observed, in principle, in several processes, e.g., WW with dilepton productions and double Drell-Yan processes (see, Refs. [1, 2, 3, 4, 5] for recent reviews). At the LHC, DPS, has been observed some years ago [6] and represents a background for the Higgs production in several channels. From a theoretical point of view, the DPS cross section is written in terms of a new quantity, the so called double parton distribution function (dPDF), $F_{ij}(x_1, x_2, \vec{z}_\perp, \mu)$, which describe the joint probability of finding two partons of flavors $i, j = q, \bar{q}, g$ with longitudinal momentum fractions x_1, x_2 and separation $\vec{z}_\perp$ in the transverse plane inside the hadron, see Ref. [7]. Here μ is the renormalization scale. However, to date, being dPDFs poorly known, in order to qualitatively estimate the magnitudo of the DPS cross section, the following approximation is usually adopted:

$$F_{ij}(x_1, x_2, \vec{z}_\perp, \mu) = q_i(x_1, \mu)\, q_j(x_2, \mu)\, T(\vec{z}_\perp, \mu)\theta(1 - x_1 - x_2)(1 - x_1 - x_2)^n , \quad (1)$$

i.e., dPDFs are evaluated in a fully factorized ansatz in terms of standard one-body parton distribution functions (PDF), $q(x)$, and $T(\vec{z}_\perp, \mu)$, the function encoding all parton correlations in the transverse plane. The spirit of the latter assumption is to neglect all possible double parton correlations (DPCs) between the two interacting partons, being the latter almost unknown, see e.g. Ref. [8] for updates. Moreover dPDFs are non perturbative quantities in QCD so that they cannot be easily evaluated from the theory. Nevertheless, as already deeply discussed in Refs. [9, 10, 11, 12], quark model calculations of dPDFs could help to grasp the basic feature of such quantities. This strategy has been largely used in the past to study unknown distributions. However, since in this scenario dPDFs are calculated at the low hadronic scale of the model, $\mu_0 \sim \Lambda_{QCD}$, in order to compare the obtained outcomes with future data taken at high energy scales, $Q > \mu_0$, it is then necessary to perform the perturbative QCD (pQCD) evolution of the model calculations, using the dPDF evolution equations, see Refs. [13, 14]. The idea supporting our analysis is that, thanks to this procedure, future data analyses of the DPS processes could be guided, in principle, by model calculations. In the first part of the present paper, we focus our attention on the study of the role of DPCs in dPDFs in order to verify, as a first step, the validity of the approximation Eq. (1), often used for data analyses. In particular, it is here recall that in all model calculations of dPDFs, see Refs. [9, 10, 11], the assumption Eq. (1) is violated. See also Ref. [15] for details on the violation of the factorized ansatz due to model independent relativistic effects in the calculated dPDFs. Moreover, as already pointed out, it is fundamental to realize to which extend DPCs survive at very high energy scales, where, due to the large population of partons, the role of DPCs could be less important than at the low scale of the used model. To this aim in Refs. [11, 16], DPCs in dPDFs have been studied at the energy scale of the experiments. In particular in Ref. [16] an extension of the approach used in Refs. [11] has been provided to include sea quarks and gluon degrees of freedom. Here, the most important results of the these analyses will be summarized. In the last section of the present paper, an investigation of the so called σ_{eff} will be discussed. Usually, indeed, DPS cross section, in processes with final state $A + B$, is written through the following ratio (see e.g. Ref. [17]):

$$\sigma_{DPS}^{A+B} = \frac{m}{2} \frac{\sigma_{SPS}^{A} \sigma_{SPS}^{B}}{\sigma_{eff}}, \tag{2}$$

where m is combinatorial factor depending on the final states A and B ($m = 1$ for $A = B$ or $m = 2$ for $A \neq B$) and $\sigma_{SPS}^{A(B)}$ is the single parton scattering cross section with final state $A(B)$. Expressing the σ_{DPS} cross section in Eq. (2) in terms of product of σ_{SPS}, one assumes that, as a first approximation, double parton distributions can be written as in Eq. (1). The present knowledge on DPS cross sections has been condensed in the experimental and model dependent extraction of σ_{eff} [17, 18, 19, 20, 21, 22, 23, 24]. To date, $\sigma_{eff} \simeq 15$ mb, compatible, within errors, with a constant, irrespective of centre-of-mass energy of the hadronic collisions, the final state and x_i. Let us stress again that the latter result is obtained considering the fully uncorrelated scenario shown in Eq. (1) where all DPCs are neglected. Also in this case non perturbative methods have been used to study σ_{eff}, through the calculation of dPDFs, in order to characterize "signals" of DPCs in σ_{eff}. On top of that, in Ref. [25], the possible dependence of the latter quantity on the longitudinal momentum fraction carried by the interacting partons has been addressed. Recent results on this topic will be here discussed, including also new intriguing outcomes addressed in Refs. [15, 26]. In Ref. [15] the contribution of relativistic effects to σ_{eff} have been estimated by means of the Light-Front approach (LF), already used for the evaluation of dPDFs in Ref. [11]. Furthermore, in Ref. [26], σ_{eff} has been calculated in a AdS/QCD framework, showing, also here, a strong x_i dependence of σ_{eff} in the valence region.

IOP Conf. Series: Journal of Physics: Conf. Series **981** (2018) 012014 doi:10.1088/1742-6596/981/1/012014

2. Calculation of dPDFs within LF CQMs

In this section, details of the calculations of dPDFs, within the LF approach (see Refs. [27, 28] for fundamental details), will be discussed verifying if the factorized ansatz Eq. (1) would work in this relativistic scenario. In this framework, among all positive features of the LF approach, let us remark that one can obtain a fully Poincaré covariant description of relativistic strongly interacting systems with a fixed number of on-shell constituents, LF boosts and the "plus" components of momenta ($a^+ = a_0 + a_3$) are kinematical operators. Furthermore, being the LF hypersurface, the one where the initial conditions of the system are fixed, tangent to the light-cone, the kinematics of DIS processes is naturally obtained. For these reasons, this strategy has been adopted in the past for the evaluation unknown distributions, see e.g. Refs. [29, 30, 31]. All the details on the present calculations are deeply discussed and reported in Ref. [11] and will not be repeated here. The final expression of the dPDF in momentum space for two unpolarized quarks of flavor u, reads:

$$F_{uu}^{\lambda_1,\lambda_2}(x_1,x_2,\vec{k}_\perp) = 2(\sqrt{3})^3 \int \left[\prod_{i=1}^{3} d\vec{k}_i \sum_{\lambda_i^f \tau_i}\right] \delta\left(\sum_{i=1}^{3}\vec{k}_i\right) \Psi^*\left(\vec{k}_1+\frac{\vec{k}_\perp}{2},\vec{k}_2-\frac{\vec{k}_\perp}{2},\vec{k}_3;\{\lambda_i^f,\tau_i\}\right)$$
$$\times\ \Psi\left(\vec{k}_1-\frac{\vec{k}_\perp}{2},\vec{k}_2+\frac{\vec{k}_\perp}{2},\vec{k}_3;\{\lambda_i^f,\tau_i\}\right)\delta\left(x_1-\frac{k_1^+}{P^+}\right)\delta\left(x_2-\frac{k_2^+}{P^+}\right),$$

being $k_\perp$ the relative transverse momentum of one of the parton in the amplitude and its complex conjugate and $\vec{k}_i$ the momentum of the $i-$ quark.

The canonical proton wave function $\psi^{[c]}$ is embedded in the function Ψ here above, which can be written as follows:

$$\Psi(\vec{k}_1,\vec{k}_2,\vec{k}_2;\{\lambda_i^f,\tau_i\}) = \prod_{i=1}^{3}\left[\sum_{\lambda_i^c} D^{*1/2}_{\lambda_i^c\lambda_i^f}(R_{cf}(\vec{k}_i))\right]\psi^{[c]}(\{\vec{k}_i,\lambda_i^c,\tau_i\}), \tag{3}$$

where λ_i^c and τ_i are the canonical parton helicity and the isospin, respectively. Here the short notation $\{\alpha_i\} = \alpha_1, \alpha_2, \alpha_3$ is introduced. In Eq. (3), the Melosh operators, which allow to rotate the LF spin into the canonical one are introduced:

$$\hat{D}_i = D^{1/2}_{\mu\lambda}(R_{cf}(\vec{k}_i)) = \langle\mu\left|\frac{m+x_iM_0-i\vec{\sigma}_i\cdot(\hat{z}_i\times\vec{k}_{i\perp})}{\sqrt{(m+x_iM_0)^2+\vec{k}_{i\perp}^2}}\right|\lambda\rangle, \tag{4}$$

being $M_0 = \sum_i\sqrt{m^2+\vec{k}_i^2}$ the total free energy mass of the system and μ and λ generic canonical spins. In this scenario, being M_0, appearing in delta functions in Eq. (3), a kinematical quantity, dPDFs are well defined in the $x_1 + x_2 < 1$ region, at variance with what happens in the instant form calculations of PDFs and dPDFs (see, e.g., Ref. [10]). Let us remark that the main ingredient in Eq. (3) is the canonical proton wave function which has been calculated by means of a relativistic CQM, the so called hyper-central CQM described in Ref. [32]. The choice of this model is motivated by its simplicity and capability to basically reproduce the spectrum of light-hadrons. Since actually for dPDFs there are not yet available data, model calculations of these quantities could provide essential information on their relevant features. Before discussing in the next section the results of the calculations of dPDFs, one should notice that in principle the role of the model independent Melosh rotations can be important. Indeed, from Eq. (4), one can realize that such operator con introduce, in the calculation of dPDFs, non trivial correlations between x_i and $k_\perp$. As deeply discussed in Ref. [15], in order to taste possible effects of these objects, it is instructive to study the following quantity:

$$DD^\dagger(\vec{k}_\perp, x_1, x_2, \vec{k}_{1\perp} = 0, \vec{k}_{2\perp} = 0) = \langle SU(6)|D_1^\dagger D_1 D_2^\dagger D_2|SU(6)\rangle \;, \tag{5}$$

being $\vec{k}_{i\perp}$ the intrinsic transverse component of the i quark momentum. In the latter quantity use has been made of the commonly adopted SU(6) symmetry in order to evaluate the spin part of the proton wave function. In this scenario, the quantity (5) is rather model independent. Here we consider two fast partons (FF) with $x_1 = 0.2$, $x_2 = 0.3$, one slow and and one fast parton (SF) with $x_1 = 0.04$, $x_2 = 0.3$ and two slow partons (SS) with $x_1 = 0.04$, $x_2 = 0.03$. The calculation of Eq. (5) in these kinematic regions is presented in Fig. (1) where one may identify three distinct regions as a function of $k_\perp$. For $k_\perp \to 0$ the Melosh's in all kinematic configurations reduce to unity. In an intermediate region of $k_\perp$ the curves show a dip whose depth depends on the chosen kinematic configuration and at larger $k_\perp$ the curves flattens out with different asymptotics. This complicated pattern, generated by Melosh's rotations, affects the calculation of dPDFs, which, in general, are distributions evaluated also at $k_\perp \neq 0$.

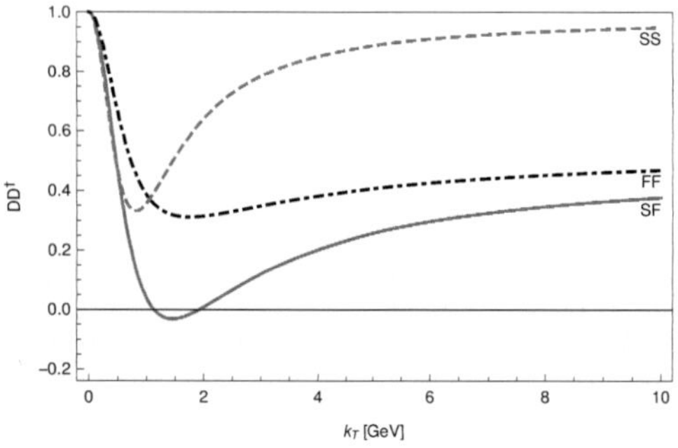

Figure 1. The quantity Eq. (5), as function of $k_T = k_\perp$, evaluated in different regions of x_1 and x_2 with $\vec{k}_{1\perp} = \vec{k}_{2\perp} = 0$.

3. Results of the calculations of dPDFs at the hadronic scale

In this section the main results of the calculations of dPDFs at the hadronic scale, $\mu_0^2 \sim 0.1$ GeV^2, will be presented. In particular, as already mentioned, here the emphasis of the analysis is focused on testing the validity of the approximation Eq. (1). As already discussed in the previous section, in general, Melosh rotations introduce correlations between $k_\perp$ and x_1, x_2 so that, as also found in Ref. [11], at the hadronic scale such factorization is violated in all model calculations of dPDFs within the LF approach, see Ref. [15]. Furthermore, in order to study the validity of the approximations Eq. (1), in the x_1 and x_2 dependence of dPDFs, the following ratio have been evaluated:

$$r_2 = \frac{2uu(x_1, x_2, k_\perp = 0)}{u(x_1)u(x_2)} \;, \tag{6}$$

As one can see on the left panel of Fig. (2), being the ratio r_2, Eq. (6), different from the unity in all the kinematical range, also the factorized form of dPDFs in terms of the product of single PDFs is not supported by the present approach. In closing, at the hadronic scale, in all model calculations of dPDFs (see also Refs. [9, 10]), the assumption Eq. (1) is violated due to DPCs.

4. pQCD evolution of the calculated dPDFs

A fundamental point, discussed e.g. in Ref. [11], is the analysis of the pQCD evolution of model calculations of dPDFs. This procedure is essential to relate CQM predictions with present and future experimental analyses. For the moment being, the pQCD evolution of dPDFs is known only for the longitudinal momentum dependence, which means $k_\perp = 0$, and using the same energy scale for both the acting partons. In these case, the evolution equations are obtained as a proper generalization of the usual Dokshitzer-Gribov-Lipatov-Altarelli-Parisi (DGLAP) ones (see Refs. [13, 14] for details), defined for the evolution of PDFs. In the first part of this section, results will be shown for the non-singlet sector. In particular, the ratio r_2, Eq. (6), has been calculated using the dPDFs evaluated at a generic high energy scale, e.g., $Q^2 = 10$ GeV^2, using pQCD evolution. As one can see on the right panel of in Fig. 2, for small values of x_i, e.g, close to the LHC kinematics, $r_2 \sim 1$. This means that dynamical correlations, for valence quarks, are suppressed after the evolution. Nevertheless, in order to complete this analysis, in Ref. [16], new ratios sensitive to longitudinal correlations have been defined and calculated at high energy scale including perturbative and non perturbative degrees of freedom. To this aim let us generalize the expression of r_2 for different partonic species:

$$ratio_{ab} = \frac{F_{ab}(x_1, x_2 = 0.01, k_\perp = 0; Q_2^2)}{a(x_1; Q_2^2)b(x_2 = 0.01; Q_2^2)} \; ; \tag{7}$$

where here $a, b = q, g, \bar{q}$, $a(x; Q^2)$ and $b(x; Q^2)$ are the single PDFs for two given partons of flavor a and b and $Q_2^2 = 250$ GeV^2. In such ratio, dPDFs in the numerator and PDFs in the denominator evolve by means of different evolution equations. Due to this feature, $ratio_{ab}$ is sensitive to non perturbative correlations, encoded in the proton wave function used to calculate dPDFs and PDFs and to the perturbative ones due to the difference in the evolution equations of dPDFs and PDFs. In order to disentangle these two different contributions to understand which among theme could affect the dPDF evaluations, the following ratios have been defined and calculated within the LF CQM approach:

$$ratio_{ab}^{P} = \frac{F_{ab}(x_1, x_2 = 0.01, k_\perp = 0; Q_2^2)|^P}{a(x_1; Q_2^2)b(x_2 = 0.01; Q_2^2)}; \qquad ratio_{ab}^{NP} = \frac{F_{ab}(x_1, x_2 = 0.01, k_\perp = 0; Q_2^2)}{F_{ab}(x_1, x_2 = 0.01, k_\perp = 0; Q_2^2)|^P} \tag{8}$$

where here:

$$F_{ab}(x_1, x_2 = 0.01, k_\perp = 0; Q_2^2)|^P = \left[a(x_1; Q_2^2)b(x_2 = 0.01; Q_2^2)\right]^{dPDFevolution} . \tag{9}$$

The latter quantity is calculated by evolving the product of PDFs by means of dPDF evolution. Due to this feature, the numerator and denominator of $ratio_{ab}^{P}$ evolve differently in pQCD while the input at the initial scale is the same, so that $ratio_{ab}^{P}$ is sensitive to perturbative correlations. On the contrary, since in ratio $ratio_{ab}^{NP}$ the numerator and denominator evolve within the same scheme, but using different input at the initial scale, it is sensitive to non perturbative correlations. Due to these features, the ratio $ratio_{ab}$, shown in Fig. 3 (left panel) for

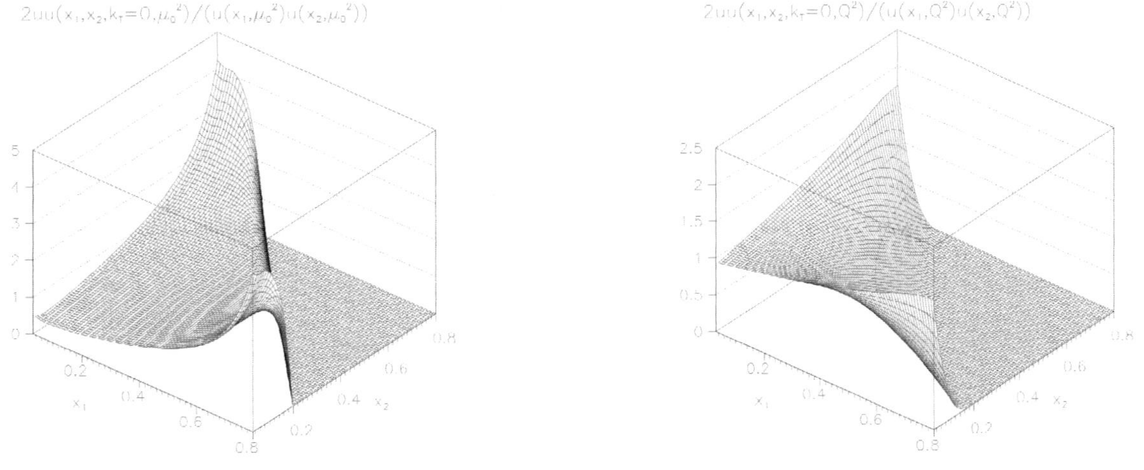

Figure 2. Left panel: the ratio r_2, Eq. (6) evaluated at the scale μ_0^2. Right panel: same of the left panel but at the scale $Q^2 = 10$ GeV^2.

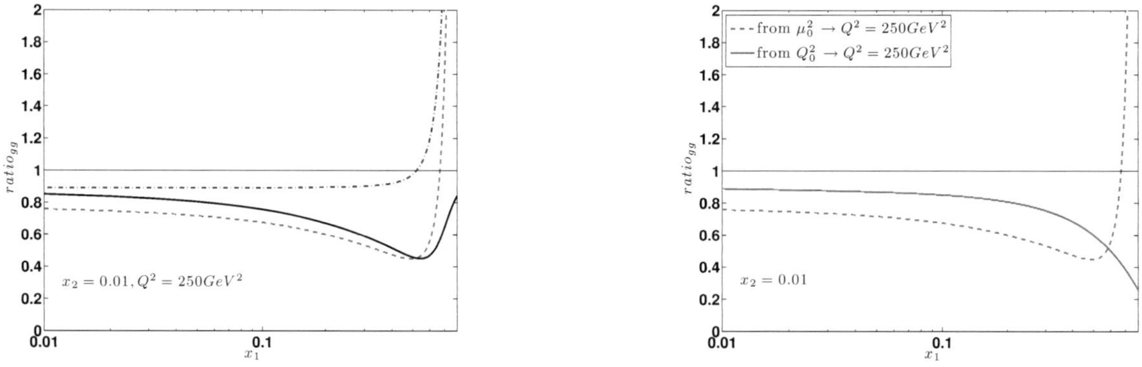

Figure 3. Left panel: gluon gluon ratios $ratio_{gg}$ (dashed line), $ratio_{gg}^{P}$ (dot-dashed) and $ratio_{gg}^{NP}$ (continuous line) evaluated at the scale $Q_2^2 = 250$ GeV^2 and for $x_2 = 0.01$. Right panel: same of the left panel but including where gluons generated non perturbatively, according to Eq. (10), are taken into account.

two $a = b = g$, is particularly emblematic. The full ratio $ratio_{ab}$, Eq. (7), (dashed line), clearly influenced by both perturbative (dot-dashed line) and non-perturbative (continuous line) effects, is compared with those where perturbative and non-perturbative correlations are disentangled, contributing to the behavior of $g - g$ dPDFs at low values of x_1 and x_2. As one can see in the left panel of Fig. 3, for the gluon-gluon distribution such components coherently interfere. Here we report results only for gluons, being the highest partonic component at LHC kinematics. Furthermore, in order to show how much these conclusions do not depend on the used model, a semi factorized ansatz has been adopted in order to include non perturbative see quarks and gluons at a given initial scale. All details of the procedure are discussed in Ref. [16]. We assume that at a given initial scale $Q_0 > \mu_0$ one has:

$$F_{uu}(x_1, x_2, k_\perp = 0, Q_0^2) \sim F_{u_V u_V}(x_1, x_2, k_\perp = 0, Q_0^2) + \{[u_V(x_1, Q_0^2)\bar{u}(x_2, Q_0^2) + \bar{u}(x_1, Q_0^2)u_V(x_2, Q_0^2)] + \bar{u}(x_1, Q_0^2)\bar{u}(x_2, Q_0^2)\}\,(1 - x_1 - x_2)^n \theta(1 - x_1 - x_2)\,. \tag{10}$$

where in the above expression, $F_{u_V u_V}(x_1, x_2, k_\perp = 0, Q_0^2)$ and $u_V(x_1, Q_0^2)$ are the dPDFs and PDFs respectively calculated by means of the LF CQM approach obtained evolving the same quantities from μ_0^2 to Q_0^2 and $\bar{u}(x_2, Q_0^2)$ is the PDF evaluated through to the MSTW2008 parametrization [33], see Ref. [16] for further details. In particular, within this choice, $Q_0^2 = 1$ GeV^2. As one can see in the right panel of Fig. 3, where $ratio_{gg}$ has been calculated starting the evolution from μ_0^2 and Q_0^2, conclusions arisen from the precedent analysis, see left panel of Fig. 3, are confirmed. For gluon-gluon the correlations induced at low-x still contain a specific sign of the correlations introduced in the valence sector and this is due to the presence of the valence component in the quark-singlet sector in the evolution procedure. The strength of the correlations seems to become smaller but they are still sizable and should not be neglected.

5. Calculation of σ_{eff}

As previously mentioned, an important quantity, relevant for the experimental analyses of DPS is the effective cross section, σ_{eff} whose expression in terms of PDFs and dPDFs, has been presented in Ref. [25]. In order to discuss the main features of σ_{eff}, we restrict the analysis to the zero rapidity region $y = 0$ and show results, within the LF approach, in the left panel of Fig. 4. Here the latter quantity has been calculated at $Q^2 = 250$ GeV^2 for a sea and a valence quarks. It is worth to notice that the three old experimental extractions of σ_{eff} from data, Refs. [34, 35, 36], lie in the obtained range of values of the calculated σ_{eff}. It is fundamental to remark that, at the hadronic scale, as demonstrated in Ref. [15], if Melosh rotations would neglected, the average value of σ_{eff} would change by a factor 2, making the

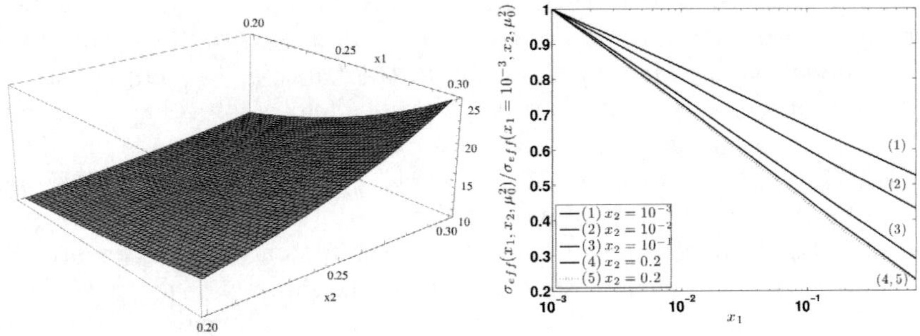

Figure 4. Left panel: $\sigma_{eff}(x_1, x_2, Q^2)$ for the values of x_1, x_2 measured in Ref. [34] evaluated for a valence quark and a sea quark. Right panel: The ratio $\sigma_{eff}(x_1, x_2, \mu_0^2)/\sigma_{eff}(x_1 = 10^{-3}, x_2, \mu_0^2)$ as a function of as a function of x_1 at fixed $x_2 = 0.001, 0.01, 0.1, 0.2$.

impact of relativistic effects in σ_{eff} very strong. In the same figure, as one can see, DPCs generate a strong x_i dependence of σ_{eff} in the valence region. Such feature, as already discussed, is related to the combined effect of pQCD evolution correlations and non perturbative ones. In order to provide a fully model independent analysis of the x dependence of σ_{eff}, in Ref. [26], such quantity has been calculated through an evaluation of dPDFs based on the AdS/QCD correspondence, see Refs. [37, 38, 39] for fundamental details. In particular the authors found that, at the hadronic scale, the x dependence of σ_{eff} is comparable with the one found within the LF approach. Moreover, in order to emphasize such feature, in the right panel of Fig. 4, the ratio $\sigma_{eff}(x_1, x_2, \mu_0^2)/\sigma_{eff}(x_1 = 10^{-3}, x_2, \mu_0^2)$ has been plotted. As one can see, such ratio strongly depends on x_2 in the valence region, at the variance of the case in which longitudinal correlations were neglected.

6. Conclusions

In this work, dPDFs have been calculated within a fully Poincaré covariant constituent quark model approach, the Light-Front one, fulfilling the essential properties of dPDFs, see Ref. [11]. The main goal of this first analysis was to establish the role of DPCs in dPDFs. In particular, the factorization of dPDFs, at the hadronic scale of the model, in the $x_1 - x_2$ and $(x_1, x_2) - k_\perp$ dependences is violated, as already found in previous analyses c.f. Refs. [9, 10]. In order to compare our results with actual and future experimental studies, the pQCD evolution of the calculated dPDFs has been deeply investigated as discussed in Refs. [11, 16] where, at very high energy scales, like the experimental ones, it was found that longitudinal correlations may survive also at small values of x_i. Furthermore, as also explained in Ref. [16], where also non perturbative degrees of freedom have been taken into account in the analysis, there are different source of longitudinal correlations, the perturbative one, induced by the pQCD evolution scheme and the non perturbative ones due to the model used to describe the proton wave function, used to calculate the dPDFs. For example, in the gluon gluon case, such correlations coherently interfere in the small x_i region at very high energy scales making theme sizable. However, since dPDFs can not be directly measured in DPS, within this model calculation approach, the so called σ_{eff}, a quantity experimentally studied to estimate the DPS cross section, has been calculated. The latter one could be very important from the theoretical and experimental point of view to obtain information on partonic structure of the proton being sensitive to DPCs. In particular, signals of DPCs in σ_{eff}, have been found in its x_i dependence in the valence region also at high energy scales, for details see Ref. [25]. Furthermore σ_{eff} has been also calculated within the AdS/QCD correspondence and also in this case, an important x_i dependence has been found, in particular in the valence region. Since, the role of correlations is crucial to grasp new information on the partonic structure of the proton, relativistic effects in the calculation of dPDFs have been studied in order to identify model independent correlations in the $(x_1, x_2) - k_\perp$ dependence of dPDFs. In particular, as deeply discussed in Ref. [15], the role of Melosh rotations, essential feature of the LF approach, introduce non negligible correlations in such dependence. Furthermore new studies for the extraction of the proton dPDFs from pA collisions are under investigations, see Ref. [40]. We conclude that the investigation of dPDFs, also trough the analysis of σ_{eff}, could unveil new and interesting information on the 3D partonic structure of the proton. I thank Sergio Scopetta, Marco Traini, Vicente Vento and Federico Alberto Ceccopieri for a nice and fruitful collaboration on this subject. This work was supported in part by the Mineco under contract FPA2013-47443-C2-1-P and SEV-2014-0398.

References

[1] Gaunt J R and Stirling W J 2010 *JHEP* **03** 005
[2] Diehl M, Ostermeier D and Schafer A 2012 *JHEP* **1203** 089
[3] Manohar A V and Waalewijn W J 2012 *Phys. Rev.* D **85** 114009
[4] Bartalini P and Fanò arXiv:1003.4220 [hep-ex]
[5] Bansal S et al., arXiv:1410.6664 [hep-ph]
[6] Aad G et al. [ATLAS Collaboration] 2013 *New J. Phys.* **15** 033038
[7] Paver N and Treleani D 1982 *Nuovo Cim.* A **70** 215
[8] Diehl M, Kasemets T and Keane S 2014 *JHEP* **05** 118
[9] Chang H M, Manohar A V and Waalewijn W J 2013 *Phys. Rev.* D **87** 034009
[10] Rinaldi M, Scopetta S and Vento V 2013 *Phys. Rev.* D **87** 11
[11] Rinaldi M, Scopetta S, Traini M and Vento V 2014 *JHEP* **12** 028
[12] Broniowski W and Arriola E R 2014 *Few Body Syst.* **55** 381
[13] Kirschner R 1979 *Phys. Lett.* B **84** 266
[14] Shelest V P, Snigirev A M and Zinovev G M 1982 *Phys. Lett.* B **113** 325
[15] Rinaldi M and Ceccopieri F A 2017 *Phys. Rev.* D **95** no. 3 034040
[16] Rinaldi M, Scopetta S, Traini M C and Vento V 2016 *JHEP* **10** 063
[17] Jung H et al. Proceedings, 7th International Workshop on Multiple Partonic Interactions at the LHC (MPI@LHC 2015) *https://bib-pubdb1.desy.de/record/297386, DESY-PROC-2016-01*
[18] Akesson T et al. [Axial Field Spectrometer Collaboration] 1987 *Z. Phys.* C **34** 163
[19] Alitti J et al. [UA2 Collaboration] 1991 *Phys. Lett.* **B268** 145
[20] Abe F et al. [CDF Collaboration] 1997 *Phys. Rev.* D **56** 3811
[21] Abazov V M et al. [D0 Collaboration] 2010 *Phys. Rev.* D **81** 052012
[22] Aad G et al. [ATLAS Collaboration] 2013 *New J. Phys.* **15** 033038
[23] Chatrchyan S et al. [CMS Collaboration] 2014 *JHEP* **03** 032
[24] Aaij R et al. [LHCb Collaboration] 2014 *JHEP* **04** 091
[25] Rinaldi M, Scopetta S, Traini M and Vento V 2016 *Phys. Lett.* B **752** 40
[26] Traini M, Scopetta S, Rinaldi M and Vento V 2017 *Phys. Lett.* B **768** 270
[27] Keister B D and Polyzou W N 1991 *Adv. Nucl. Phys.* **20** 225
[28] Brodsky S J, Pauli H C and Pinsky S S 1998 *Phys. Rept.* **301** 299
[29] Boffi S, Pasquini B and Traini M 2003 *Nucl. Phys.* B **649** 243
[30] Pasquini B, Traini M and Boffi S 2005 *Phys. Rev.* D **71** 034022
[31] Traini M 2014 *Phys. Rev.* D **89** no. 3 034021
[32] Faccioli P, Traini M and Vento V 1999 *Nucl. Phys.* A **656** 400
[33] Martin A D, Stirling W J, Thorne R S and Watt G 2009 *Eur. Phys. J.* C **63** 189
[34] Akesson T et al. [Axial Field Spectrometer Collaboration] 1987 *Z. Phys.* C **34** 163
[35] Alitti J et al. [UA2 Collaboration] 1991 *Phys. Lett.* B **268** 145
[36] Abe F et al. [CDF Collaboration] 1997 *Phys. Rev.* D **56** 3811
[37] Maldacena J M 1998 *Adv. Theor. Math. Phys.* **2** 231
Gubser S S, Klebanov I R and Polyakov A M 1998 *Phys. Lett.* B **428** 105
Witten E 1998 *Adv. Theor. Math. Phys.* **2** 253
[38] Gursoy U and Kiritsis E 2008 *JHEP* **0802** 032
Gursoy U, Kiritsis E and Nitti F 2008 *JHEP* **0802** 019
[39] Brodsky S J, de Teramond G F, Dosch H G and Erlich J 2015 *Phys. Rept.* **584** 1
[40] Salvini S, Treleani D and Calucci G 2014 *Phys. Rev.* D **89** 016020

IOP Conf. Series: Journal of Physics: Conf. Series **981** (2018) 012015 doi:10.1088/1742-6596/981/1/012015

Spin degeneracy of Hadronic molecules in the heavy quark region

Yasuhiro Yamaguchi

Istituto Nazionale di Fisica Nucleare (INFN), Sezione di Genova, via Dodecaneso 33, 16146 Genova, Italy

E-mail: yasuhiro.yamaguchi@ge.infn.it

Abstract. Hadronic molecules have been considered to appear close to the hadron-hadron threshold. For the heavy mesons, $\bar{D}$ and B, the one pion exchange potential is enhanced by the mass degeneracy of heavy pseudoscalar and vector mesons, caused by the heavy quark spin symmetry. In this study, we investigate new hadronic molecules formed by the heavy meson $P^{(*)} = \bar{D}^{(*)}, B^{(*)}$ and a nucleon N, being $P^{(*)}N$. As the interaction between $P^{(*)}$ and N, the pion and vector meson (ρ and ω) exchanges are considered. By solving the coupled-channel Schrödinger equations for PN and P^*N, we obtain the bound and resonant states in the charm and bottom sectors, and in the infinite heavy quark mass limit. In the molecular states, the $PN - P^*N$ mixing effect is important, where the tensor force of the one pion exchange potential generates the strong attraction. In the heavy quark limit, we obtain the degeneracy of the states for $J^P = 1/2^-$ and $3/2^-$.

1. Introduction

Study of the exotic hadron states attracts a great deal of interest in the hadron and nuclear physics. The constituent quark model has been successfully applied to explain the ordinary hadron spectra, describing the baryons as qqq and mesons as $q\bar{q}$ with constituent quarks q and $\bar{q}$ [1, 2]. However, recent observations of new exotic states, such as XYZ in the charm and bottom regions, cannot be explained by the ordinary heavy quarkonium picture $Q\bar{Q}$ with $Q = c, b$ [3–6]. These states motivate us to investigate exotic structures of the hadron states. These exotic states are considered to be a multiquark state being $Q\bar{Q}q\bar{q}$, a hadronic molecule being a meson-antimeson state, and the hybrid states being $Q\bar{Q}g$. For $X(3872)$ [7] and Z_b [8] have been considered to have a molecular component, $D\bar{D}^*$ and $B\bar{B}^*$, respectively, and attract our interest.

Hadronic molecule is a loosely bound state or a resonance of several hadrons appearing close to the thresholds. This is an analogous state to the deuteron or the atomic nuclei which are the bound state of protons and nucleons [9]. The deuteron is a loosely bound system with the small binding energy 2.2 MeV, and the large root mean square radius 2 fm. Similarly, the hadronic molecules in the heavy quark sector would have those properties.

In the hadron-hadron bound system, the hadron-hadron interaction plays an important role. For a heavy meson ($\bar{D}$, B) and a nucleon, the strong attraction by the one pion exchange potential (OPEP) has been discussed in Refs. [10–14]. The OPEP is enhanced by the mass degeneracy between heavy pseudoscalar mesons, $P = \bar{D}, B$, and heavy vector mesons, $P^* = \bar{D}^*, B^*$, where the mass degeneracy are induced by the heavy quark spin symmetry

(HQS) [15–17]. It has been known that the tensor force of the OPEP plays an important role to generate the strong attraction in nuclei. Therefore, the OPEP is expected to produce an attraction in heavy hadron systems. The attractive $\bar{D}N$ (BN) interaction would make it possible to give the exotic nuclear systems with heavy antiquarks ($\bar{c}$ or $\bar{b}$). These binding states are manifestly exotic states which are stable against a strong decay.

The attraction of the hadron-nucleon system shows us interesting phenomena as the impurity effects which are not present in normal atomic nuclei. To understand phenomena caused by impurities is important to investigate hadron-nucleon interactions, properties of the hadrons in nuclear medium, and impurity effects for nuclear properties. In the strangeness sector, the impurity effects such as the formation of the high density states in the $\bar{K}$ nuclei [18, 19], and the shrinking of the wave functions in the hypernuclei [20, 21] have been studied. The heavy hadron in nuclear systems have also been discussed [22–31].

In the present work, we focus on the two-body meson-nucleon system being $\bar{D}^{(*)}N$ and $B^{(*)}N$, where $\bar{D}^{(*)}$ ($B^{(*)}$) stands for $\bar{D}$ or $\bar{D}^*$ (B or B^*). The meson-baryon interaction is given by the meson exchange potential, which induces the off-diagonal components mixing the PN and P^*N channels. By solving the coupled-channel Schrödinger equations, we obtain the bound and resonant states of the genuinely exotic states. We also investigate the properties of the two-body system with an infinite heavy quark mass.

2. Interactions

As the $P^{(*)}N$ interaction, the meson exchange potential is employed. In particular, the OPEP would give a strong attraction. The OPEP appears in the process $PN \leftrightarrow P^*N \leftrightarrow PN$, while the π exchange is absent in $PN \leftrightarrow PN$, because the $PP\pi$ vertex violates parity conservation. Therefore, the OPEP is enhanced by the small mass splitting of P and P^* mesons. In the heavy quark sector, the mass differences of the mesons are small, e.g. $m_{\bar{D}^*} - m_{\bar{D}} \sim 140$ MeV for charm and $m_{B^*} - m_B \sim 45$ MeV for bottom, while $m_\rho - m_\pi \sim 600$ MeV, and $m_{K^*} - m_K \sim 400$ MeV in the up, down and strange sectors have the large mass difference. Therefore, the pion coupling will be important in the heavy meson sectors.

The π, ρ and ω exchange potentials are obtained by the meson exchange scattering amplitudes. The amplitudes are given by the effective Lagrangians for heavy mesons and light mesons, and for nucleons and light mesons [11–14, 33]. The Lagrangian for the heavy mesons and the light mesons ($\pi, \mathrm{V} = (\rho, \omega)$), based on the HQS, are

$$\mathcal{L}_{\pi HH} = ig_\pi \mathrm{Tr}\left[H_b \gamma_\mu \gamma_5 A^\mu_{ba} \bar{H}_a\right] , \tag{1}$$

$$\mathcal{L}_{VHH} = -i\beta \mathrm{Tr}\left[H_b v^\mu (\rho_\mu)_{ba} \bar{H}_a\right] + i\lambda \mathrm{Tr}\left[H_b \sigma^{\mu\nu} F_{\mu\nu}(\rho)_{ba} \bar{H}_a\right] , \tag{2}$$

where the heavy meson fields H and $\bar{H}$ are given by

$$H_a = \frac{1+\not{v}}{2}\left[P^*_{a\,\mu}\gamma^\mu - P_a\gamma^5\right] , \tag{3}$$

$$\bar{H}_a = \gamma^0 H^\dagger_a \gamma^0 , \tag{4}$$

which are expressed by the linear combination of the heavy pseudoscalar (P) and vector (P^*) meson fields. The heavy meson fields are normalized as

$$\langle 0|P|P(p_\mu)\rangle = \sqrt{p^0} , \tag{5}$$

$$\langle 0|P^*_\mu|P^*(p_\mu, \alpha)\rangle = \varepsilon(\alpha)_\mu \sqrt{p^0} , \tag{6}$$

where $\varepsilon(\alpha)_\mu$ is the polarization vector of P^* with polarization $\alpha = \pm, 0$. v^μ is the four-velocity of a heavy quark, where we take the static approximation $v^\mu = (1, \vec{0})$. The subscripts a, b are

Table 1. Mass of mesons ($\alpha = \pi, \rho, \omega$) and coupling constants in Refs. [12–14]

	m_α [MeV]	g_π	β	λ [GeV^{-1}]	$g^2_{\alpha NN}$	κ
π	137.27	0.59	—	—	13.6	—
ρ	769.9	—	0.9	0.56	0.84	6.1
ω	781.94	—	0.9	0.56	20.0	0.0

for light flavors. The axial current A^μ, and the field tensor $F^{\mu\nu}$ are

$$A^\mu = \frac{i}{2f_\pi}\partial^\mu \hat{\pi}\,, \quad \hat{\pi} = \sqrt{2}\begin{pmatrix} \pi^0/\sqrt{2} & \pi^+ \\ \pi^- & -\pi^0/\sqrt{2} \end{pmatrix}, \tag{7}$$

$$F^{\mu\nu} = \partial^\mu V^\nu - \partial^\nu V^\mu\,, \quad V^\mu = i\frac{g_V}{\sqrt{2}}\hat{V}^\mu\,, \tag{8}$$

$$\hat{V}^\mu = \sqrt{2}\begin{pmatrix} (\rho^0+\omega)/\sqrt{2} & \rho^+ \\ \rho^- & (-\rho^0+\omega)/\sqrt{2} \end{pmatrix}^\mu, \tag{9}$$

with the pion decay constant $f_\pi = 92.3$ MeV and $g_V = m_\rho/f_\pi$. The coupling constants of the Lagrangians are summarized in Table 1.

The Lagrangians for nucleons and light mesons are given by the Bonn model [34, 35] as

$$\mathcal{L}_{\pi NN} = i g_{\pi NN}\bar{N}_b\gamma^5 N_a \hat{\pi}_{ba}\,, \tag{10}$$

$$\mathcal{L}_{VNN} = g_{VNN}\bar{N}_b\left(\gamma^\mu(\hat{V}_\mu)_{ba} + \frac{\kappa}{2m_N}\sigma_{\mu\nu}\partial^\nu(\hat{V}^\mu)_{ba}\right)N_a\,. \tag{11}$$

From those Lagrangians in Eqs.(1)-(11), we obtain the $P^{(*)}N$ potentials. The π exchange potential is given by

$$V^\pi_{PN-P^*N} = -\frac{g_\pi g_{\pi NN}}{2m_N f_\pi}\frac{1}{3}\left[\vec{\varepsilon}^\dagger\cdot\vec{\sigma}C(r;m_\pi) + S_\varepsilon T(r;m_\pi)\right]\vec{\tau}_P\cdot\vec{\tau}_N\,, \tag{12}$$

$$V^\pi_{P^*N-P^*N} = \frac{g_\pi g_{\pi NN}}{2m_N f_\pi}\frac{1}{3}\left[\vec{T}\cdot\vec{\sigma}C(r;m_\pi) + S_T T(r;m_\pi)\right]\vec{\tau}_P\cdot\vec{\tau}_N\,, \tag{13}$$

and the vector meson exchange potentials are given by

$$V^{\mathrm{V}}_{PN-PN}(r) = \frac{\beta g_V g_{VNN}}{\sqrt{2}m^2_{\mathrm{V}}}C(r;m_{\mathrm{V}})\vec{\tau}_P\cdot\vec{\tau}_N\,, \tag{14}$$

$$V^{\mathrm{V}}_{PN-P^*N}(r) = \frac{\beta g_V g_{VNN}\lambda(1+\kappa)}{\sqrt{2}m^2_{\mathrm{V}}}\frac{1}{3}\left[-2\vec{\varepsilon}\cdot\vec{\sigma}C(r;m_{\mathrm{V}}) + S_\varepsilon T(r;m_{\mathrm{V}})\right]\vec{\tau}_P\cdot\vec{\tau}_N\,, \tag{15}$$

$$V^{\mathrm{V}}_{P^*N-P^*N}(r) = \frac{\beta g_V g_{VNN}}{\sqrt{2}m^2_{\mathrm{V}}}C(r;m_{\mathrm{V}})\vec{\tau}_P\cdot\vec{\tau}_N + \frac{\beta g_V g_{VNN}\lambda(1+\kappa)}{\sqrt{2}m^2_{\mathrm{V}}}\frac{1}{3}\left[2\vec{T}\cdot\vec{\sigma}C(r;m_{\mathrm{V}}) - S_T T(r;m_{\mathrm{V}})\right]\vec{\tau}_P\cdot\vec{\tau}_N\,. \tag{16}$$

The nucleon mass is given by $m_N = 940$ MeV. $\vec{T}$ is the spin-one operator, $\vec{T} = i\vec{\varepsilon}^\dagger\times\varepsilon$, and S_ε (S_T) is the tensor operator $S_{\mathcal{O}}(\hat{r}) = 3(\vec{\mathcal{O}}\cdot\hat{r})(\vec{\sigma}\cdot\hat{r}) - \vec{\mathcal{O}}\cdot\vec{\sigma}$ with $\hat{r} = \vec{r}/r$ and $r = |\vec{r}|$ for $\vec{\mathcal{O}} = \vec{\varepsilon}$ ($\vec{T}$), where $\vec{r}$ is the relative position vector between $P^{(*)}$ and N. $\vec{\sigma}$ are Pauli matrices acting on nucleon spin. $\vec{\tau}_P$ ($\vec{\tau}_N$) are isospin operators for $P^{(*)}$ (N). For the OPEP, the $PN - PN$ term is absent because the vertex of three pseudoscalar mesons violates the parity conservation.

The functions $C(r; m_\alpha)$ and $T(r; m_\alpha)$ for $\alpha = \pi, \rho, \omega$ are expressed by

$$C(r; m_\alpha) = \int \frac{d^3\vec{q}}{(2\pi)^3} \frac{m_\alpha^2}{\vec{q}^{\,2} + m_\alpha^2} e^{i\vec{q}\cdot\vec{r}} F(\vec{q}, m_\alpha), \tag{17}$$

$$S_{\mathcal{O}}(\hat{r}) T(r; m_\alpha) = \int \frac{d^3\vec{q}}{(2\pi)^3} \frac{-\vec{q}^{\,2}}{\vec{q}^{\,2} + m_\alpha^2} S_{\mathcal{O}}(\hat{q}) e^{i\vec{q}\cdot\vec{r}} F(\vec{q}, m_\alpha), \tag{18}$$

where the dipole-type form factor

$$F(\vec{q}, m_\alpha) = \frac{\Lambda_N^2 - m_\alpha^2}{\Lambda_N^2 + |\vec{q}|^2} \frac{\Lambda_P^2 - m_\alpha^2}{\Lambda_P^2 + |\vec{q}|^2} \tag{19}$$

with cutoff parameters Λ_N and Λ_P is introduced for spatially extended hadrons. From a quark model estimation, we use $\Lambda_D = 1.35\Lambda_N$ for $\bar{D}^{(*)}$ meson and $\Lambda_B = 1.29\Lambda_N$ for $B^{(*)}$ meson, and $\Lambda_{P_Q} = 1.12\Lambda_N$ for the heavy quark limit, with $\Lambda_N = 830$ MeV, as discussed in Refs. [11–13].

The Hamiltonian is given by $H = K + V_{P^{(*)}N}$, where K is the kinetic term, and $V_{P^{(*)}N}$ is the $P^{(*)}N$ potential shown above. By solving the coupled-channel Schrödinger equation, the bound state is obtained. We also obtain the resonance by analyzing the phase shift.

3. Numerical Results

In this section, we show the results for the bound and resonant states of the $\bar{D}N$ and BN molecules with $J^P = 1/2^\pm, \cdots, 5/2^\pm$, and the isospin $I = 0$. The obtained energy spectra are summarized in Table 2. The binding energy is shown as a real negative number. For the resonance, the energy is given by a complex value as $E_{\rm re} - i\Gamma/2$ with a resonance energy $E_{\rm re}$ and a decay width Γ.

We obtain the bound states both in $\bar{D}N$ and BN for $J^P = 1/2^-$. In the bound states, we find that the tensor force of the OPEP in the $PN - P^*N$ mixing component plays an important role to generate the strong attraction. For the bottom sector, the BN binding energy is larger than the $\bar{D}N$. The BN system obtains not only the suppression of the kinetic energy but also the enhancement of the mixing effect due to the small BB^* mass difference.

For $J^P = 1/2^+, 3/2^\pm$ and $5/2^+$, the resonance states are found above the PN threshold. In particular the states with $J^P = 3/2^-$ have the small decay width. We find the interesting structure of the resonances which is the Feshbach resonance. The states have the quasi bound states of the P^*N channels.

We obtain no bound state for $I = 1$. In this channel, the OPEP does not produce the strong attraction because of the small contribution of the isospin factor $\vec{\tau_P} \cdot \vec{\tau_N} = 1$ for $I = 1$, while $\vec{\tau_P} \cdot \vec{\tau_N} = -3$ for $I = 0$.

Finally, we consider the $P_Q N$ systems in the heavy quark limit, where the pseudoscalar and vector mesons, P_Q and P_Q^*, are degenerate, namely $m_{P_Q} = m_{P_Q^*}$. We study the states for $J^P = 1/2^-, 3/2^-$. We find that both states have the same binding energy, -37.4 MeV as summarized in Table 3. The obtained energies of the charm, bottom and heavy quark limit sectors are compared in Table 3. The energy difference of the $J^P = 1/2^-$ and $J^P = 3/2^-$ states decreases as the heavy quark mass increases. In the heavy quark limit, these bound states are degenerate. The results show that the $J^P = 1/2^-$ and $J^P = 3/2^-$ states belong to the heavy quark spin multiplet, caused by the HQS [38, 39].

4. Summary

We have explored the possible existence of the genuinely exotic nuclei as $\bar{D}N$ and BN. The interaction is constructed by introducing the heavy quark spin symmetry which provides the

Table 2. Energy spectra of $\bar{D}N$ and BN with $I = 0$ in Refs. [12,13]. J and P are total angular momentum and parity, respectively. Binding energy is shown by a real negative value, while the energy of the resonance states are given by the resonance energy $E_{\rm re}$ and the decay width Γ as $E_{\rm re} - i\Gamma/2$. All values are measured from the PN threshold and given in units of MeV.

$\bar{D}N$	$P = -$	$P = +$	BN	$P = -$	$P = +$
$J = 1/2$	-2.1	$26.8 - i65.7$	$J = 1/2$	-23.0	$5.8 - i3.0$
$J = 3/2$	$113.2 - i8.9$	$148.2 - i5.1$	$J = 3/2$	$6.9 - i0.05$	$31.8 - i14.4$
$J = 5/2$	—	$176.0 - i87.4$	$J = 5/2$	—	$58.4 - i24.8$

Table 3. Comparison of the energy spectra of $\bar{D}N$, BN, and P_QN for $J^P = 1/2^-, 3/2^-$ in Refs. [38, 39]. The energy difference of the $J^P = 1/2^-$ and $3/2^-$ states, $E_{3/2^-} - E_{1/2^-}$, is also shown. The values are given in the unit of MeV.

J^P	$\bar{D}N$	BN	P_QN
$1/2^-$	-2.1	-23.0	-37.4
$3/2^-$	$113.2 - i8.9$	$6.9 - i0.05$	-37.4
$E_{3/2^-} - E_{1/2^-}$	115.3	29.9	0.0

mass degeneracy of the heavy pseudoscalar and vector mesons. The meson exchange potentials for π, ρ and ω are employed as the $P^{(*)}N$ interaction, where those potentials induce the $P - P^*$ mixing. By solving the coupled-channel equations for PN and P^*N, we obtain bound states and resonances in charm and bottom sectors. The tensor force of the OPEP mixing PN and P^*N plays the important role to produce the strong attraction. This force is enhanced by the small mass splitting of mesons due to the heavy quark spin symmetry. The mechanism is unique in the heavy hadron systems with the small mass difference of the spin partner. In the heavy quark limit, we find that the bound states for $J^P = 1/2^-$ and $3/2^-$ are degenerate. The result indicates the $J^P = 1/2^-$ and $3/2^-$ states belongs to the heavy quark spin multipulet.

The obtained exotic states can be investigated in relativistic heavy ion collisions in RHIC and LHC [40, 41], J-PARC and GSI-FAIR.

References

[1] Klempt E and Richard J M 2010 *Rev. Mod. Phys.* **82** 1095 (*Preprint* 0901.2055 [hep-ph])
[2] Eichten E, Godfrey S, Mahlke H and Rosner J L 2008 *Rev. Mod. Phys.* **80** 1161 (*Preprint* hep-ph/0701208)
[3] Brambilla N, Eidelman S, Heltsley B K, Vogt R, Bodwin G T, Eichten E, Frawley A D and Meyer A B *et al.*, 2011 *Eur. Phys. J.* C **71** 1534 (*Preprint* 1010.5827 [hep-ph])
[4] Swanson E S 2006 *Phys. Rept.* **429** 243 (*Preprint* hep-ph/0601110)
[5] Voloshin M B 2008 *Prog. Part. Nucl. Phys.* **61** 455 (*Preprint* 0711.4556 [hep-ph])
[6] Godfrey S and Olsen S L 2008 *Ann. Rev. Nucl. Part. Sci.* **58** 51 (*Preprint* 0801.3867 [hep-ph])
[7] Choi S K *et al.* [Belle Collaboration] 2003 *Phys. Rev. Lett.* **91** 262001 (*Preprint* hep-ex/0309032)
[8] Bondar A *et al.* [Belle Collaboration] 2012 *Phys. Rev. Lett.* **108** 122001 (*Preprint* 1110.2251 [hep-ex])
[9] Tornqvist N A 1994 *Z. Phys.* C **61** 525 (*Preprint* hep-ph/9310247)
[10] Cohen T D, Hohler P M and Lebed R F 2005 *Phys. Rev.* D **72** 074010 (*Preprint* hep-ph/0508199)
[11] Yasui S and Sudoh K 2009 *Phys. Rev.* D **80** 034008 (*Preprint* 0906.1452 [hep-ph])
[12] Yamaguchi Y, Ohkoda S, Yasui S and Hosaka A 2011 *Phys. Rev.* D **84** 014032 (*Preprint* 1105.0734 [hep-ph])
[13] Yamaguchi Y, Ohkoda S, Yasui S and Hosaka A 2012 *Phys. Rev.* D **85** 054003 (*Preprint* 1111.2691 [hep-ph])
[14] Yamaguchi Y, Yasui S and Hosaka A 2014 *Nucl. Phys.* A **927** 110 (*Preprint* 1309.4324 [nucl-th])

[15] Isgur N and Wise M B 1989 *Phys. Lett.* B **232** 113
[16] Flynn J M and Isgur N 1992 *J. Phys.* G **18** 1627 (*Preprint* hep-ph/9207223)
[17] Manohar A V and Wise M B 2000 *Camb. Monogr. Part. Phys. Nucl. Phys. Cosmol.* **10** 1
[18] Yamazaki T, Dote A and Akaishi Y 2004 *Phys. Lett.* B **587** 167 (*Preprint* nucl-th/0310085)
[19] Dote A, Hyodo T and Weise W 2009 *Phys. Rev.* C **79** 014003 (*Preprint* 0806.4917 [nucl-th])
[20] Hashimoto O and Tamura H 2006 *Prog. Part. Nucl. Phys.* **57** 564
[21] Tamura H 2002 *Eur. Phys. J.* A **13** 181
[22] Tsushima K, Lu D H, Thomas A W, Saito K and Landau R H 1999 *Phys. Rev.* C **59** 2824 (*Preprint* nucl-th/9810016)
[23] Mishra A, Bratkovskaya E L, Schaffner-Bielich J, Schramm S and Stoecker H 2004 *Phys. Rev.* C **69** 015202 (*Preprint* nucl-th/0308082)
[24] Lutz M F M and Korpa C L 2006 *Phys. Lett.* B **633** 43 (*Preprint* nucl-th/0510006)
[25] Tolos L, Ramos A and Mizutani T 2008 *Phys. Rev.* C **77** 015207 (*Preprint* 0710.2684 [nucl-th])
[26] Mishra A and Mazumdar A 2009 *Phys. Rev.* C **79** 024908 (*Preprint* 0810.3067 [nucl-th])
[27] Kumar A and Mishra A 2010 *Phys. Rev.* C **81** 065204 (*Preprint* 1005.5018 [nucl-th])
[28] Garcia-Recio C, Nieves J, Salcedo L L and Tolos L 2012 *Phys. Rev.* C **85** 025203 (*Preprint* 1111.6535 [nucl-th])
[29] Yasui S and Sudoh K 2013 *Phys. Rev.* C **87** 015202 (*Preprint* 1207.3134 [hep-ph])
[30] Hosaka A, Hyodo T, Sudoh K, Yamaguchi Y and Yasui S 2016 *Preprint 1606.08685 [hep-ph]*
[31] Yamaguchi Y and Yasui S 2016 *Preprint* 1612.08925 [nucl-th]
[32] Bayar M, Xiao C W, Hyodo T, Dote A, Oka M and Oset E 2012 *Phys. Rev.* C **86** 044004 (*Preprint* 1205.2275 [hep-ph])
[33] Casalbuoni R, Deandrea A, Di Bartolomeo N, Gatto R, Feruglio F and Nardulli G 1997 *Phys. Rept.* **281** 145 (*Preprint* hep-ph/9605342)
[34] Machleidt R, Holinde K and Elster C 1987 *Phys. Rept.* **149** 1
[35] Machleidt R 2001 *Phys. Rev.* C **63** 024001 (*Preprint* nucl-th/0006014)
[36] Pudliner B S, Pandharipande V R, Carlson J, Pieper S C and Wiringa R B 1997 *Phys. Rev.* C **56** 1720 (*Preprint* nucl-th/9705009)
[37] Hiyama E, Kino Y and Kamimura M 2003 *Prog. Part. Nucl. Phys.* **51** 223
[38] Yasui S, Sudoh K, Yamaguchi Y, Ohkoda S, Hosaka A and Hyodo T 2013 *Phys. Lett.* B **727** 185 (*Preprint* 1304.5293 [hep-ph])
[39] Yamaguchi Y, Ohkoda S, Hosaka A, Hyodo T and Yasui S 2015 *Phys. Rev.* D **91** 034034 (*Preprint* 1402.5222 [hep-ph])
[40] Cho S *et al.* [ExHIC Collaboration] 2011 *Phys. Rev. Lett.* **106** 212001 (*Preprint* 1011.0852 [nucl-th])
[41] Cho S *et al.* [ExHIC Collaboration] 2011 *Phys. Rev.* C **84** 064910 (*Preprint* 1107.1302 [nucl-th])

IOP Conf. Series: Journal of Physics: Conf. Series **981** (2018) 012016 doi:10.1088/1742-6596/981/1/012016

Heavy flavour in high-energy nuclear collisions: a theoretical overview

Andrea Beraudo
INFN - Sezione di Torino, Via Pietro Giuria 1, I-10125 Torino
E-mail: beraudo@to.infn.it

Abstract. The peculiar role of heavy-flavour observables in relativistic heavy-ion collisions is discussed. Produced in the early stage, c and b quarks cross the hot deconfined plasma arising from the collision, interacting strongly with the latter, until they hadronize. Transport calculations are the tools to follow their propagation in the medium: their formulation as well as their conceptual basis are briefly reviewed. Depending on the strength of the interaction heavy quarks may or not approach kinetic equilibrium with the plasma, tending in the first case to follow the collective flow of the expanding fireball. The presence of a hot deconfined medium may also affect heavy-quark hadronization, being possible for them to recombine with the surrounding light thermal partons, so that the final heavy-flavour hadrons inherit part of the flow of the medium. Here we show how it is possible to develop a complete transport setup allowing one to describe heavy-flavour production in high-energy nuclear collisions. The ultimate goal will be to extract from the experimental data the heavy-flavour transport coefficients in the Quark-Gluon Plasma: we will comment on how far we are from this achievement. Information coming from recent lattice-QCD simulations concerning both the heavy-flavour transport coefficients in the hot QCD plasma and the nature of the charmed degrees around the deconfinement transition is also presented. Finally, the possibility that the formation of a hot deconfined medium even in small systems (high-multiplicity p-Au and d-Au collisions, so far) may affect also heavy-flavour observables is investigated.

1. Introduction

Quantum Chromo-Dynamics is characterized by a non trivial phase-diagram in which, depending on the temperature and the baryon density, the active degrees of freedom may be coloured quarks and gluons or "white" hadrons. Relativistic heavy-ion collisions are the only experimental tool to study, under controlled conditions, such a phase diagram. In particular, at the highest energies, nuclear collisions allows one to explore the transition from a plasma of deconfined quarks and gluons (QGP) to a gas of hadrons and resonances in the region of high temperature and (almost) vanishing baryon density. In such a regime, experienced also by the universe during its thermal evolution, the transition is actually a smooth crossover and is accompanied by the spontaneous breaking of chiral symmetry (responsible for most of the baryonic mass of our universe).

The matter produced in heavy-ion collisions undergoes the following evolution. There is an initial pre-equilibrium stage, during which the energy stored in the strong colour-fields is converted into particles (quarks and gluons). There is evidence that, quite rapidly ($\tau_0 \lesssim 1$ fm/c), the system reaches local thermodynamical equilibrium, spending some time ($\sim 5-10$ fm/c) in the QGP phase, during which it expands and cools until getting to a regime in which the active degrees of freedom are again mesons and baryons. When the collision rate becomes too small to maintain kinetic equilibrium hadrons decouple and follow a free-streaming until being detected. Hence, information on the possible onset of deconfinement and on the properties of the produced medium has to be obtained quite indirectly, starting from hadronic observables.

Soft probes, i.e. low-p_T hadrons (the bulk of particle production), provide evidence for the collective behaviour of the produced system, which displays a hydrodynamics expansion driven by pressure gradients. The success of hydrodynamics in explaining peculiar features of soft-particle production (in particular the azimuthal asymmetry of the spectra) suggests that the medium formed in heavy-ion collisions is characterized by a mean-free-path much smaller than its size, $\lambda_{\rm mfp} \ll L$, and that during most of its expansion it remains close to local thermal equilibrium. *Hard probes*, on the contrary, are those particles (high-p_T partons, heavy quarks and quarkonia) produced in hard pQCD processes in the very first instants and which, before reaching the detectors, cross the fireball formed in the collision of the two nuclei, performing tomography of the latter. In particular, the suppression of high-p_T particle production, loosely speaking referred to as jet-quenching, provides evidence that the medium formed in heavy-ion collisions is very opaque: partons propagating through it lose a non-negligible fraction of their energy. Within this framework, heavy-flavour (HF) particles play a peculiar role. Soft observables are described by hydrodynamics, assuming to deal with a system close to local thermal equilibrium as a working hypothesis. The study of jet quenching involves modeling the energy-degradation of *external* probes. On the other hand, the description of HF observables requires dealing with a more complex situation, developing a setup (based on *transport theory*) able to describe how particles, initially shot into the medium, asymptotically approach kinetic equilibrium with the latter. Notice that, at high-p_T, the interest in HF particles is no longer related to their possible thermalization, but to the mass and colour-charge dependence of parton energy-loss and jet quenching: we will not address this issue here.

Why are c and b-quarks considered heavy, based on their mass M? First of all because $M \gg \Lambda_{\rm QCD}$, setting a hard scale which ensures that their initial production is well described by pQCD. Secondly, because $M \gg T$, entailing that their thermal abundance in the plasma would be negligible: in heavy-ion collisions, with an expanding fireball with a quite short lifetime ~ 10 fm/c, the final multiplicity is set by the initial hard production. Finally, because $M \gg gT$, the latter being the typical momentum exchange in the collisions with the plasma particles, implying that many soft scatterings are necessary to change significantly the momentum and trajectory of a heavy quark: this will be of relevance in order to developed a simplified transport setup.

The present contribution is organized as follows. In Sec. 2 we discuss what lattice-QCD calculations tell us about the behaviour of charm around deconfinement. In Sec. 3 we give an overview on medium modification of HF observables in heavy-ion collisions, illustrating the complex setup needed to get to a meaningful theory-to-experiment comparison and discussing the information one can obtain from the present experimental data. In Sec. 4 we move to smaller systems, discussing whether also in high-multiplicity p-A or d-A collisions one can expect final-state effects associated to the formation of a hot deconfined medium and whether this can affect also HF observables. Finally, in Sec. 5, we draw our conclusions and discuss the perspectives of future studies.

2. Charm degrees of freedom around deconfinement: lattice-QCD results

Lattice-QCD calculations show that, in the limit of vanishing baryon density, the deconfinement transition is actually a crossover, i.e. a rapid but smooth change in the nature of the active degrees of freedom, carriers of conserved charges: something similar to what occurred to the electromagnetic plasma in the primordial universe, with the Saha equation describing the evolution of the ionization fraction, i.e. the fraction of charged protons not bound to electrons in neutral hydrogen atoms. Actually, there is an important difference with respect to such a case. In the early universe, at recombination, the Debye radius of the electric interaction

IOP Conf. Series: Journal of Physics: Conf. Series **981** (2018) 012016 doi:10.1088/1742-6596/981/1/012016

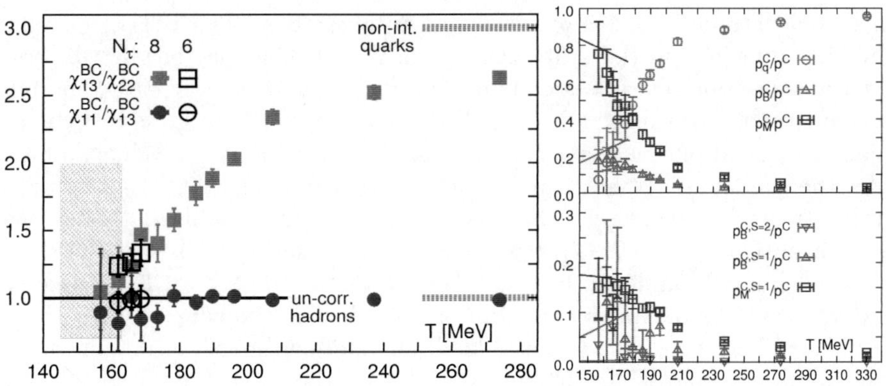

Figure 1. Left panel: correlators of charm and baryon-number fluctuations around deconfinement [1]. Right panel: the partial pressures of charmed quarks, mesons and baryons around deconfinement [2].

$r_D \equiv (T/n_e e^2)^{1/2} \approx 24$ cm was orders of magnitude larger than the typical atomic radius $a_0 \sim 10^{-10}$ m, leaving atomic properties unaffected. On the contrary, in the QGP for the Debye radius of the colour interaction $r_D^{-1} \equiv gT(N_c/3 + N_f/6)^{1/2}$ at temperatures $T \approx 0.2$ GeV around the deconfinement transition one has $r_D \approx 0.4\,\text{fm} \sim r_h$, comparable to the typical size of a hadron: the QCD crossover occurs in a dense, strongly interacting regime, in which hadron properties are sizably modified.

Due to its large mass compared to the other quarks, it is interesting to study the behaviour of charm across the deconfinement transition. QCD interactions do not change flavour, so that charm can be considered a conserved charge in heavy-ion collisions, as baryon number. Based on the fact that charm quarks have $B = 1/3$ while charmed hadrons have $B = 0, 1$, from the correlations of C and B fluctuations one can get information on the nature of the charm-carrying degrees of freedom, i.e. whether they are mostly partonic or hadronic (C excess being always associated to B excess in the partonic phase). Taking derivatives of the pressure with respect to the chemical potentials one obtains the various higher-order cumulants of the above fluctuations

$$\chi_{kl}^{BC} = \left.\frac{\partial^{k+l}[P/T^4]}{\partial \hat{\mu}_B^k \partial \hat{\mu}_C^l}\right|_{\hat{\mu}_i=0} \quad \text{where} \quad \hat{\mu}_i \equiv \mu_i/T, \tag{1}$$

with a very different behaviour in the two phases. For instance one has

$$\frac{\chi_{mn}^{BC}}{\chi_{m+1,n-1}^{BC}} = B^{-1} \quad (1 \text{ for hadrons, } 3 \text{ for quarks}), \tag{2}$$

allowing one to identify the nature of charm-carrying degrees of freedom. Such an analysis was carried out in [1] and results are displayed in the left panel of Fig. 1. It is also possible to separate, as done in [2], the various partial pressures of charmed particles (quarks and hadrons) around the deconfinement crossover: results are shown in the right panel of Fig. 1. Overall, lattice-QCD calculations show that also for charm the deconfinement transition is actually a smooth crossover, with part of the charmed degrees of freedom being still/already hadronic states at temperatures slightly above T_c.

3. Heavy Flavour in heavy-ion collisions

A realistic study of HF observables in high-energy nuclear collisions requires developing a multi-step setup. First of all, one needs to simulate the initial hard production of the $c\bar{c}$ and $b\bar{b}$ pairs in the elementary nucleon-nucleon collisions: for this automated pQCD tools are available, which one can generalize to include initial-state effects, such as nuclear parton distribution functions (nPDF's) and transverse-momentum broadening in nuclear matter. Secondly, one must have at his disposal a realistic description of the fireball in which the heavy quarks propagate, provided by hydrodynamic calculations validated against soft-hadron data. The core of the setup is represented by the modeling of the heavy-quark propagation in the hot plasma: one has to develop a transport code working in the case of an evolving medium, taking advantage of the previous information. Once the heavy quarks reach a fluid cell below the deconfinement temperature they hadronize and one has to model also this stage: can it be described simply via fragmentation functions like in the vacuum or are other mechanisms, such as recombination with light thermal partons, possible? Clearly, medium modification of hadronization in heavy-ion collisions is an item of interest in itself; however, it is at the same time a source of systematic uncertainty if one wants to extract information on the properties of the deconfined medium. Finally, since experimental data sometimes does not refer directly to D and B-mesons, one must also simulate the decay into the relevant channels (e.g. $D \to X\,\nu_e\,e$, $D \to X\,\nu_\mu\,\mu$, $B \to X\,J/\Psi ...$) accessible to the experiments.

3.1. *Heavy-quark production*

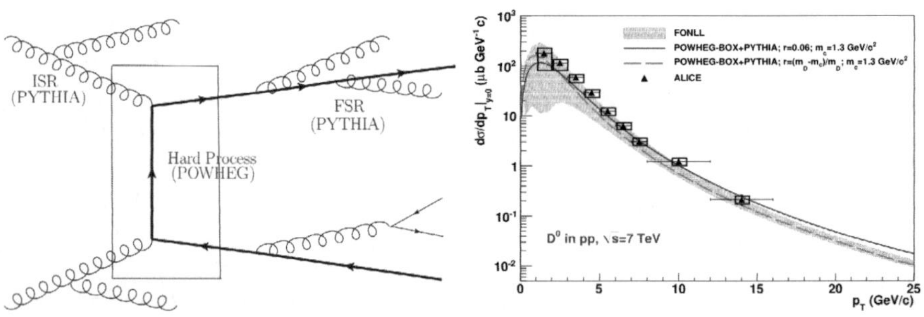

Figure 2. Left panel: $Q\overline{Q}$ event generated by POWHEG-BOX [4], with the hard process interfaced with a parton-shower stage. Right panel: FONLL [3] and POWHEG-BOX predictions (with different fragmentation functions) for D^0 production in p-p collisions at 7 TeV compared to ALICE data [6].

Due to the large mass, heavy-quark production is a hard short-distance process occurring at the level of the individual binary nucleon-nucleon collisions. It is then necessary to validate the simulation of the initial hard production against p-p data. The state of the art for its theoretical description is represented by the Fixed Order Next-to-Leading Log (FONLL) calculation, developed to provide accurate results both at low and at high-p_T through a proper resummation of large transverse-momentum logarithms [3]. However, FONLL simply provides an inclusive single-particle spectrum, while for practical applications it is more convenient to work with an event generator. In this regard, the POWHEG-BOX package [4] represents a convenient tool. The latter generates hard events at NLO accuracy and interface them with a parton-shower stage, simulating initial and final-state radiation. A comparison of FONLL and POWHEG-BOX results (with $m_c = 1.3$ GeV, $\mu_R = \mu_F = m_T$ and HQET fragmentation functions [5]) with ALICE data for D^0 production in p-p collisions at the LHC [6] is displayed in the right panel of Fig. 2. Overall, the data tend to stay at the upper edge of the (quite large, in the case of charm) theoretical uncertainty band, arising from the choice of the quark mass and of the factorization and renormalization scales. For a systematic comparison of the various automated pQCD tools for the calculation/simulation of heavy-quark production see Ref. [7]. Once validated in the p-p case and supplemented with nPDF's [8], the above pQCD tools can be employed to simulated the initial $Q\overline{Q}$ production in heavy-ion collisions. For each event the full kinematic information on the produced $Q\overline{Q}$ pairs can be stored and used in the subsequent steps of the simulations.

3.2. Heavy-quark transport

As a next step, one needs to implement a numerical transport calculation describing the heavy-quark propagation in the plasma. For pioneering work in this fields see for instance Refs. [9, 10]. The starting point of any transport calculation is the Boltzmann equation for the evolution of the heavy-quark phase-space distribution

$$\frac{d}{dt}f_Q(t,\vec{x},\vec{p}) = C[f_Q] \quad \text{with} \quad C[f_Q] = \int d\vec{q}[w(\vec{p}+\vec{q},\vec{q})f_Q(t,\vec{x},\vec{p}+\vec{q}) - w(\vec{p},\vec{q})f_Q(t,\vec{x},\vec{p})], \tag{3}$$

where the collision integral $C[f_Q]$ is expressed in terms of the $\vec{p}\to\vec{p}-\vec{q}$ transition rate $w(\vec{p},\vec{q})$. The direct solution of the Boltzmann equation is numerically demanding; however, as long as $q \ll p$ (q being typically of order gT), one can expand the collision integrand in powers of the momentum exchange. Truncating the expansion to second order

$$C[f_Q] \approx \int d\vec{q}\left[q^i\frac{\partial}{\partial p^i} + \frac{1}{2}q^iq^j\frac{\partial^2}{\partial p^i\partial p^j}\right][w(\vec{p},\vec{q})f_Q(t,\vec{p})] \tag{4}$$

leads to the Fokker-Planck (FP) equation

$$\frac{\partial}{\partial t}f_Q(t,\vec{p}) = \frac{\partial}{\partial p^i}\left\{A^i(\vec{p})f_Q(t,\vec{p}) + \frac{\partial}{\partial p^j}[B^{ij}(\vec{p})f_Q(t,\vec{p})]\right\}, \tag{5}$$

having defined

$$A^i(\vec{p}) = \int d\vec{q}\,q^i w(\vec{p},\vec{q}) \quad \underset{\text{isotropic medium}}{\longrightarrow} \quad A^i(\vec{p}) = A(p)\,p^i$$

$$B^{ij}(\vec{p}) = \frac{1}{2}\int d\vec{q}\,q^iq^j w(\vec{p},\vec{q}) \quad \underset{\text{isotropic medium}}{\longrightarrow} \quad B^{ij}(\vec{p}) = (\delta^{ij} - \hat{p}^i\hat{p}^j)B_0(p) + \hat{p}^i\hat{p}^jB_1(p).$$

In the case of an isotropic system the whole information on the particle-medium interaction is encoded in three quantities: the friction and transverse/longitudinal momentum-broadening coefficients $A(p)$ and $B_{0/1}(p)$. Actually, the above coefficients cannot be taken as independent, but they have to satisfy the Einstein fluctuation-dissipation relation

$$A(p) = \frac{B_1(p)}{TE_p} - \left[\frac{1}{p}\frac{\partial B_1(p)}{\partial p} + \frac{d-1}{p^2}(B_1(p) - B_0(p))\right], \tag{6}$$

which ensures that the FP equation admits the steady equilibrium solution $f_{\rm eq}(p) \equiv e^{-E_p/T}$.

The FP equation can be recast in a form more suited to a numerical implementation interfaced to an event generator: the Langevin equation. The latter in its discretized form

$$\Delta\vec{p}/\Delta t = -\eta_D(p)\vec{p} + \vec{\xi}(t) \tag{7}$$

provides a recipe to update the heavy quark momentum through the sum of a deterministic friction force and a random noise term specified by its temporal correlator

$$\langle\xi^i(\vec{p}_t)\xi^j(\vec{p}_{t'})\rangle = b^{ij}(\vec{p}_t)\delta_{tt'}/\Delta t \qquad b^{ij}(\vec{p}) \equiv \kappa_\perp(p)(\delta^{ij} - \hat{p}^i\hat{p}^j) + \kappa_\parallel(p)\hat{p}^i\hat{p}^j. \tag{8}$$

IOP Conf. Series: Journal of Physics: Conf. Series **981** (2018) 012016 doi:10.1088/1742-6596/981/1/012016

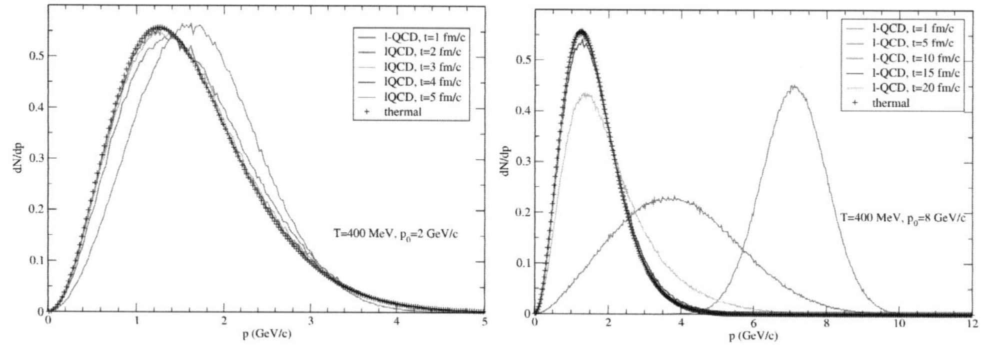

Figure 3. The evolution of a sample of c quarks in a static medium, starting from different initial momenta. Non-perturbative transport coefficients are employed.

The transport coefficients $\kappa_{\perp/\|}(p)$ (representing the average transverse/longitudinal squared momentum exchanged with the plasma per unit time) are directly related to the FP ones by $\kappa_{\perp/\|}(p) = 2B_{0/1}(p)$, while $\eta_D(p)$ depends on the discretization scheme. In the Ito *pre-point* scheme it coincides with the friction coefficient in the FP equation: $\eta_D(p) = A(p)$ [11, 12]. One has then to solve Eq. (7) throughout the whole medium evolution [13, 14], described by hydrodynamic calculations [15, 16]. As a validation of the tool, in Fig. 3 we show how, in the case of a static medium, the momenta of a sample of heavy-quarks actually approach a relativistic Maxwell distribution, at the same equilibrium temperature as the surrounding plasma.

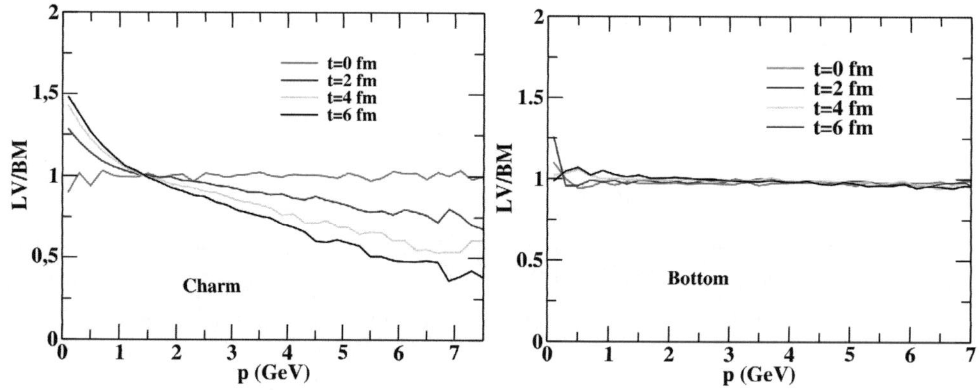

Figure 4. Langevin-to-Boltzmann ratio of c (left panel) and b-quark (right panel) spectra at different times [17]. In the case of beauty the Langevin and the Boltzmann equations provide an equivalent description of the heavy-quark propagation in the medium.

If the expansion of the collision integral allows one to develop a more tractable transport setup with respect to the original Boltzmann equation, one may wonder whether this soft-scattering approximation is fully justified in situations in which the typical momentum exchange $\sim gT$ is not so small compared to the heavy-quark mass. The question was addressed in [17], where the authors performed a systematic comparison of the Boltzmann and Langevin approaches: as shown in Fig. 4, the two schemes provides identical results in the case of beauty, while for charm discrepancies (dependent on the values of medium-parameters) are visible.

In the Langevin approach physics enters through the transport coefficients, which one aims at deriving through a first-principle calculation. In the left panel of Fig. 5, referring to charm, we

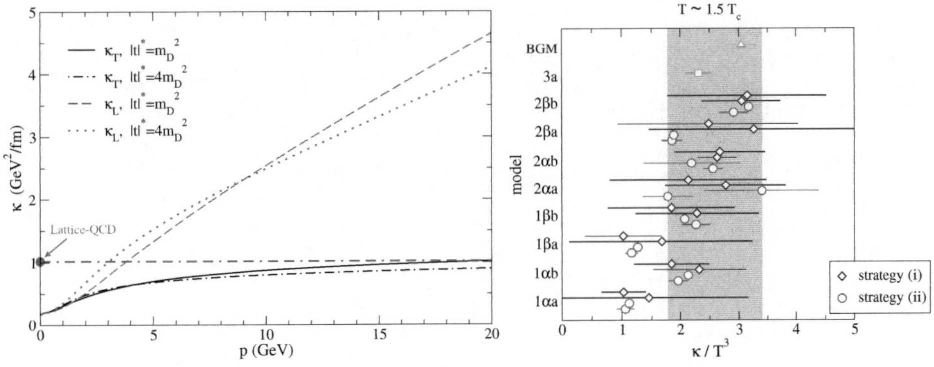

Figure 5. Heavy-quark momentum diffusion coefficients. Left panel: weak-coupling results for charm at $T=400$ MeV, with different choices of the cutoff separating hard and soft collisions [13]. Right panel: lattice-QCD result for a static quark ($\kappa_T=\kappa_L\equiv\kappa$) in a gluonic plasma [18]. The uncertainty band arises from extracting real-time information from imaginary-time correlators.

show the results of a weak-coupling evaluation, accounting for $2\to 2$ collisions, with resummation of medium effects in the case of soft-gluon exchange [13]. Instead, a non-perturbative estimate can be provided by lattice-QCD simulations, so far limited to the case of a static, infinitely heavy , quark. In this limit one can neglect the momentum dependence of the strength of the noise, so that

$$\langle\xi^i(t)\xi^j(t')\rangle=\delta^{ij}\delta(t-t')\kappa \quad\longrightarrow\quad \kappa=\frac{1}{3}\int_{-\infty}^{+\infty}dt\langle\xi^i(t)\xi^i(0)\rangle_{\rm HQ}\underset{p\to 0}{\sim}\frac{1}{3}\int_{-\infty}^{+\infty}dt\langle F^i(t)F^i(0)\rangle_{\rm HQ}. \tag{9}$$

The momentum-diffusion coefficient κ can be extracted from a force-force correlator in an ensemble of states containing one heavy quark: for a static colour source the only force playing a role is the chromo-electric field. Notice that κ, as any other transport coefficient, is a real-time quantity, to be reconstructed from lattice correlators, which, however, can be evaluated only for imaginary-times: this leads to the large theoretical uncertainty band [18] in the right panel of Fig. 5.

Finally, heavy quarks, when reaching a fluid cell below a critical temperature, are allowed to hadronize. Various models have been developed to describe how the medium can modify this stage, all based on the idea that a heavy quark can easily find one or more light partons nearby to recombine with, producing a colour-singlet object, which can be taken to be directly a HF hadron (coalescence [19, 20]), a heavy resonance [21] or a string to further fragment [14].

3.3. Results

Let us now present some selected outcomes of a specific transport setup [14], pointing out some quite general qualitative findings common with the analysis performed by other groups [19, 20, 21]. In Fig. 6 we consider the *nuclear modification factor* of the p_T-spectra and the *elliptic flow*

$$R_{\rm AA}\equiv\frac{(dN/dp_T)^{\rm AA}}{\langle N_{\rm coll}\rangle(dN/dp_T)^{\rm pp}}\quad\text{and}\quad v_2\equiv\langle\cos[2(\phi-\psi_{\rm RP})]\rangle \tag{10}$$

of D^0 mesons in Au-Au and Pb-Pb collisions at RHIC and LHC ($\langle N_{\rm coll}\rangle$ being the average number of binary nucleon-nucleon collisions). It is interesting to start addressing the limit of very (infinitely) large transport coefficients. In this case HF particles would approach local kinetic

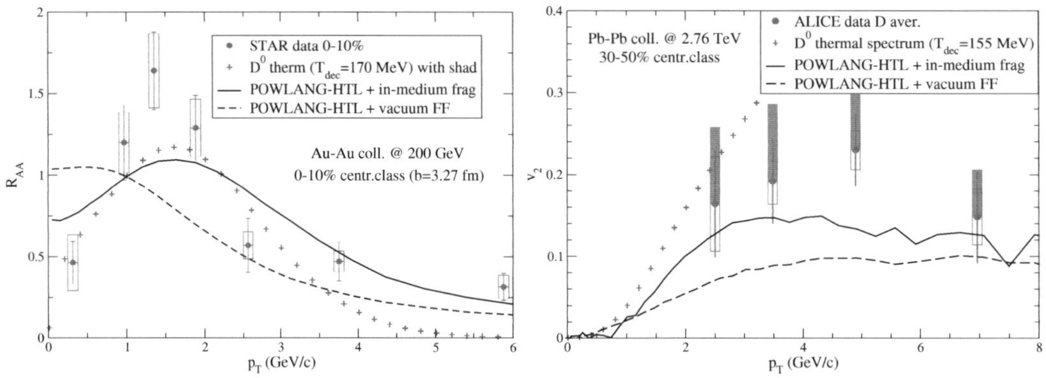

Figure 6. Left panel: the nuclear modification factor of D^0 spectra in 0-10% Au-Au collision at $\sqrt{s_{\rm NN}} = 200$ GeV compared to STAR data [22]. Right panel: the elliptic flow of D^0 mesons in 30-50% Pb-Pb collisions at $\sqrt{s_{\rm NN}} = 2.76$ TeV compared to ALICE data [23]. Theory results include the full kinetic-equilibrium limit (green crosses), transport calculations interfaced with vacuum fragmentation functions (dashed curves) and transport calculations followed by in-medium hadronization (continuous curves).

equilibrium with the fireball and would be thermally emitted from a decoupling hypersurface, identified (rather schematically) by a temperature below which interactions are supposed to become too rare to further modify the spectra. The usual Cooper-Frye formula gives:

$$E\frac{dN}{d\vec{p}} = \frac{1}{(2\pi)^3}\int_{\Sigma_{\rm dec}} p^\mu d\Sigma_\mu \exp\left[-\frac{p\cdot u(x)}{T_{\rm dec}}\right]. \tag{11}$$

As a consequence, high-energy particles would be thermally suppressed; on the other hand, the collective flow of the fluid u^μ would tend to drag very soft particles to larger p_T. This gives rise to the bump in the $R_{\rm AA}$ seen in the left panel of Fig. 6 (green crosses) and displayed also by the experimental data. At the same time, in non-central collision the flow of the fluid is characterized by an azimuthal anisotropy, due to the initial larger pressure gradient along the reaction-plane direction $\psi_{\rm RP}$: the azimuthal distribution of particles flowing with the fluid would inherit this anisotropy, displaying a strong elliptic flow, as shown in the right panel of Fig. 6 and supported by the data. Does the above considerations entail that in heavy-ion collisions charm actually reaches kinetic equilibrium with the medium during its partonic phase? This is not necessarily the case, as shown by the curves in Fig. 6 referring to Langevin calculations with finite (weak-coupling) transport coefficients. Transport in the QGP phase alone does not appear able to transfer to the charmed particles enough radial and elliptic flow (dashed curves with vacuum fragmentation functions). On the other hand, modeling HF hadronization through recombination with thermal partons from the medium leads to results in better agreement with the data, thanks to the additional flow acquired from the light companion (continuous curves).

In Fig. 7 we compare the nuclear modification factors of D^0 mesons in non-central Pb-Pb events at the LHC emitted along (in-plane) or orthogonally (out-of-plane) to the direction of the impact parameter of the collision. We explore the sensitivity to different decoupling temperatures and hadronization schemes. Curves including in-medium hadronization are again characterized by a bump at moderate p_T: experimental data at lower p_T are clearly necessary in order to confirm such an occurrence.

Besides affecting the momentum and angular distribution of the produced particles, in-medium hadronization may also change the heavy-flavour hadrochemistry. The issue was addressed, for instance, in [25], where the authors, implementing a mechanism based on the

IOP Conf. Series: Journal of Physics: Conf. Series **981** (2018) 012016 doi:10.1088/1742-6596/981/1/012016

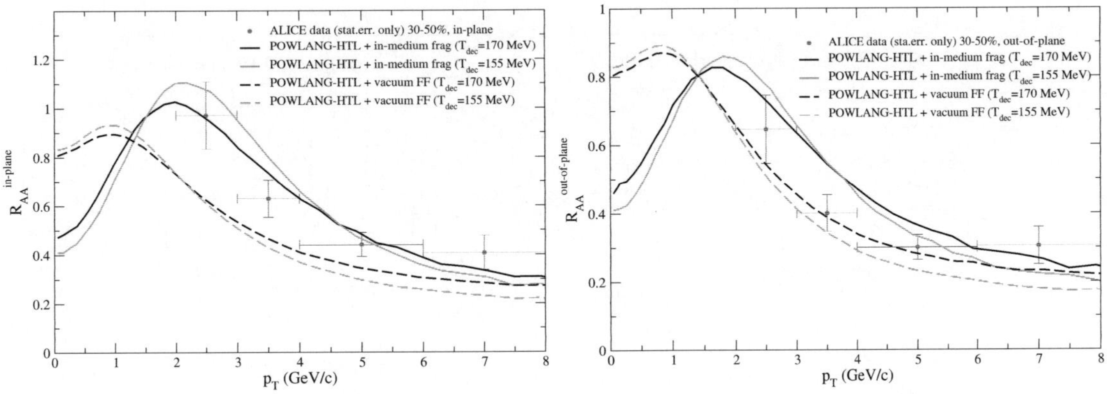

Figure 7. The in-plane (left panel) and out-of-plane (right panel) nuclear modification factor of D^0 mesons in 30-50% Pb-Pb collisions at $\sqrt{s_{\rm NN}}=2.76$ TeV. Transport results with different hadronization schemes and decoupling temperatures are compared to ALICE data [24].

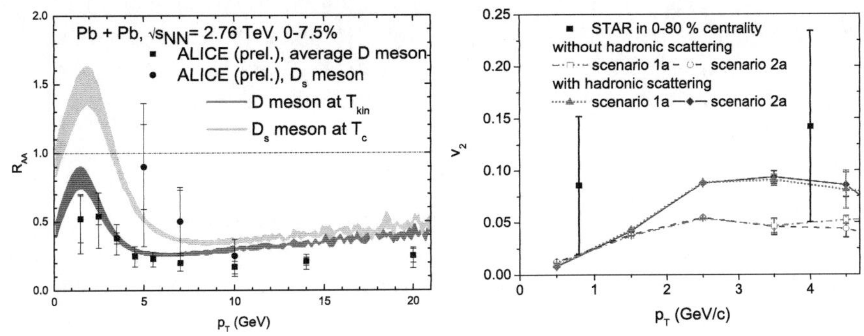

Figure 8. Left panel: The nuclear modification factor of D and D_s mesons in central Pb-Pb collisions at $\sqrt{s_{\rm NN}}=2.76$ TeV; transport results with in-medium hadronization [25] are compared to ALICE data [26]. Right panel: the effect of rescattering in the hadronic phase on the D-meson elliptic flow in Au-Au collisions at $\sqrt{s_{\rm NN}}=200$ GeV; results of model calculations [27] are compared to STAR data.

formation/decay of heavy-light resonances, studied the production of D and D_s mesons in nucleus-nucleus collisions. Results for their nuclear modification factor are displayed in the left panel of Fig. 8, compared to preliminary ALICE data [26]: a relative enhancement of D_s mesons, due to recombination with strange quarks at hadronization, is visible. Finally, also the interactions in the hadronic phase may affect the final particle distributions. In particular, this can be of relevance for the D-meson v_2. In fact, although transport coefficients are much smaller, in the hadronic phase the medium is characterized by the largest elliptic flow: results obtained in [27] modeling $D-\pi$ interactions with and effective chiral Lagrangian are displayed in the right panel of Fig. 8.

4. Heavy Flavour in small systems: room for medium effects?

One of the most surprising findings in the experimental search for the Quark-Gluon Plasma is certainly the signature of possible collective effects, suggestive of the formation of a hot strongly-interacting medium, recently observed in collisions involving small projectiles, like p-Pb at the LHC and d-Au (and now also ^{3}He-Au) at RHIC, in particular when selecting events

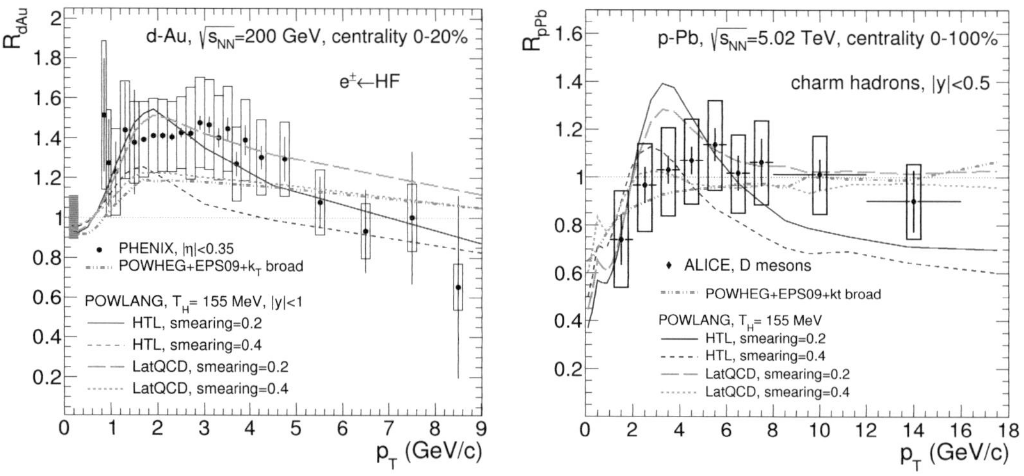

Figure 9. Left panel: the nuclear modification factor of HF decay electrons in 0-20% most central d-Au collisions at RHIC measured by PHENIX [28]. Right panel: the nuclear modification factor of D-mesons in 0-100% p-Pb collisions at the LHC measured by the ALICE collaboration [29]. The various curve refers to different choices of transport coefficients and initial conditions.

characterized by a high multiplicity of produced particles. Various observables support the above picture, in particular transverse-momentum spectra and azimuthal correlations, suggesting that soft hadrons take part in a common collective expansion, with a radial, elliptic and triangular flow arising from the response of a strongly-interacting medium to the initial pressure gradients and fluctuating geometry. On the contrary, high-p_T hadrons and jets look essentially unmodified with respect to the p-p benchmark. Due to the peculiar role of HF particles, produced in initial hard events but tending to (at least partially) thermalize with the medium in heavy-ion collisions, it is clearly of interest to study their behaviour also in small systems, looking for possible modifications in the final spectra introduced by a hot deconfined plasma. Due to the large systematic error bands, current experimental data are not able to provide a clear answer: the nuclear modification factor of HF electrons (from charm and beauty decays) in d-Au collisions at RHIC measured by PHENIX [28] seems to display a moderate enhancement in the intermediate-p_T region; on the other hand, within the present uncertainties, the D-meson $R_{\rm pA}$ measured by ALICE in p-Pb collisions at the LHC looks compatible with unity. In [30] the HF transport setup developed in [14] was applied to such small systems. At variance with the nucleus-nucleus case, the initial energy-density profile is not related to the impact parameter of the collision, but to event-by-event fluctuations. Hence, one starts the hydrodynamic evolution of the medium with Glauber Monte-Carlo initial conditions, so to capture at least the fluctuations arising from the random event-by-event nucleon positions. As discussed in detail in [30], the final results arise from the interplay of several initial/final-state effects: nuclear PDF's, with gluon shadowing tending to reduce charm production at low-p_T; transverse-momentum broadening in nuclear matter (before the initial $Q\overline{Q}$ production), tending to move particles to moderate p_T; interaction with the deconfined plasma produced in the collision, tending to suppress heavy-quark spectra at high-p_T and making them inherit part of the flow of the fireball; in-medium hadronization, which also transfer part of the (radial and elliptic) flow of the fireball to the final HF hadrons. Overall, the final outcomes are the ones shown in Fig. 9, in which the various curves reflect the different initial conditions and transport coefficients, giving an assessment of the theoretical uncertainty.

5. Conclusions and perspectives

The comparison between transport calculations and experimental data provides robust indications that heavy quarks strongly interact with the hot deconfined medium formed in high-energy nuclear collisions. Due to such interactions, besides soft hadrons, also heavy-flavour particles carry signatures of the collective flow of the background medium, although the expansion and limited lifetime of the latter may prevent them to reach full kinetic equilibrium with the plasma. A number of theoretical questions and experimental challenges remain to be addressed. Charm measurements down to $p_T \to 0$ will be important to get information on the flow and thermalization of charm and its total cross-section in heavy-ion collisions [31]. D_s and Λ_c measurements [32] will give a definite answer on possible changes of HF hadronization due to the recombination of the heavy quarks with light thermal partons, which seems currently supported by D-meson $R_{\rm AA}$ and v_2 data. Furthermore, D_s and Λ_c reconstruction will contribute to a quantitative estimate of the total charm cross-section, relevant for the study charmonium suppression in heavy-ion collisions on solid ground. In the near future the issue of HF production in small systems (e.g. high-multiplicity p-A collisions), already started in the last few years, will certainly deserve further investigation in order to point out and give a quantitative assessment of initial/final-state effects arising e.g. from the nuclear PDF's and from the possible formation of a hot deconfined medium. Besides the interest for the heavy-ion community, theoretical results for charm production in high-energy p-A collisions validated against experimental data will be of relevance for experiments like IceCube [33], to get a robust estimate of the charm-decay contribution to the background of high-energy atmospheric neutrinos [34].

Concerning forthcoming improvements of our transport approach [13, 14, 30], we are planning to interface our Langevin evolution to a full $(3+1)$-D hydrodynamic background, dropping the approximation of longitudinal boost-invariance: this will allow us to face also observables at forward rapidity (e.g. HF-decay muons measured by ALICE, J/Ψ from B-decays measured by CMS...) and to rely on a more realistic description of the medium formed in light-heavy collisions, like p-A and d-A, of interest also for the above mentioned cosmic-ray studies.

The most important challenge, which would represent a major experimental achievement, remains however the extraction of the HF transport coefficients in hot-QCD: are current experimental data sufficient to put tight constraints on them? In case so, how do they compare with first-principle theoretical calculations? Notice that, for what concerns heavy-quark momentum diffusion, the definition itself of the associated transport coefficient relies on the ansatz of a Langevin dynamics of the heavy particle. Although current experimental results, once compared to theory calculations, already gives some qualitative guidance, a solid answer to this issue requires waiting for beauty measurements at low p_T via exclusive B-meson hadronic decays. Beauty is in fact a better probe of the medium due to its large mass: the latter makes the initial hard production (described by pQCD) under better control, ensures the effective equivalence between the Langevin and the Boltzmann equation and, furthermore, makes the extraction of HF transport coefficients from lattice-QCD simulations conceived for static quarks more meaningful.

References

[1] Bazavov A *et al.* 2014 *Phys. Lett. B* **737** 210
[2] Mukherjee S *et al.* 2016 *Phys. Rev. D* **93** 014502
[3] Cacciari M *et al.* 1998 *J. High Energy Phys* JHEP05(1998)007; 2001 *J. High Energy Phys* JHEP03(2001)006
[4] Alioli S, Nason P, Oleari C and Re E, 2010 *J. High Energy Phys* JHEP06(2010)043
[5] Braaten E, Cheung K M, Fleming S and Yuan T C 1995 *Phys. Rev. D* **51** 4819
[6] ALICE Collaboration (Abelev B *et al.*) 2012 *J. High Energy Phys* JHEP01(2012)128
[7] Cacciari M *et al.* 2012 *J. High Energy Phys* JHEP10(2012)137

[8] Eskola K J, Paukkunen H and Salgado C A 2009 *J. High Energy Phys* JHEP04(2009)065
[9] Svetitsky B 1988 *Phys.Rev. D* **37** 2484
[10] Moore G D and Teaney D 2005 *Phys.Rev. C* **71** 064904
[11] Lau A W C and Lubensky T C 2007 *Phys. Rev. E* **76** 011123
[12] Beraudo A, De Pace A, Alberico W M and Molinari A 2009 *Nucl. Phys. A* **831** 59
[13] Alberico W M *et al.* 2011 *Eur. Phys. J. C* **71** 1666; 2013 *Eur. Phys. J. C* **73** 2481
[14] Beraudo A *et al.* 2015 *Eur. Phys. J. C* **75** 121
[15] Luzum M and Romatschke P 2008 *Phys. Rev. C* **78** 034915
[16] Del Zanna L *et al.* 2013 *Eur. Phys. J. C* **73** 2524
[17] Das S K *et al.* 2014 *Phys. Rev. C* **90** 044901
[18] Francis A *et al.* 2015 *Phys. Rev. D* **92** 116003
[19] Gossiaux P B, Bierkandt R and Aichelin J 2009 *Phys. Rev. C* **79** 044906
[20] van Hees H et al. 2008 *Phys. Rev. Lett.* **100** 192301
[21] He M, Fries R J and Rapp R 2012 *Phys. Rev. C* **86** 014903
[22] STAR Collaboration (Adamczyk L *et al.*) 2014 *Phys. Rev. Lett.* **113** 142301
[23] ALICE Collaboration (Abelev B *et al.*) 2013 *Phys. Rev. Lett.* **111** 102301
[24] ALICE Collaboration (Abelev B *et al.*) 2014 *Phys. Rev. C* **90** 034904
[25] He M, Fries R J and Rapp R 2014 *Phys. Lett. B* **735** 445
[26] Barbano A (for the ALICE Collaboration) 2016 *J. Phys. Conf. Ser.* **668** 012040
[27] Song T *et al.* 2015 *Phys. Rev. C* **92** 014910
[28] PHENIX Collaboration (Adare A *et al.*) 2012 *Phys. Rev. Lett.* **109** 242301
[29] ALICE Collaboration (Abelev B *et al.*) 2014 *Phys. Rev. Lett.* **113** 232301
[30] Beraudo A *et al.* 2016 *J. High Energy Phys* JHEP03(2016)123
[31] ALICE Collaboration (Abelev B *et al.*) 2014 *J. Phys. G* **41** 087002
[32] Zhou L (for the STAR Collaboration) 2017 arXiv:1704.04364
[33] IceCube Collaboration (Aartsen M G *et al.*) 2013 *Science* **342** 1242856
[34] Garzelli M V, Moch S and Sigl G 2015 *J. High Energy Phys* JHEP10(2015)115

Initial State fluctuations from midperipheral to ultracentral collisions in a transport approach

Salvatore Plumari[1,2], Francesco Scardina[1,2], Vincenzo Minissale[1,2], Lucia Oliva[1,2] and Vincenzo Greco[1,2]

[1] Department of Physics and Astronomy, University of Catania, Via S. Sofia 64, I-95125 Catania (Italy)

[2] Laboratorio Nazionale del Sud, INFN-LNS, Via S. Sofia 63, I-95125 Catania (Italy)

E-mail: salvatore.plumari@ct.infn.it

Abstract. We study the build up of the anisotropic flows v_n for a fluid at fixed $\eta/s(T)$ by means of an event-by-event transport approach. Usually in the partonic approach are used massless partons implying a $\epsilon = 3p$ Equation of State (EoS). In this paper we extend previous studies to finite partonic masses tuned to simulate a fluid with an EoS close to the recent lQCD results. In particular we study the role of the equation of state and the effect of the η/s ratio and its T dependence on the build up of the $v_n(p_T)$ for Au+Au collisions at $\sqrt{s} = 200\,GeV$ and for $Pb + Pb$ collisions at $\sqrt{s} = 2.76\,TeV$. We find that the sensitivity to the EoS of the $v_n(p_T)$ increase with the order of the harmonics n. In particular we find a mass ordering for the elliptic flow at low p_T. We find that for the two different beam energies considered the suppression of the $v_n(p_T)$ due to the viscosity of the medium have different contributions coming from the cross over or QGP phase depending on the collision energies. The study reveals that in ultra-central collisions $(0 - 0.2\%)$ the $v_n(p_T)$ have a stronger sensitivity to the T dependence of η/s in the QGP phase and this sensitivity increases with the order of the harmonic n.

1. Introduction

Ultra-relativistic heavy-ion collisions (uRHICs) experiments, conducted first at the Relativistic Heavy-Ion Collider (RHIC) and in the recent years at Large Hadron Collider (LHC), are the only way to access experimentally the properties of hot strongly interacting matter. In the last decades it has been reached a general consensus confirmed by several experimental signatures that the matter created in these collisions consists of a strongly interacting quark-gluon plasma (QGP). One of the most important discoverties was the large elliptic flow $v_2 = \langle (p_x^2 - p_y^2)/(p_x^2 + p_y^2) \rangle$ observed [1, 2]. The elliptic flow is a measurement of the anisotropy in momentum space of the emitted particles and it is an observable that encodes information about the Equation of State (EoS) and transport properties of the matter created in these collisions[3, 4]. Theoretical calculations first within the viscous hydrodynamics framework [5, 6] and in the recent years also within kinetic transport approach [7, 8] have shown that the large value of v_2 is consistent with a matter with a very low shear viscosity to entropy density ratio $4\pi\eta/s \sim 1 - 2$ close to the conjectured lower bound for a strongly interacting system [9].

Moreover, in the recent years experimentally it has been possible to measure the event-by-event angular distribution of emitted particles and it has been possible to extend this analysis to higher order harmonics $v_n = \langle cos(n\,\varphi_p) \rangle$ [10, 11]. The origin of these high order harmonics

IOP Conf. Series: Journal of Physics: Conf. Series **981** (2018) 012017 doi:10.1088/1742-6596/981/1/012017

flows can be attributed to the fluctuations in the initial geometry [12, 13, 14]. In the recent years most of the theoretical efforts has been focused on the development of the event-by-event viscous hydrodynamics simulations. The comparison with the experimental results for $v_n(p_T)$ seems to confirm a finite but not too large average value of η/s with $4\pi\langle\eta/s\rangle \sim 1-3$ [14]. However, a small value of $\langle\eta/s\rangle$ is not an evidence of the creation of a QGP phase. Moreover it is well known that the η/s of the QGP is expected to have a temperature dependence with a minimum close to the cross over region. The collection of several different theoretical calculations show the evidence of this picture [15, 16, 17, 18, 19], see Fig.1. A phenomenological estimation of $\eta/s(T)$ could give further information if the matter created in these collisions undergoes a phase transition [20, 21, 22]. However due to the large error bars in the lQCD results for η/s at moment it is not possible to infer a clear temperature dependence in the QGP phase. Recently, the CMS, ATLAS, and ALICE collaborations have been able to measure the anisotropic flow coefficients $\langle v_n\rangle$ in ultra-central heavy ion collisions [23, 11, 24]. In particular the experimental results have shown that the triangular flow $n = 3$ appears to be comparable or even larger of the elliptic flow $n = 2$. Such a result poses a problem to hydrodynamical approach that predict a significantly larger $\langle v_2\rangle$ w.r.t. $\langle v_3\rangle$. Very recently it has been pointed-out that at LHC energies and for ultra-central collisions a quite large correlation between $\langle\epsilon_n\rangle$ and $\langle v_n\rangle$ is present up to $n = 5$ at variance with RHIC energies and/or other centralities [25]. At the same time a much larger sensitivity to $\eta/s(T)$ has been spot. Therefore the study of the anisotropic flows in ultra-central collision offer a powerfull tool infer information about the initial geometry of the fireball.

In this paper we extend the analysis to high order harmonics studying the role of the Equation of State and $\eta/s(T)$ on the build up of $v_n(p_T)$ using a kinetic transport approach with initial state fluctuations. In the transport calculations shown in this paper we use massive particles which supply the possibility of having a soften equation of state with a decreasing speed of sound when the crossover region is approached similar to the one of lQCD calculations [26, 27]. We will show results on $v_n(p_T)$ for $n = 2, .., 5$ for two different systems: $Au + Au$ collisions at $\sqrt{s} = 200\,GeV$ and $Pb + Pb$ collisions at $\sqrt{s} = 2.76\,TeV$ at different centralities.

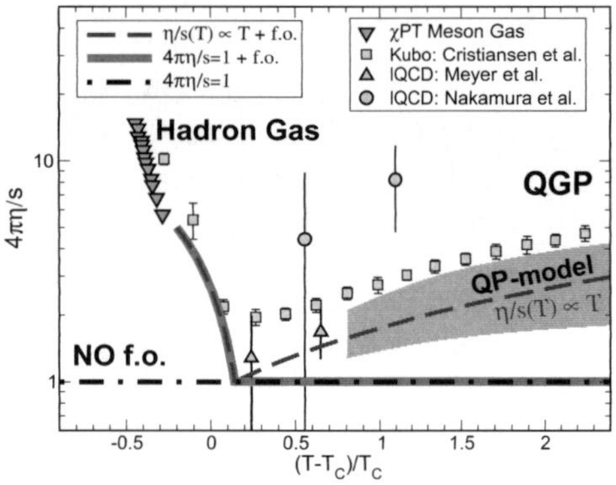

Figure 1. Different parametrizations for the temperature dependence of η/s. The orange area refers to the quasi-particle model predictions for η/s [28] while up-triangles, circles and squares are lQCD calculation on pure gauge [17, 18, 19]. The orange down-triangles are calculation in chiral perturbation theory for a meson gas [15, 16]. Finally the three different lines indicate different possible T dependencies used in this paper.

2. Transport approach at fixed η/s

In this work the fireball evolution is studied employing a transport code developed to perform studies of the dynamics of heavy-ion collisions for different energies [7, 29, 30, 31, 32, 33]. Recently this transport code has been extended to include the initial state fluctuations in order to study the harmonics $v_n(p_T)$ with $n \geq 2$. For a more detailed discussion about the implementation of the initial state fluctuations in transport approach see [25]. In this paper we extend the solution of the Relativistic Boltzmann Transport (RBT) equation to finite mass partons which allows to simulate a fluid with an EoS similar to the recent lQCD calculations

[26, 27]. This implies that the fluid has an equation of state that exhibit a large trace anomaly ($T^\mu_\mu = \epsilon - 3p \neq 0$) and consequently a sound velocity $c_S^2(T)$ that is significantly smaller than 1/3 and reaching about 1/10 at $T \approx T_C = 155\, MeV$.

The evolution of the phase-space distribution function $f(x,p,t)$ is given by solving the RBT equation:

$$p^\mu\, \partial_\mu f(x,p) = \int_{2,1',2'} (f_{1'} f_{2'} - f f_2) |\mathcal{M}|^2 \delta^4(p + p_2 - p_{1'} - p_{2'}) \tag{1}$$

with $\int_{2,1',2'} = \int \Pi_{k=2,1',2'} d^3p_k / 2E_k (2\pi)^3$ while $\mathcal{M}$ denotes the transition amplitude for the elastic processes. $\mathcal{M}$ is directly related to the differential cross section $|\mathcal{M}|^2 = 16\pi\, s\, (s - 4M^2) d\sigma/dt$ with s the Mandelstam invariant. For the results shown in this paper we have considered an isotropic cross section. To study the expansion dynamics of the fireball with a certain $\eta/s(T)$, we determine locally in space and time the total cross section σ_{tot} according to the Chapmann-Enskog theory [34]. The motivation of this approach is inspired by the success of the hydrodynamical approach that has shown the key role played by the η/s to describe the experimental data [35, 14]. Therefore on one hand we use the RBT equation as an approach converging to hydrodynamics for small scattering relaxation time $\tau \sim \sigma\rho$ (or small η/s) [36, 37]. Furthermore it permits to study directly the impact of a T dependence of η/s on observables like the anisotropic flows $v_2(p_T)$ and $v_3(p_T)$ which is the main focus of this paper. On the other hand, this approach it is not based on an ansatz for the viscous corrections for the phase-space distribution function δf and it is naturally valid also at large η/s or for $p_T >> T$ and it is possible to include initial non equilibrium effects to the distribution function $f(x,p)$ (see [31, 32]). As shown in [34] for a massive gas in the Chapmann-Enskog approximation at the first order η is given by

$$\eta = f(z) \frac{T}{\sigma_{tot}} \tag{2}$$

with $z = m/T$ while the function $f(z)$ is given by

$$f(z) = \frac{15}{16} \frac{\left[z^2 K_3(z)\right]^2}{(15z^2 + 2)K_2(2z) + (3z^3 + 49z)K_3(2z)} \tag{3}$$

where K_n-s are the modified Bessel functions. While the entropy density for a massive system is given by $s = \rho(4 + zK_1(z)/K_2(z))$. Notice that in the ultra-relativistic limit $z \to 0$ the $f(z) \to 1.2$ and we recover the massless limit for the η. As shown in [30] the expression for η in Eq.(2) is in quite good agreement at level of $3-5\%$ with the Green-Kubo formula and it describes correctly the η/s of the system in the range of interest of HIC.

In Fig.1 we show the three different T dependencies of η/s studied: one of constant $4\pi\eta/s = 1$ during all the evolution of the system shown by dot dashed line another one of $4\pi\eta/s = 1$ at higher temperature in the QGP phase and an increasing η/s in the cross over region towards the estimated value for hadronic matter $4\pi\eta/s \approx 6$ [16, 38] and shown by solid line. Finally the third one shown by the dashed line in Fig.1 where we consider the increase of η/s at higher temperature with a linear temperature dependence and a minimum close to the critical temperature with a temperature dependence similar to that expected from general considerations as suggested by quasi particle models [28] or by recent calculations on pure gauge [19]. Notice that in our approach an increase of η/s at lower temperature $0.8T_C \leq T \leq 1.2T_C$ permits to account for a smooth realistic kinetic freeze-out (f.o.) because at lower T the mean free path λ goes like $\lambda \propto \frac{\eta}{s}\frac{1}{T}$. In the following discussion the term f.o. means to take into account the increase of η/s at $T < 1.2T_C$.

3. Role of the Equation of State on the v_n

In the following we will consider two systems at different centralities: $Au + Au$ collisions at $\sqrt{s_{NN}} = 200\,GeV$ produced at RHIC and $Pb + Pb$ collisions at $\sqrt{s_{NN}} = 2.76\,TeV$ at LHC. To generate an event by event initial profile we use the Monte-Carlo Glauber model. We employ the geometrical method to determine if the initial nucleons distribution. Within this method two nucleons collide each other if the relative distance in the transverse plane is $d_T \leq \sqrt{\sigma_{NN}/\pi}$ where σ_{NN} is the nucleon-nucleon cross section. In our calculation $\sigma_{NN} = 4.2\,fm^2$ is taken for RHIC at $\sqrt{S_{NN}} = 200\,GeV$ and $\sigma_{NN} = 7.0\,fm^2$ for LHC at $\sqrt{s_{NN}} = 2.76\,TeV$. The resulting initial nucleon distribution is converted in a smooth one by assuming a Gaussion distribution centered in the nucleon position as done in several hydrodynamics calculations. In the results shown in this paper we have fixed the Gaussian width of the fluctuations to $\sigma_{xy} = 0.5\,fm$. In our calculation we assume a longitudinal boost invariant distribution from $y = -2.5$ to $y = 2.5$. For the initialization in momentum space at RHIC (LHC) energies we have considered for partons with transverse momentum $p_T \leq p_0 = 2\,GeV$ $(3\,GeV)$ a thermalized spectrum in the transverse plane. The initial local temperature in the transverse plane $T(x, y)$ is related to the local densisty in the transverse plane $\rho_T(x, y)$ by using the standard thermodynamical relation for massive particles $\rho_T(x, y) = \gamma T^3 z^2 K_2(z)/(2\pi^2)$ with $z = m/T$ and $\gamma = 40$ for $N_c = 3$ colors and $N_f = 2$ flavours. While for partons with $p_T > p_0$ we have assumed the spectrum of non-quenched minijets according to standard NLO-pQCD calculations with a power law shape [39]. The initial transverse momentum of the particles is distributed uniformly in the azimuthal angle. We fix the starting time of the simulation to $\tau_0 = 0.6\,fm/c$ for RHIC and $\tau_0 = 0.3\,fm/c$ for LHC as commonly done also in hydrodynamical approaches.

As well known the anisotropic flow coefficients are observable sensitive to the bulk properties of the created matter. In particular the $v_n(p_T)$ encode information about the Equation of State (EoS) and transport properties of the medium like the η/s ratio. To extract information about the EoS we have evaluated the average energy-momentum tensor $\langle T^{\mu\nu}\rangle$.

$$\langle T^{\mu\nu}\rangle = \frac{1}{V N_{test}} \sum_{i=1}^{N_{tot}} \frac{p_i^\mu p_i^\nu}{p_i^0} \tag{4}$$

$\langle T^{\mu\nu}\rangle$ is evaluated in a cylinder of radius $r_T = 1.5\,fm$ and in longitudinal direction $|\eta| < 0.3$. We evaluate the energy density by $\epsilon = \langle T^{00}\rangle$ and the pressure by assuming that $p = \frac{1}{3}\sum_i \langle T^{ii}\rangle$.

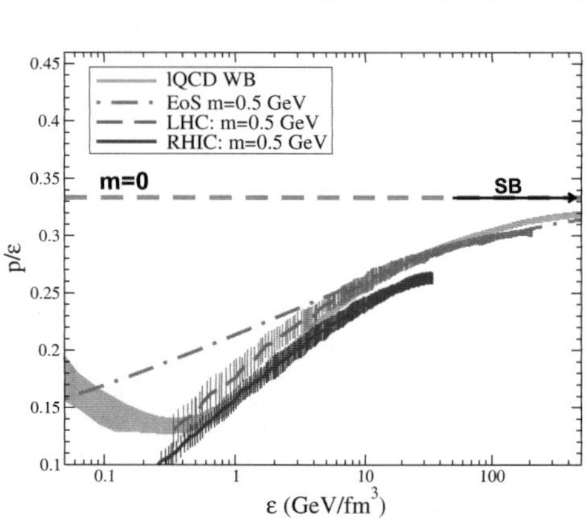

Figure 2. p/ϵ ratio as a function of the energy density. The orange band are the lQCD data taken from the Wuppertal-Budapest collaboration [40]. The solid green and dashed blue lines are respectively for LHC and RHIC energies for central collisions and mid rapidity.

In Fig.2 we compare the p/ϵ ratio obtained during the dynamical evolution in our simulations with the one of lQCD for both $Au + Au$ collisions at $\sqrt{s_{NN}} = 200\,GeV$ and $Pb + Pb$ collisions at $\sqrt{s_{NN}} = 2.76\,TeV$. The results shown are for a fireball in central collisions. As shown we get a reasonably agreement with lQCD data for both energies. In these calculation we have fixed

$m = 0.5\,GeV$ and these values of masses are in agreement with the masses extracted within quasi-particle models [28]. Due to the initial fluctuations in density and therefore in energy density the corresponding value of p/ϵ can fluctuate event-by-event the blue and green bands show the possible range of trajectories covered in our simulations. This is quite interesting that such an effect is small and the bands follows closely the lQCD calculations. The band is due to the fact that locally the relation between pressure and energy density can be modified by non equilibrium effects.

The elliptic flow $v_2(p_T)$ and the high order harmonics $v_3(p_t)$, $v_4(p_T)$ and $v_5(p_T)$ have been calculated using the event plane method as

$$v_n = \langle \cos\left[n(\phi - \Psi_n)\right] \rangle \tag{5}$$

where the momentum space angles Ψ_n are given by

$$\Psi_n = \frac{1}{n} \arctan \frac{\langle \sin(n\phi) \rangle}{\langle \cos(n\phi) \rangle} \tag{6}$$

As a first step in our analysis we study the time evolution of the final anisotropic flows $\langle v_n \rangle$. Where the average $\langle ... \rangle$ is over all the events. In our simulation we have performed 1000 events for each centrality class.

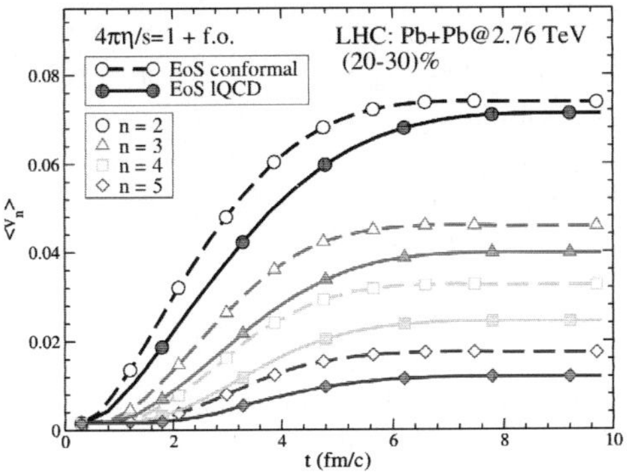

Figure 3. Time evolution of $\langle v_n \rangle$ at mid rapidity respectively for $(20-30)\%$ centrality collisions. Results for $Pb+Pb$ collisions at $\sqrt{s_{NN}} = 2.76\,TeV$ and for the case $4\pi\eta/s = 1$ and the kinetic freeze out (f.o.). Different symbols are for different harmonics, solid lines are for the massive case (lQCD EoS) while dashed lines are for massless case (conformal limit).

In Fig.3 we plot the time evolution of the final anisotropic flow coefficients $\langle v_n \rangle$ for both cases massless (dashed lines) and massive case (solid lines). These calculation are for LHC energies at mid rapidity and for $(20-30)\%$ centrality collisions. Results show that in general for both massless and massive case for all harmonics the $v_n(p_T)$ are built-up within the first few fm/c and for both cases different harmonics have a different formation time with the following ordering $t_{v_2} < t_{v_3} < t_{v_4}$.... This was noted in Ref.[25], but here we consider the impact of the EoS. Moreover we observe that for massless case the system is more efficient to convert the initial anisotropy in coordinate space into final anisotropy in momentum space where the effect of a realistic EoS is to reduce the $\langle v_n \rangle$. This effect as shown in Fig.3 increase with the increasing of the order of the harmonics. The increasing sensitivity to the EoS for larger n is related to the different formation time of the harmonics where for example the second coefficient $\langle v_2 \rangle$ is almost completely developed within $3 - 4 fm/c$ where the $p/\epsilon \approx 0.3$ and close to the conformal limit. The effect increases for large n because higher harmonics develop later at lower temperature where $p/\epsilon < 0.3$ and the system is less efficient in converting the initial anisotropy in coordinate in the final $\langle v_n \rangle$.

In Fig.4 we compare the final $v_n(p_T)$ for both massless (dashed lines) and massive case (solid lines). These results are for $Pb+Pb$ collisions at $\sqrt{s_{NN}} = 2.76\,TeV$ and for $(20-30)\%$ centrality collisions. We observe for the elliptic flow v_2 a mass ordering where larger is the mass

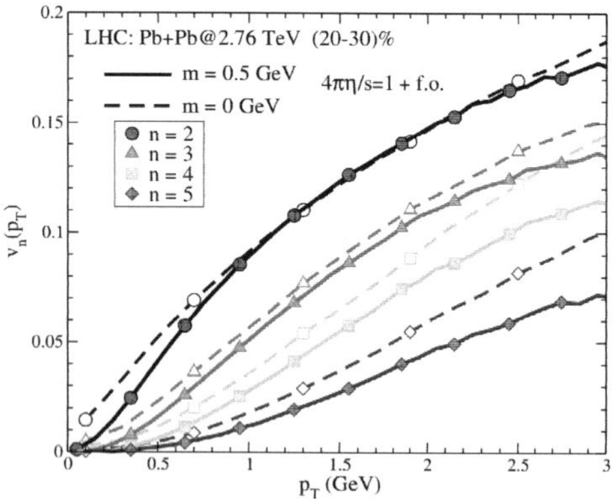

Figure 4. Differential $v_n(p_T)$ for $Pb+Pb$ collisions at $\sqrt{s_{NN}} = 2.76\,TeV$ for $(20-30)\%$ centrality collisions and for the case $4\pi\eta/s = 1$ and the kinetic freeze out (f.o.). Different symbols are for different harmonics n. Solid lines are for the massive case while the dashed lines are for the massless case.

and smaller is the corresponding elliptic flow and this is typical of hydrodynamic expansion that boost strongly massive particles [41, 3]. This ordering can be explained by the fact that at low p_T the elliptic flow is $v_2(p_T) \propto p_T - \langle\beta_T\rangle m_T$ therefore for light particles $m_T = \sqrt{p_T^2 + m^2} \approx p_T$ and $v_2(p_T) \propto p_T$ while for massive particle the coupling of m_T with the radial flow gives at the same p_T a smaller $v_2(p_T)$ and it is translated in a non linear behavior at low p_T. A similar mass ordering we observe also for higher harmonics $v_n(p_T)$. In particular at low p_T the effect of the mass is to give a non linear behavior for the second and third coefficients. Moreover we observe that the sensitivity to the EoS of the $v_n(p_T)$ increases with the increasing of the harmonic n where for example a reduction of about 15% for $v_4(p_T)$ and of about 25% for $v_5(p_T)$ is observed. The enhancement of the sensitivity with the increase of the order of the harmonic is again related to the different formation time of the harmonics. In fact for large n the corresponding $v_n(p_T)$ develops later where the speed of sound c_s^2 or the p/ϵ is smaller then the one of conformal limit, see Fig.2.

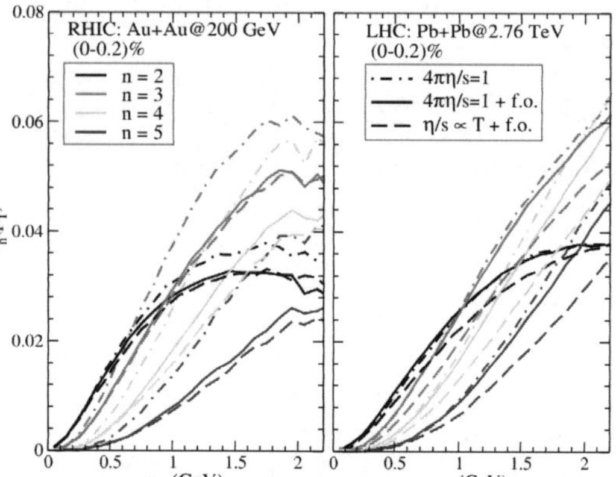

Figure 5. Differential $v_n(p_T)$ at mid rapidity and for ultra-central collisions. The comparison is between the two systems: $Au+Au$ at $\sqrt{s} = 200\,GeV$ (left panel) and $Pb+Pb$ at $\sqrt{s} = 2.76\,TeV$ (right panel). Different colors are for different harmonics n. The dot dashed lines refer to the case with a constant $\eta/s = (4\pi)^{-1}$ during all the evolution. The solid lines refer to the case with $\eta/s = (4\pi)^{-1}$ at higher temperature and kinetic f.o. while the dashed lines refer to the case with $\eta/s \propto T$ at higher temperature and the kinetic f.o. at lower temperature.

Recently it has been possible to access experimentally the ultra-central collisions. In ultra-central collisions the initial asymmetry in coordinate space measured by ϵ_n comes completely from the fluctuations in the initial geometry and there is no coupling with the global overlap shape that has a circular symmetry. Moreover as shown by the event-by-event ideal and viscous hydrodynamics calculations [42, 43] and recently also in event-by-event transport approach [25] for these collision centralities the final integrated anisotropic flows coefficient $\langle v_n\rangle$ are strongly correlated to the initial eccentricities $\langle\epsilon_n\rangle$ with a linear correlation coefficient $C(\epsilon_n, v_n) \approx 1$. In

Fig.5 it is shown the comparison of $v_n(p_T)$ produced in $Au+Au$ collisions at $\sqrt{s} = 200\,GeV$ (left panel) with the one produced in $Pb+Pb$ collisions at $\sqrt{s} = 2.76\,TeV$ (right panel) for ultra-central collisions. Comparing both panel of Fig.5 we observe that at low p_T for both energies the $v_n(p_T) \propto p_T^n$. On the other hand at hight p_T we observe that the elliptic flow $v_2(p_T)$ shows a saturation while $v_n(p_T)$ for $n \geq 3$ have an almost linear p_T dependence. This is in qualitative agreement with what has been observed experimentally.

Similarly to what has been obtained in viscous hydrodynamical calculations the increase of the viscosity of the medium has the effect to reduce the produced $v_n(p_T)$. These results are qualitatively in agreement with previous results within kinetic transport approach and in the conformal limit [25]. Comparing the dot dashed lines with the solid lines in the left panel of Fig.5, we observe that at RHIC energies the $v_n(p_T)$ are sensitive to the increase of the η/s at lower temperature close to the cross over region. This sensitivity increase with the order of the harmonics. In particular for $n \geq 3$ the reduction for central collisions is about $30-35\%$. On contrary at LHC energies right panel of Fig.5 the $v_n(p_T)$ are more sensitive to the increase of η/s at high temperature as shown by a larger suppression of the $v_n(p_T)$ for the case where $\eta/s \propto T$ in the QGP phase (dashed lines). Notice that the 30% effect is determined by a slowly linear rising of η/s with T and with a minimum close T_C of $\eta/s \approx 1/(4\pi)$ as the dashed lines in Fig.1.

4. Conclusions

We have investigated the build up of the anisotropic flows $v_n(p_T)$ for $n = 2, 3, 4$ and 5 within an event-by-event transport approach. In particular we have studied the role of the EoS (by a finite mass) and the the role of a temperature dependent η/s, on the time evolution of $\langle v_n \rangle$ and on the differential $v_n(p_T)$. Our calculations show that the effect of the mass is to reduce the $v_n(p_T)$ with a sensitivity that increase with the order of the harmonics. Finally, we have studied the effect of η/s ratio on $v_n(p_T)$ for two different beam energies: at RHIC for $Au+Au$ collisions at $\sqrt{s} = 200\,GeV$ and at LHC for $Pb+Pb$ collisions at $\sqrt{s} = 2.76\,TeV$. We have found that at RHIC the $v_n(p_T)$ are more affected by the value of η/s in the cross over region ($T < 1.2T_C$) with an enhancement of the sensitivity for larger harmonics. On contrary, at LHC energies we get that almost all the $v_n(p_T)$ are more affected by the η/s of the QGP. However the sensitivity to the T dependence of the η/s is quite weak. These results are in qualitative agreement with the results obtained for the massless case. In general we found an enhancement of the sensitivity of the $v_n(p_T)$ that for $n = 2, ..., 5$ reaches about a $30-35\%$.

5. Acknowledgments

V. Greco, S. Plumari, F. Scardina acknowledge the support of the ERC-StG Grant under the QGPDyn project.

References

[1] Adams J *et al.* (STAR Collaboration) 2005 *Nucl.Phys.* **A757** 102–183
[2] Aamodt K *et al.* (ALICE Collaboration) 2010 *Phys.Rev.Lett.* **105** 252302
[3] Ollitrault J Y 2008 *Eur. J. Phys.* **29** 275–302
[4] Gale C, Jeon S and Schenke B 2013 *Int. J. Mod. Phys.* **A28** 1340011
[5] Romatschke P and Romatschke U 2007 *Phys.Rev.Lett.* **99** 172301
[6] Song H and Heinz U W 2008 *Phys.Rev.* **C78** 024902
[7] Ferini G, Colonna M, Di Toro M and Greco V 2009 *Phys.Lett.* **B670** 325–329
[8] Xu Z and Greiner C 2009 *Phys.Rev.* **C79** 014904
[9] Kovtun P, Son D and Starinets A 2005 *Phys.Rev.Lett.* **94** 111601
[10] Adare A *et al.* (PHENIX Collaboration) 2011 *Phys.Rev.Lett.* **107** 252301
[11] Aad G *et al.* (ATLAS) 2012 *Phys. Rev.* **C86** 014907
[12] Petersen H, Qin G Y, Bass S A and Muller B 2010 *Phys.Rev.* **C82** 041901
[13] Holopainen H, Niemi H and Eskola K J 2011 *Phys.Rev.* **C83** 034901
[14] Schenke B, Jeon S and Gale C 2012 *Phys.Rev.* **C85** 024901

[15] Prakash M, Prakash M, Venugopalan R and Welke G 1993 *Phys.Rept.* **227** 321–366
[16] Chen J W, Li Y H, Liu Y F and Nakano E 2007 *Phys.Rev.* **D76** 114011
[17] Meyer H B 2007 *Phys.Rev.* **D76** 101701
[18] Nakamura A and Sakai S 2005 *Phys.Rev.Lett.* **94** 072305
[19] Christiansen N, Haas M, Pawlowski J M and Strodthoff N 2015 *Phys. Rev. Lett.* **115** 112002
[20] Csernai L P, Kapusta J and McLerran L D 2006 *Phys.Rev.Lett.* **97** 152303
[21] Lacey R A, Ajitanand N, Alexander J, Chung P, Holzmann W *et al.* 2007 *Phys.Rev.Lett.* **98** 092301
[22] Plumari S, Greco V and Csernai L 2014 *Nuovo Cim.* **C037** 68–72
[23] Abelev B *et al.* (ALICE) 2013 *Phys. Lett.* **B719** 18–28
[24] Chatrchyan S *et al.* (CMS) 2014 *JHEP* **02** 088
[25] Plumari S, Guardo G L, Scardina F and Greco V 2015 *Phys. Rev.* **C92** 054902
[26] Borsanyi S, Fodor Z, Hoelbling C, Katz S D, Krieg S and Szabo K K 2014 *Phys. Lett.* **B730** 99–104
[27] Bazavov A *et al.* (HotQCD) 2014 *Phys. Rev.* **D90** 094503
[28] Plumari S, Alberico W M, Greco V and Ratti C 2011 *Phys.Rev.* **D84** 094004
[29] Plumari S and Greco V 2012 *AIP Conf.Proc.* **1422** 56–61
[30] Plumari S, Puglisi A, Scardina F and Greco V 2012 *Phys.Rev.* **C86** 054902
[31] Ruggieri M, Scardina F, Plumari S and Greco V 2013 *Phys.Lett.* **B727** 177–181
[32] Ruggieri M, Scardina F, Plumari S and Greco V 2014 *Phys.Rev.* **C89** 054914
[33] Puglisi A, Plumari S and Greco V 2014 *Phys. Rev.* **D90** 114009
[34] Wiranata A and Prakash M 2012 *Phys.Rev.* **C85** 054908
[35] Gardim F G, Grassi F, Luzum M and Ollitrault J Y 2012 *Phys. Rev. Lett.* **109** 202302
[36] Huovinen P and Molnar D 2009 *Phys.Rev.* **C79** 014906
[37] Plumari S, Guardo G L, Greco V and Ollitrault J Y 2015 *Nucl. Phys.* **A941** 87–96
[38] Demir N and Bass S A 2009 *Phys.Rev.Lett.* **102** 172302
[39] Greco V, Ko C and Levai P 2003 *Phys.Rev.Lett.* **90** 202302
[40] Borsanyi S, Endrodi G, Fodor Z, Jakovac A, Katz S D, Krieg S, Ratti C and Szabo K K 2010 *JHEP* **11** 077
[41] Huovinen P, Kolb P F, Heinz U W, Ruuskanen P V and Voloshin S A 2001 *Phys. Lett.* **B503** 58–64
[42] Gardim F G, Grassi F, Luzum M and Ollitrault J Y 2012 *Phys.Rev.* **C85** 024908
[43] Niemi H, Denicol G, Holopainen H and Huovinen P 2013 *Phys.Rev.* **C87** 054901

Ab initio calculations for non-strange and strange few-baryon systems

Winfried Leidemann
Dipartimento di Fisica, Università di Trento, I-38123 Trento, Italy
and INFN, TIFPA, I-38123 Trento, Italy
E-mail: winfried.leidemann@unitn.it

Abstract. Concerning the non-strange particle systems the low-energy excitation spectra of the three- and four-body helium isotopes are studied. Objects of the study are the astrophysical S-factor S_{12} of the radiative proton deuteron capture $d(p,\gamma)^3$He and the width of the ^{4}He isoscalar monopole resonance. Both observables are calculated using the Lorentz integral transform (LIT) method. The LIT equations are solved via expansions of the LIT states on a specifically modified hyperspherical harmonics (HH) basis. It is illustrated that at low energies such a modification allows to work with much higher LIT resolutions than with an unmodified HH basis. It is discussed that this opens up the possibility to determine astrophysical S-factors as well as the width of low-lying resonances with the LIT method. In the sector of strange baryon systems binding energies of the hypernucleus $^3_\Lambda$H are calculated using a nonsymmetrized HH basis. The results are compared with those calculated by various other groups with different methods. For all the considered non-strange and strange baryon systems it is shown that high-precision results are obtained.

1. Introduction

Non-strange and strange few-baryon systems are particularly interesting particle systems in the hadronic sector. On the one hand they serve for parametrization and test of potential models for nucleon-nucleon (NN), nucleon-hyperon (NY), hyperon-hyperon (YY) interactions and the various analogous three-body interactions. On the other hand they play an important role in testing the quality of ab initio methods, for example in benchmark calculations. In the present work the second of these two aspects is of relevance. In fact our aim is to test the quality of two different ab initio techniques for specific physical questions as is explained in greater detail in the following.

One of the tested ab initio methods is the LIT. The LIT approach is well-established [1] and allows to determine observables involving the many-body continuum without the necessity to calculate continuum wave functions. In the present work it is investigated to what extent specific features in the low-energy electromagnetic response of nuclei can be determined with the LIT method. We consider two examples: (i) the width of the ^{4}He isoscalar monopole resonance 0^+ and (ii) the threshold cross section in ^{3}He photodisintegration. The inverse reaction of the latter, the radiative proton-deuteron capture, is of relevance for the nucleosynthesis and usually parametrized via the astrophysical S-factor S_{12}. As explained in section 2 the crucial point for an exact description of the observables mentioned above lies in the question whether a sufficiently high density of LIT states can be obtained in the low-energy region. In a rather recent LIT calculation based on HH expansions [2], where the ^{4}He inelastic isoscalar monopole response function was computed with realistic nuclear forces, this aim could not be achieved even though the HH basis was quite large. Therefore it was not possible to determine the width of the 0^+ resonance in this calculation. In [3] it was then shown that the problem is due to the employed HH basis and that a somewhat modified many-body basis solves the problem of a too low low-energy density of LIT states. The modification consists in using for an A-body system, instead of an A-body HH basis, an $(A-1)$-body HH basis times a basis set for the relative motion of the A-*th* particle with respect to the center of mass of the $(A-1)$-body system.

For the strange particle systems responses to external probes have not yet been determined in experiment. In fact the knowledge of such systems is still rather scarce. One source of experimental information are binding energies of hypernuclei. One of the future aims of our group are ab initio calculations of such binding energies with realistic forces. However, first we want to test the reliability and the precision of the ab initio approach chosen by us. Different from the basis systems, which is used for the above mentioned LIT calculations, where the various HH basis states have a well-defined permutational symmetry, we take for the bound-state calculations of strange few-body baryon systems a nonsymmetrized HH (NSHH) basis. The results are then compared with results coming from other ab initio approaches. We switch from a symmetrized to a nonsymmetrized HH basis because we are confident that calculations for A-body baryon systems with $A \geq 6$ can be carried out with less computational effort.

Since in the present work it is the aim to test the precision of various theoretical ab initio approaches we do not employ realistic interaction models, but use instead simpler potential models, which will be defined in the following sections.

This work is organized as follows. In section 2 the LIT method and the used many-body basis systems are briefly described. Furthermore, the LIT results for the above mentioned S-factor S_{12} as well as for the 0^+ resonance of ^{4}He are discussed. In both cases results with the HH and the new basis are compared. In section 3 it is described how nonsymmetrized basis systems can nonetheless be used to determine ground states of systems which obey a specific permutational symmetry. Subsequently the results for binding and Λ separation energies of $^3_\Lambda$H are illustrated in comparison to results from other authors with different ab initio few-body techniques. Finally, in section 4 a summary is given.

2. The LIT method

Nuclear cross sections of inclusive reactions with electromagnetic probes are expressed in terms of inclusive response functions, which contain the information about the dynamics of the nucleus under investigation. Inclusive response functions are in general of the following form

$$R(\omega) = \int df |\langle f|\hat{O}|0\rangle|^2 \delta(E_f - E_0 - \omega)\,, \tag{1}$$

where $|0\rangle$ and $|f\rangle$ are nuclear ground and final states, E_0 and E_f are the corresponding eigenenergies and ω is the energy of the exchanged real (photoabsorption) or virtual photon (electron scattering). Finally, $\hat{O}$ denotes the operator inducing the reaction.

A calculation of $R(\omega)$ can be become very difficult or even impossible for cases where $|f\rangle$ is a many-body continuum state. However an explicit calculation of $|f\rangle$ can be avoided by the use of the LIT, which is an integral transform defined as follows

$$L(\sigma) = \int d\omega \, \frac{R(\omega)}{(\omega - \sigma_R)^2 + \sigma_I^2} \tag{2}$$

with $\sigma = \sigma_R + i\sigma_I$. Due to the variable width of $2\sigma_I$ of the Lorentzian kernel the LIT is an integral transform with a controlled resolution. But it is important to realize that in a given calculation one cannot simply increase the resolution by choosing smaller and smaller σ_I values. In fact one has to make sure that the precision of the LIT calculation allows the choice of a smaller σ_I value. How this can be achieved becomes clearer in the discussion that follows next.

Since the aim is to determine the response function without the knowledge of the continuum wave function it is useless to calculate the LIT via its definition of eq. (1). Fortunately, the LIT can be determined in an alternative way, namely by solving an equation, the LIT equation, given by

$$(\hat{H} - E_0 - \sigma)\,|\tilde{\Psi}(\sigma)\rangle = \hat{O}|0\rangle\,, \tag{3}$$

where $\hat{H}$ is the nuclear Hamiltonian. The important feature of the solution $\tilde{\Psi}(\sigma)$ is that it is a localized function. Therefore one can compute $\tilde{\Psi}(\sigma)$ using bound-state methods.

After having determined $\tilde{\Psi}(\sigma)$ one calculates the LIT from the following expression

$$L(\sigma) = \langle\tilde{\Psi}(\sigma)|\tilde{\Psi}(\sigma)\rangle\,. \tag{4}$$

In order to obtain the response function $R(\omega)$ one has to invert the LIT. Details about inversion methods are described in [1, 4].

As already mentioned in the introduction an expansion on a complete many-body basis is used for the solution of the LIT equation (3). First, the Hamiltonian matrix for such a basis is determined, then, in a subsequent diagonalization of this matrix, N eigenvalues E_n and eigenstates ϕ_n (LIT states) ($n = 1, 2, ..., N$) are obtained, where N is the dimension of the basis. The LIT can then be expressed in terms of the energy eigenvalues and the LIT states. One obtains

$$L(\sigma) = \sum_{n=1}^{N} \frac{S_n}{(\sigma_R - (E_n - E_0))^2 + \sigma_I^2}\,. \tag{5}$$

with

$$S_n = |\langle\phi_n|\hat{O}|0\rangle|^2\,. \tag{6}$$

Coming back to the question which resolution or in other words which value of σ_I can be sustained in a given LIT calculation it is already pointed out in the introduction that the density of LIT states plays a crucial rule. In fact the higher the density of LIT states the higher is also the resolution. This point will be better illustrated in section 2.2.

2.1. Many-body basis systems

In most of the LIT applications the HH basis has been used for the expansions of ground-state wave functions and LIT states of the considered A-body system. This is mainly attributed to its property of being a complete A-body basis for localized states.

An HH basis has the following form

$$HH_{[K]n}(\Omega_A, \rho_A) = \mathcal{Y}_{[K]}(\Omega_A) R_n(\rho_A)\,. \tag{7}$$

It consists of an hyperangular part $\mathcal{Y}_{[K]}(\Omega_A)$ and a hyperradial part $R_n(\rho_A)$, where Ω_A is a set of $3A-4$ hyperangles, ρ_A denotes the hyperradius and $[K]$ stands for a set of hyperspherical quantum numbers. For the hyperradial basis functions Laguerre polynomials $L_n^{(\beta)}(\rho_A)$ times an exponential factor $\exp(-\rho_A/2b)$ are used, where β and b are a free parameters. In addition one can also introduce as a multiplicative factor NN short-range correlation functions of Jastrow type, which can be purely central or can also become spin and/or isospin dependent (see e.g. [5]).

Usually the hyperangular states $\mathcal{Y}_{[K]}$ are constructed with a well-defined permutational symmetry. With a complementary permutational symmetry of the spin-isospin part of the nuclear wave function one then obtains an antisymmetric basis. In case of $A = 2$ the

IOP Conf. Series: Journal of Physics: Conf. Series **981** (2018) 012018 doi:10.1088/1742-6596/981/1/012018

hyperangular basis functions reduce to the well-known spherical harmonics Y_{lm}. A detailed description of the HH expansion technique is given e.g. in [1].

In the LIT applications of the present work a new basis $\Phi_{[K]nn'l}$ is employed in addition. It consists of a separation of the A-body basis in a $(A-1)$-part with HH basis functions $R_n(\rho_{A-1})\mathcal{Y}_{[K]}(\Omega_{A-1})$ and a single-particle part with basis functions $R^{(2)}_{n'}(r'_A)Y_{lm}(\Omega_{r'_A})$:

$$\Phi_{[K]nn'l} = \mathcal{Y}_{[K]}(\Omega_{A-1})\, R_n(\rho_{A-1})\, R^{(2)}_{n'}(r'_A)Y_{l(m)}(\Omega_{r'_A}) \tag{8}$$

with $\mathbf{r}'_A = \mathbf{r}_A - \mathbf{R}^{(A-1)}_{cm}$, where $\mathbf{r}_A$ and $\mathbf{R}^{(A-1)}_{cm}$ is the position of the A-th particle and the center of mass of the $(A-1)$-particle system, respectively. For $R^{(2)}_{n'}(r'_A)$ a similar expansion as for the hyperradial part is taken, namely a Laguerre polynomial $L^{(2)}_{n'}(r'_A)$ times an exponential factor $\exp(-r'_A/2b_A)$. Also in this case one may use the NN correlation functions, discussed above, in addition.

Of course, as for the HH basis given in eq. (7), one has to multiply the basis functions of eq. (8) with appropriate A-body spin-isospin wave functions and then one has to care for an antisymmetric state by making a proper antisymmetrization of the basis states.

2.2. Photodisintegration of ^{3}He and astrophysical S-factor S_{12}

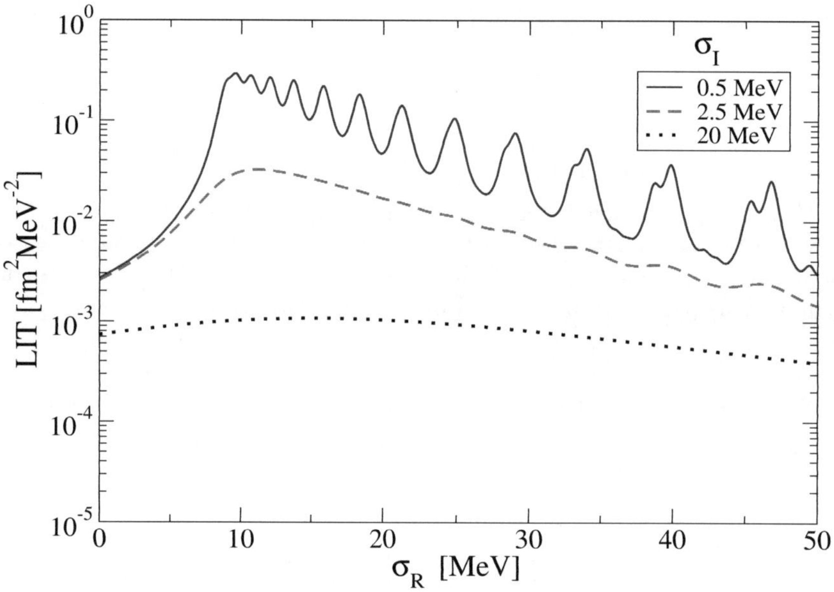

Figure 1. LIT of the ^{3}He response function $R_{\rm E1}(\omega)$ using an HH basis with 31 hyperradial ($b = 0.3$ fm) and 30 hyperangular basis states.

As pointed out in the introduction, in this work the S-factor of the reaction $d(p,\gamma)^3$He is determined via the inverse reaction, the ^{3}He photodisintegration, then time reversal invariance is applied to obtain S_{12}.

We take the unretarded dipole approximation for the calculation of the ^{3}He photodisintegration cross section, which is given by

$$\sigma_{\rm E1}(\omega) = 4\pi^2\alpha\omega R_{\rm E1}(\omega)\,, \tag{9}$$

where α is the fine structure constant and $R_{\rm E1}(\omega)$ denotes the dipole response function. In this case the components $\hat{O}$ and $|0\rangle$ of eq. (1) become equal to D_z, the third component of the nuclear dipole operator $\mathbf{D}$, and the ^{3}He ground-state wave function, respectively.

In order to determine the S-factor S_{12} one only needs to take into account the low-energy ^{3}He photoabsorption cross section, which is exclusively due to the two-body breakup channel ^{3}He+$\gamma \rightarrow p + d$. Since the pd channel has isospin $T = 1/2$, only the $T = 1/2$ channel is considered for the LIT equation (3). As NN potential the MT-I/III potential [6] is employed. To speed up the convergence of the expansions, both with HH basis and new basis, the already mentioned central NN short-range correlation functions are taken into account in addition.

First LIT results with a three-body HH basis are considered. As mentioned in section 2.1 the hyperradial basis functions contain Laguerre polynomials $L_n^{(\beta)}$, here $\beta = 5$ is taken. In fig. 1 the LIT for the case with 30 hyperangular and 31 hyperradial states ($b = 0.3$ fm) is shown. One sees that a smooth LIT is obtained with $\sigma_I = 20$ MeV, while with $\sigma_I = 2.5$ MeV the contributions of single LIT states becomes visible at higher energies. Such contributions due to single LIT states become even the dominant feature for $\sigma_I = 0.5$ MeV. This is a clear sign that the LIT-state density is too low to support a resolution with a σ_I value of 0.5 MeV. In fact the resolution of strength encoded in the LIT depends on the relative distance ΔE of two neighbouring LIT states. Structures with a width smaller than ΔE can hardly be resolved by an inversion of the LIT. In other words the higher the density of LIT states the finer the details that can be resolved inverting the LIT.

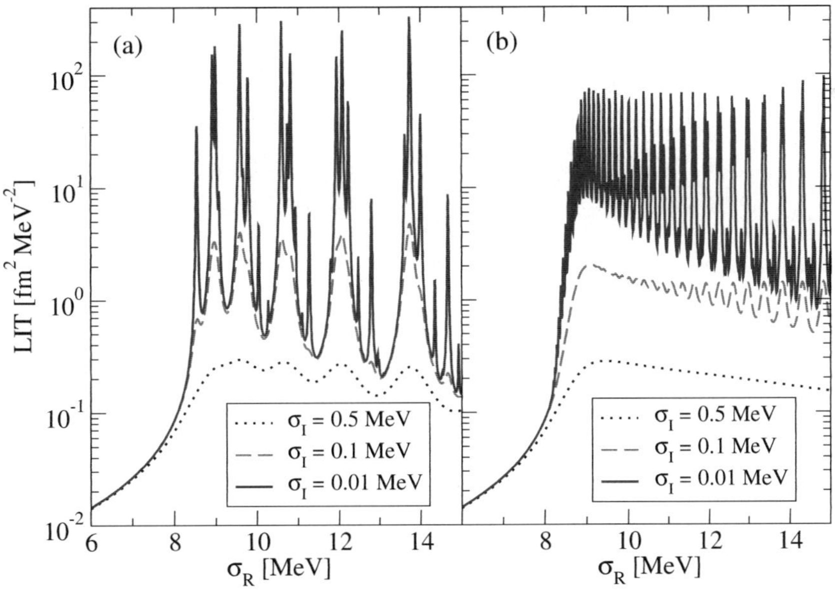

Figure 2. LIT of the ^{3}He response function $R_{\rm E1}(\omega)$ using an HH basis as in fig. 1 (a) and with 76 hyperradial ($b = 1$ fm) and 40 hyperangular basis states (b).

In order to enhance the LIT-state density one can increase the number of HH basis states taking more hyperangular and/or hyperradial states. In addition one can use a larger spatial extension of the basis by taking a greater value for the hyperradial parameter b, which then leads to a shift of LIT states towards lower energies. The resulting effects on the LIT are discussed in greater detail in [7]. Here, in fig. 2, we only compare the LITs of fig. 1 at low energies with those obtained with an HH basis of 40 hyperangular and 76 hyperradial states with $b = 1$ fm. From the LIT results with $\sigma_I = 0.01$ MeV it is readily seen that the density of LIT states becomes much larger with the increased HH basis. Accordingly one finds for the lower resolutions of σ_I equal to 0.1 and 0.5 MeV much smoother LIT results with the increased HH basis. However, one also notes a very important point: even in fig. 2b there is not a single LIT state below the three-body breakup threshold at about 8 MeV (^{3}He binding energy with MT potential). Thus the information about the response function is only rather scarce in the energy range between the two-body breakup threshold at about 5.8 MeV and the three-body breakup threshold. Also with regard to the results following in section 2.3 and to those of [2] one may conclude that for an HH basis a systematic increase of the LIT-state density cannot easily be achieved in an energy region where only two-body breakup channels are open. Therefore the HH basis is not very suitable to obtain precise results for such cases, as for example the present one of low-energy ^{3}He photodisintegration.

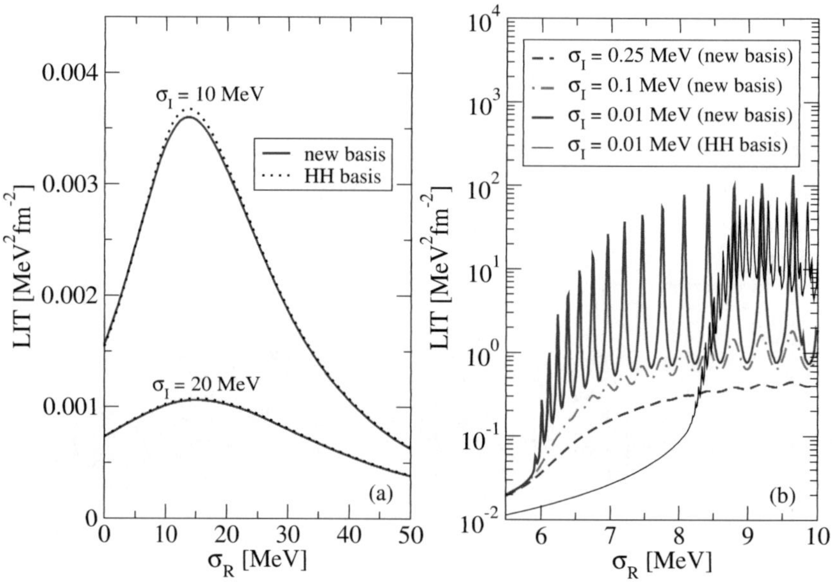

Figure 3. LIT of the ^{3}He response function $R_{\mathrm{E1}}(\omega)$ using an HH basis as in fig. 2b (a) and new basis as described in the text (b).

Now we turn to the results with the new basis described in section 2.1. The $(A-1)$-basis corresponds in the present three-body case to a two-body basis with basis states $Y_{lm}(\Omega_r)\, L^{(2)}_{n'}(r)\, \exp(-r/b_2)$ with $\mathbf{r} = \mathbf{r}_2 - \mathbf{r}_1$, where $\mathbf{r}_i$ is the position of the *i-th* particle. A basis is used with 25 and 80 radial states for the two-body and single-particle basis, respectively ($b_2 = 0.75$ fm, $b_3 = 0.5$ fm). Since in the present case only the low-energy part of the response function is relevant, it is sufficient to take into account only s-states for the two-body basis.

In fig. 3 LITs resulting from the increased HH basis of fig. 2b and those obtained with the new basis are shown. Figure 3a illustrates that both results are very similar for $\sigma_I = 20$ MeV, while with $\sigma_I = 10$ MeV one finds some differences in the region of the maximum. For a much smaller σ_I of 0.01 MeV, shown in fig. 3b, strong differences become evident. In fact, only with the new basis LIT states are present right above the two-body breakup threshold at about 5.8 MeV. Moreover these states have a rather high density. More details of the LIT calculation with the new basis, like for example the convergence behaviour with respect to the two-body and single-particle basis systems, are discussed in [7].

The results presented in fig. 3 show that the use of a proper many-body basis can become important for specific questions. There are two conditions which should be considered: (i) is the density of LIT states sufficiently high in order to extract specific structures in the response function and (ii) is the LIT-state density sufficiently regular in order to work with a single σ_I value. If the second condition is not fulfilled one should take in a region of lower LIT-state density a different, more suitable, value for σ_I, otherwise one risks to misplace strength in the inversion. In fact in [7] quite a number of different σ_I values were used in order to take into account a lower LIT-state density with growing energy.

For the LIT with the present HH basis one can conclude that the completely missing LIT states in the two-body breakup region do not only prevent to resolve the correct threshold behaviour of $R_{\rm E1}(\omega)$, but that one would also obtain an overestimation of the peak height of the response function if for the inversion one uses a LIT with a σ_I value much smaller than 20 MeV. As illustrated in [7] (see fig. 7 therein) one can still obtain a rather reasonable inversion result with $\sigma_I = 20$ MeV using the standard inversion method, described in [1], where the correct threshold behaviour of the response function is implemented.

Concerning the LIT results with the new basis one can certainly say that the low-energy density of LIT states is quite high and that the pattern is very regular. Thus one may expect that the low-energy response, and thus the astrophysical S-factor S_{12}, can be determined very precisely by the inversion. In fig. 4 we show the result for S_{12} obtained with the new basis in comparison to a calculation of S_{12} with explicit wave functions for the proton-deuteron continuum states (for details of the continuum state calculation see [7]). As anticipated one observes an excellent agreement between both results. Note that the figure also contains an error estimate due to the LIT inversion, also here we refer to [7] for a more detailed description of the error estimate.

2.3. The ^{4}He isoscalar monopole resonance

The isoscalar monopole resonance 0^+ of the α-particle leaves a strong signal in inclusive inelastic electron scattering experiments [8, 9, 10]. The corresponding transition form factor was studied in a LIT calculation, where an HH basis and modern realistic forces were used [2]. A rather strong potential model dependence was found, but the experimental data were overestimated quite a bit. The present work, however, is not devoted to determine the strength of the transition form factor, but rather to a different aspect of the resonance, namely its rather small width of 270(50) keV as determined in the ^{4}He(e,e') experiments mentioned above. In [2] this question could not be addressed because the density of LIT states was not sufficiently high in the region of the 0^+ resonance, which is located closely above the lowest ^{4}He two-body breakup threshold.

The isoscalar monopole response function $R_{\rm C0}(q,\omega)$ depends on energy transfer ω and momentum transfer q mediated in electron scattering by the exchanged virtual photon. Thus the corresponding transition operator $\hat{O}$ of eq. (1) becomes q-dependent:

$$\hat{O}(q) = \frac{G_E^s(q^2)}{2} \sum_{i=1}^{A} j_0(qr_i)\,. \tag{10}$$

In the equation above $G_E^s(q^2)$ is the nucleon isoscalar electric form factor, $\mathbf{r}_i$ is the position of nucleon i, and j_0 is the spherical Bessel function of 0th order.

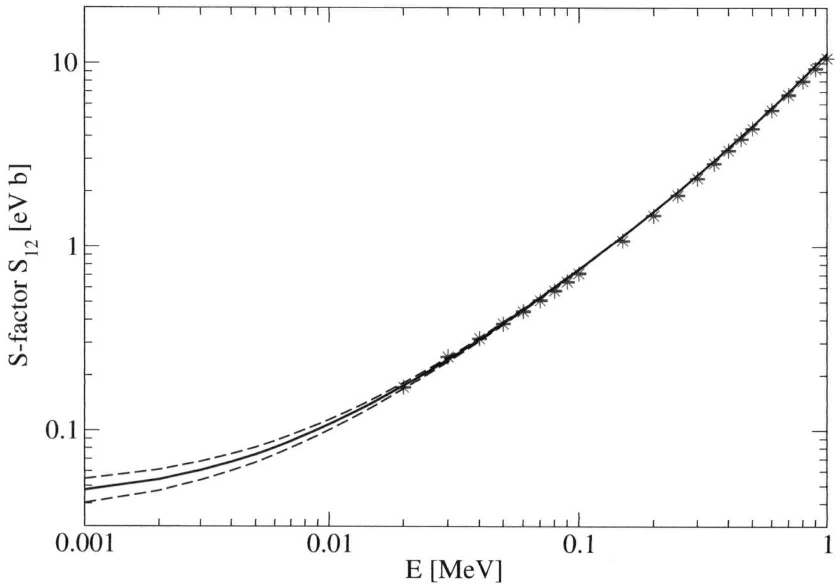

Figure 4. Astrophysical S-factor S_{12} calculated with the LIT method (full curve) with additional error estimate due to inversion (dashed curves); in addition results of a calculation with explicit *pd* continuum wave functions (stars).

For the present study the LIT of $R_{C0}(q,\omega)$ is taken at $q = 300$ MeV/c, a q value, which lies in the momentum transfer range of maximal strength of the 0^+ transition form factor. Here we consider results with an HH basis for the four-body system and in addition the new basis as described in section 2.1 (three-body HH basis plus single-particle basis). For details of the used basis states we refer to [3]. As NN potential model the central TN potential is taken, it has been used in the very first LIT applications for the α-particle (see for example [1]). Like in the previous case in section 2.2 central NN correlation functions are used in order to accelerate the convergence of HH and new basis.

In fig. 5 the LIT results for the response function are shown. For the HH basis in fig. 5a one sees a very similar picture as in fig. 2a. There is only one essential difference, in fig. 2a there are no LIT states below the many-body breakup threshold, whereas in fig. 5a one finds just one LIT state below the many-body breakup threshold at about 30 MeV (note ^{4}He binding energy with TN potential is 31.4 MeV). The isolated low-energy LIT state for the ^{4}He case is due to the 0^+ resonance. It is evident that with a single LIT state it is impossible to determine a resonance width. In the already mentioned LIT calculation for $R_{C0}(q,\omega)$ with realistic nuclear forces of [2] the situation was somewhat better, but the LIT-state density could not be systematically improved in order to have sufficient information to compute the 0^+ resonance width.

As illustrated in fig. 5b the situation is much better in case of the new basis. It is interesting to study the results with the various σ_I values. With a low resolution of $\sigma_I = 5$ MeV one does not realize that there is a resonance (note the logarithmic scale). If one increases the resolution using smaller σ_I values the resonance becomes more and more distinct from the background. For the highest resolution of $\sigma_I = 0.01$ MeV one sees that there are various LIT states in the region of the resonance at about 26 MeV. In fact in [3] it was possible to determine the width using besides the results of fig. 5b also LIT results with a different basis size (for details see [3]). The obtained width of 180(70) keV agrees quite well with the experimental value of 270(50) keV.

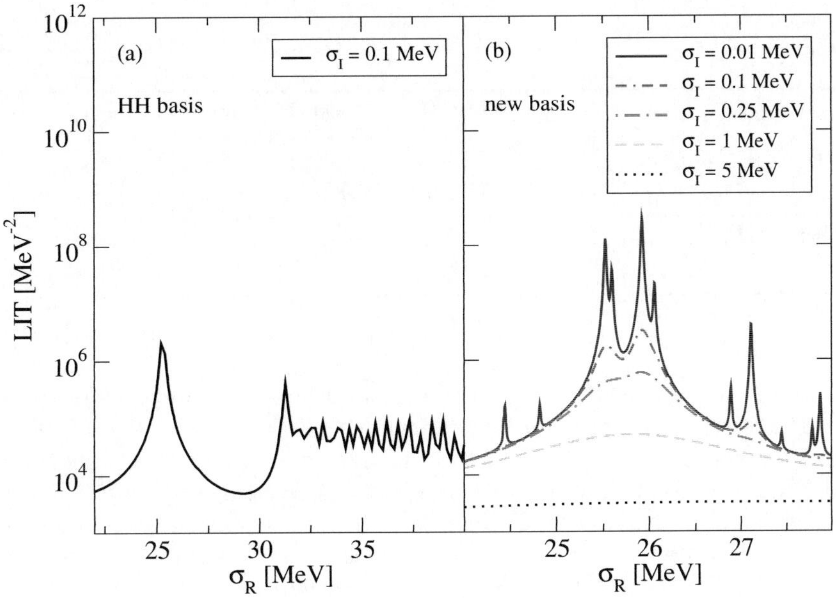

Figure 5. LIT of the ^{4}He response function $R_{C0}(q,\omega)$ at $q = 300$ MeV/c using a four-body HH basis (a) and the new basis $\Phi_{[K]nn'l}$ of eq. (8) (b) (a detailed description of the considered basis states is given in [3]).

3. Strange baryon systems

In the benchmark calculation of [11] bound baryon systems from three up to five particles are considered, where one of the baryons is the Λ hyperon which has strangeness $S = -1$. As already mentioned in the introduction the experimental information about the YN interaction is still rather scarce. On the other hand, our present aim is not yet a realistic calculation of the binding energy of hypernuclei, but rather a check of the precision of the ab initio method used by us. Therefore in [11] calculations with non-fully realistic interactions models are made. Before coming to some of these results in section 3.2, first, a short description of the ab initio method of our choice is given in the following section.

3.1. The nonsymmetrized hyperspherical harmonics (NSHH) expansion

The NSHH expansion relies on the HH expansion given in eq. (7), but the hyperspherical functions $\mathcal{Y}_{[K]}(\Omega_A)$ are not constructed with any permutational symmetry. Also the spin-isospin part of a hypernuclear basis state is taken without imposing a permutational symmetry. On the other hand it is clear that a hypernuclear wave function has to be antisymmetric under the exchange of two identical fermions. At this point it is helpful to consider the Casimir operator $\hat{C}$ of the particle system. Taking an A-body baryon system with N nucleons ($n = 1, 2, ..., N$) and L Λ hyperons ($n = N+1, N+2, ..., N+L = A$) one has

$$\hat{C} = \hat{C}_N + \hat{C}_\Lambda = \sum_{j>i=1}^{N} \hat{P}_{ij} + \sum_{j>i=N+1}^{A} \hat{P}_{ij} \,, \qquad (11)$$

where the operator $\hat{P}_{ij}$ exchanges particles i and j. The eigenvalues $\lambda_{[I]}$ of the Casimir operator depend on the specific permutational symmetry of the eigen functions.

Table 1. Binding energy BE and Λ separation energy S_Λ for the hypernucleus ${}^3_\Lambda$H.

Potential model	Energy	NSHH	AFDMC	FY	GEM
AV4'+Bodmer-Usmani	BE	2.530(3)	2.42(6)	2.537(1)	-
	S_Λ	0.290(3)	0.18(6)	0.292(1)	-
AV8'+NSC97f	BE	2.41(2)	-	2.415(1)	-
	S_Λ	0.17(2)	-	0.189(1)	0.19(1)

In our case with just one Λ particle only the permutational symmetry of the nucleons is relevant. One has the lowest eigenvalue for the antisymmetric case ($\lambda_{[A]} = -N(N-1)/2$) and the highest for the symmetric case ($\lambda_{[S]} = +N(N-1)/2$). Thus diagonalizing the Hamiltonian matrix for an NSHH basis, one can find out the symmetry of a given eigenstate by calculating the corresponding eigenvalue $\lambda_{[f]}$ applying the Casimir operator. This is the strategy which has been put forward in [12]. If, however, the NSHH basis is very large and one wants to find the lowest state being antisymmetric for the nucleonic part, which in general is not the lowest energy state, it is more convenient concerning the computational resources to apply the strategy of [13]. In fact in case of large basis systems it is not advisable to perform a complete diagonalization of H. It is much better to use the Lanczos technique, which saves computational resources and leads to a fast determination of the lowest energy state. In order to bring the lowest antisymmetric state for the nucleonic part to the lowest overall state the following fictitious Hamiltonian $\hat{H}'$ has been introduced in [13]:

$$\hat{H}' = \hat{H} + \gamma \hat{C}_N \,. \tag{12}$$

Thus, for a sufficiently large γ, such a lowest antisymmetric state will become the absolute ground state of the particle system with ground-state energy E_0. Therefore, taking $\hat{H}'$ instead of $\hat{H}$, one can apply the Lanczos technique to find the proper ground state for an A-baryon system with $A-1$ nucleons and one Λ hyperon. In order to have the correct bound-state energy one needs to correct E_0 only by $\gamma N(N-1)/2$.

3.2. The hypernucleus ${}^3_\Lambda H$

Here the benchmark results for one of the strange baryon systems discussed in [11] are illustrated, namely those for ${}^3_\Lambda$H. Two different potential sets were used in these calculations: (i) the AV4' NN potential [14] together with the Bodmer-Usmani YN potential [15] and (ii) the AV8' NN potential [14] together with a parametrization [16] of the meson-theoretical NSC97f YN potential [17]. In order to accelerate the NSHH expansion an effective interaction is used as described in [18]. Besides the NSHH results new results with the auxiliary field diffusion Monte Carlo (AFDMC) technique [19] have been obtained. Also results due to the Faddeev approach (FY) and due to the Gaussian expansion method [16] (GEM) are included in [11].

In table 1 the binding energy and the Λ separation energy of ${}^3_\Lambda$H are listed for the two different potential models defined above. One observes a rather good agreement between the various methods. Only the AFDMC results are a bit different, but this is not a real surprise since the AFDMC is an ab initio method more suitable for systems with more than three particles and with some preference for closed shell nuclei. In fact further calculations discussed in [11] show that the comparison of AFDMC results with those of the other ab initio methods become decisively better for ${}^4_\Lambda$H and ${}^5_\Lambda$He.

With the results of table 1 and the further ones given in [11] one can conclude that the NSHH method is very well suited to give precise results for observables of hypernuclei. A

IOP Conf. Series: Journal of Physics: Conf. Series **981** (2018) 012018 doi:10.1088/1742-6596/981/1/012018

further benchmark with the AFDMC method, where also three-body interactions are taken into account, will be published in the near future. Thus, more ambitious calculations with more realistic interaction models can be tackled with the NSHH method in future.

4. Summary

The purpose of this work is twofold. Firstly, it is a check of the applicability of the LIT method for a precise determination of specific details in nuclear low-energy cross sections that are induced by external electromagnetic probes. To this end the reactions ^{3}He(γ) and ^{4}He(e, e') have been considered. The ^{3}He photodisintegration has been calculated in order to obtain the astrophysical S-factor S_{12} of its inverse reaction, i.e. $d(p,\gamma)^3$He, by applying time reversal invariance. Thus the actual aim has been a precise determination of S_{12} via the LIT method. Comparing to results of a calculation with explicit proton-deuteron continuum wave functions it has been shown that the LIT leads to excellent results for S_{12}. The calculation has been carried out with a simple central NN potential, but the real importance of the calculation does not lie in a realistic calculation of S_{12}, more important is the fact that the LIT method could serve to calculate astrophysical S-factors of reactions involving more than three nucleons. Also the inclusive electrodisintegration of ^{4}He has been computed here with a central NN interaction. Again, the essential aim has not been to obtain realistic results, but to test the possibility to determine the width of a narrow resonance with the LIT method. In fact the ^{4}He continuum exhibits a rather narrow resonance, the so-called isoscalar monopole resonance 0^+. Therefore the LIT calculation has been performed for the ^{4}He isoscalar monopole response function $R_{C0}(q,\omega)$. It has been shown that the resonance width can be determined with the LIT method and that a value of 180(70) keV is obtained. This agrees quite well with the experimental result of 270(50) keV. Thus, there is cause for hope that the 0^+ resonance width can be determined with the LIT method also for the case of a realistic nuclear force.

It has been pointed out that the decisive point for precise determinations of S-factor S_{12} and of ^{4}He 0^+ resonance width is a sufficiently high density of LIT states. Unfortunately, this seems to be very difficult to achieve if one uses as A-body basis an HH A-body basis. It is much better to take an HH hybrid basis consisting of an $(A-1)$-body HH basis and an additional single-particle basis for the $A-th$ nucleon. It has been illustrated that with such a basis it is possible to systematically increase the density of LIT states in the two-body breakup region at low-energies.

The second purpose of the present work has been a benchmark calculation for $^3_\Lambda$H. The ab initio method of our choice, expansion of the hypernuclear ground state on a nonsymmetrized HH basis, has been discussed, in particular, how such a basis can serve to obtain a ground state with a proper permutational symmetry. Two different potential model sets have been employed for NN and NΛ interactions. In comparison to results from other ab initio approaches it has been found that the present calculation leads to reliable results for the $^3_\Lambda$H binding energy and the corresponding Λ separation energy.

Finally, it should be mentioned that more details of the various calculations are given in [3, 7, 11].

References

[1] Efros V D, Leidemann W, Orlandini G and Barnea N 2007 *J. Phys. G* **34** R459
[2] Bacca S, Barnea N, Leidemann W and Orlandini G 2013 *Phys. Rev. Lett.* **110** 042503
[3] Leidemann W 2015 *Phys. Rev. C* **91** 054001
[4] Leidemann W 2008 *Few-Body Syst.* **42** 139
[5] Efros V D, Leidemann W, Orlandini G and Tomusiak E L 2000 *Phys. Lett. B* **484** 223
[6] Malfliet R A and Tjon J A 1969 *Nucl. Phys. A* **127** 161
[7] Deflorian S, Efros V D and Leidemann W 2017 *Few-Body Syst.* **58** 3
[8] Walcher Th 1970 *Phys. Lett. B* **31** 442

[9] Frosch R F, Rand R E, Yearian M R, Crannell H and Suelzle L R 1965 *Phys. Lett.* **19**, 155
Frosch R F, Rand R E, Crannell H, Mccarthy J S and L R Suelzle L R 1968 *Nucl. Phys. A* **110** 657
[10] Köbschall G, Ottermann C, Maurer K, Röhrich K, Schmitt Ch and Walther V H 1983 *Nucl. Phys. A* **405** 648
[11] Ferrari Ruffino F, Barnea N, Deflorian S, Leidemann W, Lonardoni D, Orlandini G and Pederiva F 2017 *preprint* arXiv:1701.06399
[12] Gattobigio M, Kievsky A and Viviani M 2011 *Phys. Rev. C* 83 024001
[13] Deflorian S, Barnea N, Leidemann W and Orlandini G 2013 *Few-Body Syst.* 54 1879
[14] Wiringa R B and Pieper S C 2002 *Phys. Rev. Lett.* **89** 182501
[15] Bodmer A R and Usmani Q N 1988 *Nucl. Phys. A* **477** 621
Usmani A A and Khanna F C 2008 *J. Phys. G* **35** 025105
[16] Hiyama E, Ohnishi S, Gibson B F and Rijken Th A 2014 *Phys. Rev. C* **89** 061302
[17] Rijken Th A, Stoks V G J and Yamamoto Y 1999 *Phys. Rev. C* **59** 21
[18] Barnea N, Leidemann W and Orlandini G 2000 *Phys. Rev. C* **61** 054001
[19] Lonardoni D, Gandolfi S, Pederiva F 2013 *Phys. Rev. C* **87** 041303R
Lonardoni D, Gandolfi S, Pederiva F 2014 *Phys. Rev. C* **89** 014314

IOP Conf. Series: Journal of Physics: Conf. Series **981** (2018) 012019 doi:10.1088/1742-6596/981/1/012019

Recent progress in *ab-initio* studies of nuclear reactions of astrophysical interest with $A \leq 3$

Laura E Marcucci

Department of Physics "E. Fermi", University of Pisa, and INFN-Pisa, 56127 Pisa, Italy

E-mail: laura.elisa.marcucci@unipi.it

Abstract. We review the most recent theoretical studies of nuclear reactions of astrophysical interest involving few-nucleon systems. In particular, we focus on the consequences for the solar neutrino fluxes of the recent determination for the astrophysical S-factor of the proton weak capture by proton, and on the radiative capture of protons by deuterons in the energy range of interest for Big Bang Nucleosynthesis.

1. Introduction

The weak proton capture on protons, *i.e.*, the reaction $p + p \to d + e^+ + \nu_e$ (hereafter labelled *pp*), is the most fundamental process in stellar nucleosynthesis: it is the first reaction in the *pp* chain, which converts hydrogen into helium in main sequence stars like the Sun. Also the proton radiative capture by deuteron, *i.e.*, the reaction $p + d \to^3 \mathrm{He} + \gamma$ (hereafter labelled *pd*), is among the nuclear fusion which are part of the *pp* chain. However, being the only possible reaction after the initial step, it has no interest for the solar neutrino fluxes. Nevertheless, this reaction is of interest for the theory of Big Bang Nucleosynthesis (BBN), since it is the main process through which deuterons are destroyed, and therefore it strongly affects the primordial deuterium abundance.

For both reaction, it is customary to express the reaction rates in terms of the astrophysical S-factor, $S(E)$, where E is the initial center-of-mass (c.m.) energy of the two particles (*pp* or *pd*), by the relation

$$S(E) = E \exp(2\pi\,\eta)\,\sigma(E) \ . \tag{1}$$

Here $\eta = \alpha/v_{\rm rel}$ is the Sommerfeld parameter, α being the fine structure constant and $v_{\rm rel}$ the *pp* or *pd* relative velocity, and $\sigma(E)$ is the capture cross section.

The most recent studies of the *pp* reaction are those of Refs. [1, 2]. The study of Ref. [1] is the first one performed within the so-called chiral effective field theory (χEFT) approach. The goal of this work was to apply χEFT to the *pp* fusion, comparing the results for the S-factor with what was obtained within the conventional phenomenological approach, and studying the effects to the S-factor arising from higher order electromagnetic contributions in the Hamiltonian and the P-wave components of the initial *pp* scattering state. A first attempt to estimate the theoretical uncertainty was also performed, basically varying consistently in the nuclear Hamiltonian and nuclear currents the Λ cutoff, present in the regularization momentum-cutoff function needed to take care of the power-law behavior for large momenta of potentials and currents. It should be noticed, though, that some inconsistencies in the study of Ref. [3] were present, as, for example, the different chiral order for the two- and three-nucleon potential and also the axial current. These inconsistencies have been overcome in the study of Ref. [2], where the same chiral order is kept in the nuclear potentials and currents. This allows to quantify the theoretical uncertainty for the *pp* fusion in a more consistent way. However, the contributions beyond S-waves in the initial *pp* scattering state and the higher order terms in the electromagnetic part of the Hamiltonian were not included. The conclusions of that study are that the theoretical uncertainty for the zero-energy astrophysical S-factor cannot be reduced more than 0.7 %. Furthermore, the central values of $S(0)$ is in very nice agreement with the results of Ref. [1].

We will review in Section 2 the χEFT framework used to perform the *pp* fusion calculation, and we will consider the consequences of these most recent determinations of $S(E)$ on the solar neutrino fluxes, as they have been investigated in Ref. [4].

The interest in the *pd* radiative capture has received a boost recently within the BBN context, since, in order to achieve agreement between the measured primordial deuterium abundance [5] and the BBN prediction, it has been questioned the accuracy of the only available experimental data in the BBN region [6]. Therefore, an *ab-initio* calculation of the astrophysical S-factor, as well as a more precise direct measurement, have become very urgent. The most recent theoretical study has been performed in Ref. [3], within the so-called conventional approach, which uses realistic phenomenological models for nuclear potentials and currents. Before, the S-factor was studied in Refs. [7, 8], still within the conventional approach, and it was found an excellent agreement between the theoretical results and the the available experimental data [9] around the solar Gamow peak [8, 10]. In the energy range of interest for BBN, on the other hand, the theoretical predictions of Ref. [8] were found to be 2–10 % higher than the central value of the polynomial best fit of Ref. [10]. Furthermore, the only available experimental data in the BBN energy range are those of Ref. [6], which are quite in disagreement with the polynomial best fit of Ref. [10]. Therefore, a new experimental determination is highly necessary and this is the main motivation behind the experiment recently proposed by the LUNA Collaboration at the Gran Sasso National Laboratories (Italy), with the goal of measuring the *pd* S-factor in the BBN energy range with a 3 % accuracy.

We will review in Section 3 the conventional phenomenological framework used to study the *pd* astrophysical S-factor, as performed in Ref. [3], and we will review its implications for BBN in Subsection 3.2.

2. The χEFT framework and the *pp* fusion

We consider here the study of the *pp* fusion within the χEFT framework of Ref. [1]. We review in the following subsection the main ingredients of the calculation, and in Subsection 2.2 we discuss the results for the astrophysical S-factor and their consequences on the solar neutrino fluxes.

2.1. The χEFT theoretical framework

In a very schematic view, the χEFT framework can be seen as a formulation of Quantum Chromodynamics (QCD) in terms of effective degrees of freedom suitable for low-energy nuclear physics: pions and nucleons. The symmetries of QCD, in particular its (spontaneously broken) chiral symmetry, severely restrict the form of the interactions of nucleons and pions among themselves and with external electroweak fields, and make it possible to expand the Lagrangian describing these interactions in powers of Q/Λ_χ, Q being the pion momentum and Λ_χ being the chiral-symmetry-breaking scale (here $\Lambda_\chi \sim 700$ MeV). As a consequence, classes of Lagrangians emerge, each of the order $(Q/\Lambda_\chi)^n$ and each involving a certain number of unknown coefficients, the so-called low-energy constants (LECs), which arise when the high-energy degrees of freedom are integrated out. These LECs are in practice constrained by fits to experimental data. The potentials and currents derived within this framework have power-law behavior for large momenta, and are regularized by introducing a momentum-cutoff function. In Ref. [1], this function is taken of the type $\exp(-q^4/\Lambda^4)$ (q is the three-momentum transfer), and it is expected that increasing the chiral order n, the dependence on Λ will become weaker. In χEFT, two- and three-nucleon interactions (TNIs) arise naturally and are used to describe the few-nucleon systems. Of course, in the case of the pp fusion, TNIs are not present. However, they are needed to fix the LECs entering the axial current (see below). In Ref. [1], the two-nucleon (NN) interaction is the one derived by Entem and Machleidt in Ref. [11] at next-to-next-to-next-to leading order (N3LO) and the TNI is derived at next-to-next-to leading order (N2LO), in the local form of Ref. [12]. To be noticed that, in the pp sector, the nuclear NN potential has been augmented by the Coulomb interaction and the higher-order electromagnetic (EM) terms, due to two-photon exchange and vacuum polarization. These higher-order terms are the same as those of the Argonne v_{18} (AV18) NN potential [13], and therefore also retain short-range corrections associated with the finite size of the proton charge distribution. The effect of the additional distortion of the pp wave function, induced primarily by vacuum polarization, on the astrophysical S-factor will be discussed below.

The charge-changing weak current has been first derived in χEFT by Park *et al.* in the late nineties, using the so-called heavy-baryon chiral perturbation theory (HBχPT) approach, where the baryons are treated as heavy static sources, and the perturbative expansion is performed in terms of the involved momenta over the baryon mass (see Refs. [14, 15] and references therein). In particular, the polar-vector part is related, via the conserved-vector-current constraint, to the (isovector) EM current, and at N3LO includes in the model of Park *et al.* [16], apart from one- (OPE) and two-pion-exchange (TPE) terms, two contact terms—one isoscalar and the other isovector—whose strengths are parametrized by the LECs g_{4S} and g_{4V}. Few years ago, the problem of deriving the EM current and charge operators in χEFT has been revisited by Pastore *et al.* [17] and, in parallel, by Kölling *et al.* [18]. Pastore *et al.* used time-ordered perturbation theory (TOPT) to calculate the EM transition amplitudes, which allows for an easier treatment of the so-called reducible diagrams than the HBχPT approach. On the other hand, Kölling *et al.* used the method of unitary transformation, the same one used to derive the chiral potentials mentioned above. The EM operators of Pastore *et al.* have been found in good agreement with those of Kölling *et al.* but significantly different from those of Park *et al.*, especially for the structure of the contact terms. This might give rise to significant differences in the study of those reactions where the polar-vector current gives significant contributions, as muon capture on light nuclei. Studies to verify whether this is really the case are currently underway.

The axial-vector current are diagrammatically represented in figure 1 up to N3LO, where they are listed according to their scaling in Q, the pions' and nucleons' momenta. The leading-order (LO) contribution consists of the well known single-nucleon axial current, and is of order Q^{-3}. At order Q^{-2} it turns out that there are no contributions, and therefore the next-to-leading (NLO) contribution is of order Q^{-1}, and arises from the $(Q/m)^2$ relativistic corrections to the LO contribution (m is the nucleon mass). The N2LO currents, of order Q^0, consist of the OPE term and a contact term with one LEC, denoted with d_R. The N3LO contributions arise from loop and TPE terms, and they have been calculated in Refs. [19, 20] using the TOPT approach mentioned above. They are not all represented in figure 1. The diagrammatic representation and a full discussion of these contributions can be found in the original references. These N3LO contributions at the moment have been used only to calculate the Gamow-Teller matrix element of tritium β-decay [20]. In fact, in the study of the pp fusion of Ref. [1], as well as Ref. [2], the N3LO axial currents are not retained. Therefore, in principle, a new study of this reaction with the N3LO terms in the axial current is required. However, considering that the two-body contributions to the astrophysical S-factor are of the order of a 1–2 %, we do not expect to see a significant change in the S-factor, going to N3LO.

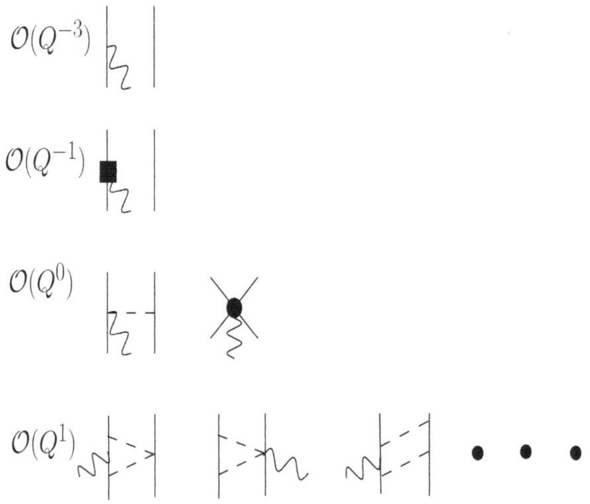

Figure 1. Diagrams illustrating one- and two-body χEFT axial currents entering at LO (Q^{-3}), NLO (Q^{-1}), and N2LO (Q^0), and N3LO (Q^1). Nucleons, pions, and weak probes are denoted by solid, dashed, and wavy lines, respectively. The solid square represents the relativistic corrections to the one-body current, while the solid circles in the N2LO contribution represent the contact terms. Only three of all possible N3LO contributions are shown.

Some remarks on the fitting procedure of the LEC d_R are in order. As first shown in Ref. [21], d_R can be related to the LEC c_D entering one of the two contact terms present in the TNI at N2LO, via the relation

$$d_R = \frac{m}{\Lambda_\chi g_A} c_D + \frac{1}{3} m (c_3 + 2c_4) + \frac{1}{6} , \tag{2}$$

where g_A is the single-nucleon axial coupling constant, c_3 and c_4 are LECs of the πN Lagrangian, already part of the chiral NN potential at NLO, and Λ_χ is the the chiral-symmetry-breaking scale. Therefore, it has become common practice since the work of Ref. [22], to fit c_D (and c_E—the other LEC entering the N2LO TNI) to the triton binding energy and the Gamow-Teller (GT) matrix element in tritium β-decay. The values obtained in this way for c_D and c_E are listed, for different values of the cutoff Λ, in Refs. [15, 23, 24, 25], where they have been used to study muon capture on deuteron and ^{3}He, the $A = 3$ and 4 elastic scattering observables, and the nuclear matter equation of state in many-body perturbation theory, respectively. In the study of the pp fusion of Ref. [1], the values obtained in Ref. [15] were used. However, it should be noticed that these values are strongly dependent on the chiral order retained in the calculation, as shown in Ref. [20]. Such a dependence should not be considered a problem of the theory, since LECs are not observables. It is expected from the theory, though, that the physical observables do not significantly change, if the calculation is performed consistently. Whether this is the case is currently under investigation.

2.2. Results for the pp fusion

We recall the main characteristics of the calculation of the pp fusion performed in Ref. [1]: (i) the astrophysical S-factor is studied in the wide energy range, from 2 to 100 keV. The solar Gamow peak is at $\simeq 6$ keV, but in larger-mass stars the Gamow peak becomes 15 keV. (ii) The χEFT N3LO NN potential of Ref. [11] is augmented not only of the Coulomb interaction but also of the higher-order EM interaction contributions, like those due to two-photon exchange,

Darwin-Foldy and vacuum polarization. (iii) All the $L \leq 1$ pp partial waves are considered. To our knowledge, this is the first time that the P-wave contributions are retained. The results for the pp astrophysical S-factor can be summarized as follows: (i) the zero-energy S-factor $S(0)$ obtained retaining all the $L \leq 1$ partial waves and the full one- plus two-body weak current contributions is $S(0) = (4.030 \pm 0.006) \times 10^{-23}$ MeV fm^2, where the theoretical uncertainty is due to the fitting procedure (the uncertainty on the value for the experimental GT matrix element reflects on an uncertainty for the LECs). The theoretical uncertainty arising from the cutoff dependence is extremely small. (ii) The P-wave contribution is $\simeq 0.02 \times 10^{-23}$ MeV fm^2. Therefore, the S-factor obtained retaining only the $^1S_0$ contributions in the initial pp wave function results in $S(0) \simeq 4.01$ MeV fm^2, in very nice agreement with the value quoted in Ref. [10]. (iii) The higher-order EM contributions to the initial pp wave functions are of the order of 1% or less, as already obtained in Ref. [26]. (iv) The study of Ref. [1] has shown that the results obtained in χEFT or within the conventional phenomenological approach are in excellent agreement.

To be noticed that the most recent study for the pp astrophysical S-factor at zero energy is given in Ref. [2], where $S(0) = (4.047^{+0.024}_{-0.032}) \times 10^{-23}$ MeV fm^2, in nice agreement with the results of Ref. [1]. As already mentioned above, the 7 ‰ theoretical uncertainty has been determined within a more robust approach. See Ref. [2] for more details.

From the astrophysical S-factor, it is possible to obtain the pp reaction rate R via the relation

$$\begin{aligned} R &= N_A\langle\sigma v\rangle = \frac{3.73 \times 10^{10}}{\sqrt{\hat{\mu}\, T_9^3}} \times \\ &\times \int_0^\infty S(E) \exp\Big(-2\pi\eta - 11.605 \frac{E}{T_9} \Big)\, \mathrm{d}E\,, \end{aligned} \tag{3}$$

where N_A is the Avogadro number, $\langle\sigma v\rangle$ is the Maxwellian-average rate, $\hat{\mu} = 0.504$ is the pp reduced mass in atomic mass units, T_9 is the temperature in units of 10^9 K, and η is the Sommerfeld parameter as defined above. The integration over the center-of-mass energy E in Eq. (3) can be performed numerically with standard techniques. This has been done in Ref. [4], where we have analyzed the effect of the adoption of different pp reaction rates on stellar models, focusing, in particular, on the age of mid and old stellar clusters (1-12 Gyr) and on standard solar model predictions. Here we consider only the standard solar model predictions, and, in particular, the results for the neutrino fluxes. In table 1 we report the values for the central solar temperature and the neutrino fluxes obtained using the astrophysical S-factor of Ref. [1], and that of Refs. [27, 10, 28]. By inspection of the table we can conclude that: (i) the largest difference in the central temperature (for the standard solar model with the pp rate from Ref. [27]) is of the order of 3‰. (ii) Due to the high temperature dependence for all the neutrino fluxes except for the pp one, the effect on solar neutrinos is not totally negligible, even if small, and it can reach the order of few %. (iii) The maximum difference for the solar neutrino fluxes corresponds to the adoption of the pp reaction rate of Ref. [27], reaching in some cases a maximum of about 8%. (iv) The $\sim$ 1% effect of the P-partial waves to the pp astrophysical S-factor, and therefore on the reaction rate, turns out into a maximum of 3% effect on the ^{8}B, ^{15}O and ^{17}F neutrino fluxes.

3. The conventional approach and the *pd* fusion

We consider here the pd fusion reaction, studied within the conventional approach. We review this approach in the next Subsection, and we summarize the results for the pd fusion and their implications for BBN in Subsection 3.2.

Table 1. The central solar temperature (K) and neutrino fluxes ($s^{-1}\,cm^{-2}$) as obtained with the *pp* S-factor of Ref. [1] with both S- and P-waves included, considered as our reference model and labelled $(S+P)$, are given in the first column. The others columns list the relative difference between our reference model and the results obtained from the S-factor of Ref. [1], but with only the ${}^1S_0$ component, labelled (S), Ref. [27], Ref. [10], and Ref. [28].

	Ref. [1]$(S+P)$ reference model	Ref. [1](S)	Ref. [27]	Ref. [10]	Ref. [28]
		relative differences			
$T_c\,[10^7\,K]$	1.54794	$-1‰$	$-3‰$	$< 1‰$	$-1‰$
$\Phi^\nu_{pp}\,[10^{10}]$	6.020	$1‰$	$2‰$	$< 1‰$	$1‰$
$\Phi^\nu_{pep}\,[10^8]$	1.446	$-2‰$	$-6‰$	$1‰$	$-1‰$
$\Phi^\nu_{hep}\,[10^3]$	8.584	$-1‰$	$-3‰$	$2‰$	$2‰$
$\Phi^\nu_{Be-7}\,[10^9]$	4.503	-1%	-3%	$< 1‰$	$-9‰$
$\Phi^\nu_{B-8}\,[10^6]$	3.694	-3%	-7%	$4‰$	-2%
$\Phi^\nu_{N-13}\,[10^8]$	2.417	-2%	-6%	$6‰$	-1%
$\Phi^\nu_{O-15}\,[10^8]$	1.811	-3%	-8%	$7‰$	-2%
$\Phi^\nu_{F-17}\,[10^6]$	3.373	-3%	-8%	$7‰$	-2%

3.1. The conventional approach

The conventional approach has been developed well before the χEFT approach described above, and, although very successful in reproducing a large body of experimental data, it is affected by the "original sin" of having no connection (or no clear connection) with QCD. In fact, this approach is purely phenomenological, and is based on realistic models for the nuclear interactions and currents. The nuclear interaction includes both two- and three-nucleon potentials. These are constructed to reproduce the $A = 2$ large body of experimental data with a χ^2/datum ~ 1 , and the $A = 3$ binding energies. In our study of the *pd* fusion, the Argonne v_{18} (AV18) model [13] for the NN interaction, and the Urbana IX (UIX) model [29] for the TNI are used. These models are quite complex, and therefore it is necessary to have at hand an *ab-initio* technique to solve the $A = 3$ bound and scattering problem. To this aim, we use the hyperspherical harmonics (HH) technique, a variational approach which adopts a basis of HH functions to expand the bound or the scattering wave functions (in this second case, for relative interparticle distances small compared with the interaction range). We will not describe this method in details and refer the reader to the review of Ref. [30]. Here, we only remark that the HH method is the only one which can calculate the nuclear wave function for the *pd* scattering state at low relative energies, as the ones of interest for BBN, including the Coulomb interaction between the charged initial particles. The $A = 3$ bound and scattering properties, as the triton and ^{3}He binding energies, the *nd* scattering length and elastic scattering observables as cross sections and analyzing powers, are in general nicely reproduced by the AV18/UIX potential model, used in conjunction with the HH technique. Some discrepancies appear only for few delicate polarization observables (see for instance Ref. [31]).

The interaction between the considered nuclear systems and the external electromagnetic probe implies the construction of the electromagnetic current and charge operators. Here we consider only the current operator. A recent review of also the electromagnetic charge operator can be found in Ref. [32]. The nuclear electromagnetic current operator is constructed so as to satisfy the current conservation relation (CCR) exactly with the adopted Hamiltonian. The one-body term, the so-called impulse approximation (IA), obtained performing a $1/m$ expansion (m is the nucleon mass) of the single-nucleon covariant current, satisfies CCR with the kinetic energy operator. The two-body electromagnetic current consists of two contributions, the so-

IOP Conf. Series: Journal of Physics: Conf. Series **981** (2018) 012019 doi:10.1088/1742-6596/981/1/012019

called model-independent (MI) and model-dependent (MD) terms. The MI contributions have been constructed in Ref. [8], where particular care has been put to verify CCR with the AV18 potential. In particular, the two-body currents related to the momentum-space components of the two-body potential (spin-orbit $L \cdot S$, L^2 and $(L \cdot S)^2$ operators) are difficult to treat in the meson-exchange scheme and a new procedure based on minimal substitution has been devised. In addition, in order to satisfy CCR with the "full" Hamiltonian operator, which also includes a three-body interaction term (here the UIX model), three-body electromagnetic currents need to be considered. They have been constructed in Ref. [8] both in the meson-exchange and minimal substitution scheme. The three-body currents, although they have found to give small contribution, are essential to describe correctly those observables sensitive to current conservation, as the T_{20} and T_{21} polarization observables in radiative capture of polarized deuterons on protons. The MD terms of the electromagnetic current are due to the $\rho\pi\gamma$ and $\omega\pi\gamma$ transitions and to the current associated with the excitation of one intermediate Δ resonance. These terms are transverse, and therefore they do not affect CCR. In the calculation for the *pd* radiative capture of Ref. [3], a new one-body term is included in the nuclear electromagnetic current, which is a relativistic correction of the order $1/m^3$ to the leading $1/m$ term. It was first derived in Refs. [17, 16] in the context of χEFT, and it was found in Ref. [33] that it reduces the *nd* radiative capture total cross section at thermal energies of about 4-5 %, bringing the theoretical prediction in a much better agreement with the experimental datum (within 4 %).

3.2. Results for the pd fusion

The astrophysical S-factor for the *pd* radiative capture is crucial to determine the consistency of BBN theoretical prediction for deuterium abundance, the new Planck results, and the most recent experimental determination of such abundance. In the absence of an accurate experimental determination in the energy range of interest for BBN, 30-300 keV, the *pd* reaction has been studied within the conventional approach described above in Ref. [3]. The predicted astrophysical S-factor in the energy range of interest for BBN is shown in figure 2, where it is compared with the previous *ab-initio* calculation of Ref. [8], with the best fit of Ref. [10], and with the available experimental data of Refs. [34, 35, 6, 9]. By inspection of the figure we can conclude that the results of Ref. [3] are systematically larger than those of Ref. [8] as well as the polynomial fit of Ref. [10]. The origin of this difference can be traced back in part to the one-body $1/m^3$ term, responsible for an increase of 1–3 % over the whole energy range, but mostly to the more accurate new solutions for the $A = 3$ scattering problem. The difference with the available experimental data in the BBN region, *i.e.* those of Ref. [6], is even larger, which is the main motivation for a new experimental determination, as the one under progress by the LUNA Collaboration. Finally, the numerical uncertainty relative to the solution of the $A = 3$ scattering problem is lower than 1 %.

The effect of this new *ab-initio* determination of the astrophysical S-factor on the primordial deuterium abundance has been studied in Ref. [3] using the numerical code `PArthENoPE` (see Ref. [36]). In particular, the deuterium to hydrogen density ratio $^2\mathrm{H/H}_{th}$ has been computed as function of two parameters, the baryon density $\Omega_b h^2$ and the effective neutrino number N_{eff}, and has been compared with the experimental determination $^2\mathrm{H/H}_{exp}$ of Ref. [5]. It has been found in Ref. [3] that for the Planck 2015 value of $\Omega_b h^2$ [37], and for standard value for N_{eff}, the ratio between the deuterium and hydrogen abundances is $^2\mathrm{H/H}_{th} = (2.46 \pm 0.03 \pm 0.03) \times 10^{-5}$, where the two errors are due to nuclear rate and $\Omega_b h^2$ uncertainties, respectively. This result is indeed in nice agreement with the experimental value of Ref. [5], $(2.53 \pm 0.04) \times 10^{-5}$. Of course, these results ought to be confirmed by direct measurement of the *pd* S-factor.

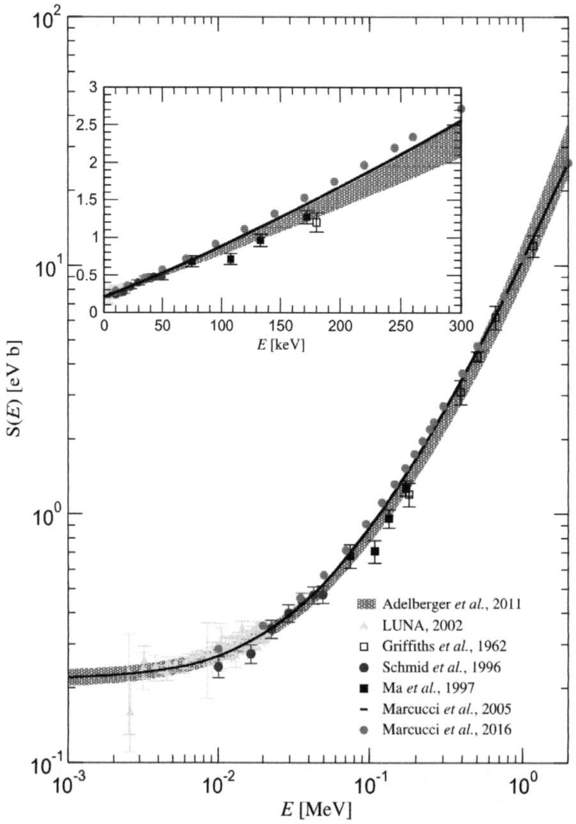

Figure 2. The astrophysical S-factor as calculated in Ref. [3] (red points) is plotted together with the available experimental data of Refs. [34, 35, 6, 9], the calculation of Ref. [8] (solid black line), and the quadratic best fit to the data of Ref. [10] (green band). The inset shows the astrophysical S-factor in the 0-300 keV energy range, since the relevant BBN energy range is 30-300 keV.

4. Summary and Conclusions

The two low-energy reactions $p+p \to d+e^+ +\nu_e$ and $p+d \to^3\text{He}+\gamma$ have been studied within an *ab-initio* approach, the former using χEFT, the latter within the conventional approach. The observable of interest, the astrophysical S-factor, has been calculated, and the implications for the astrophysical environments where these reactions are important have been discussed. In particular, the S-factor for the *pp* fusion is by now known with such an accuracy that it does not represent a source of uncertainty for stellar evolution studies. In fact, the χEFT framework allows for a robust determination of the theoretical uncertainty, quantified in Ref. [2] at the 7–8 ‰ level. The determination for the *pp* S-factor of Ref. [1] implies a change in the central solar temperature of at most 3 ‰, which however reflects to relative differences in the solar neutrino fluxes from the ^{8}B, ^{13}N, ^{15}O and ^{17}F of 2–8 %, depending on which older database is used. To be noticed that the most commonly used database of Ref. [27] is the one which gives rise to the largest differences.

The *pd* fusion has been studied in the energy range relevant for BBN, 30–300 keV, within the conventional approach, since it allows to derive electromagnetic currents which do satisfy *exactly* the current conservation relation. The *pd* S-factor has been found systematically larger than the available experimental data in the BBN range [6]. Furthermore, the S-factor predictions are larger also than the older *ab-initio* determination of Ref. [8] and than the polynomial best fit of Ref. [10]. The origin of this increase respect to the results of Ref. [8] can be mostly traced back to the use of more accurate solutions for the $A = 3$ scattering problem. The primordial deuterium abundance which derives from this new determination of the *pd* S-factor is in very nice agreement with the most recent experimental determination of Ref. [5], when the Planck 2015 value for $\Omega_b h^2$ [37] and the standard value for $N_{\rm eff}$ are used.

Further studies in this sector can move into two directions: first, we would like to apply the χEFT framework also to the electromagnetic processes. This would allow to determine the theoretical uncertainty to be assigned to our predictions. In order to do this, it is necessary to construct a set of χEFT potential and currents, such that the current conservation relation is *exactly* satisfied. A first step is represented by the new set of χEFT potentials derived in Ref. [38] and first used in Ref. [39]. These new potentials include Δ-isobar degrees of freedom in the two-pion exchange term up to N3LO. Furthermore, being minimally non-local and having an operatorial structure similar to the one of the AV18 potential, they are simpler to be used than the other χEFT potentials. Finally, it is of interest to extend the *ab-initio* framework, which seems to be so successful, also to reactions which involve a larger number of nucleons, in particular $A = 4, 5$. Among the reactions which will be studied soon there are the $d + d$ transfer reactions, as well as the $d + d$ radiative capture. These processes are again of interest for BBN, since they are the source at the moment of the largest uncertainties to determine deuterium production and destruction during BBN.

References

[1] Marcucci L E, Schiavilla R, Viviani M 2013 *Phys. Rev. Lett.* **110** 192503
[2] Acharya B, Carlsson B D, Ekström A, Forssèn C, Platter L 2016 *Phys. Lett. B* **760** 584
[3] Marcucci L E, Mangano G, Kievsky A, Viviani M 2016 *Phys. Rev. Lett.* **116** 102501; Erratum *Phys. Rev. Lett.* **117** 049901
[4] Tognelli E, Degl'Innocenti S, Marcucci L E, Prada Moroni P G 2015 *Phys. Lett. B* **742** 189
[5] Cooke R *et al.* 2014 *Astrophys. J.* **781** 31
[6] Ma L *et al.* 1997 *Phys. Rev. C* **55** 588
[7] Viviani M *et al.* 2000 *Phys. Rev.* C **61** 064001
[8] Marcucci L E, Viviani M, Schiavilla R, Kievsky A, Rosati S 2005 *Phys. Rev. C* **72** 014001
[9] Casella C *et al.* (LUNA Collaboration) 2002 *Nucl. Phys. A* **706** 203
[10] Adelberger E G *et al.* 2011 *Rev. Mod. Phys.* **83** 195
[11] Entem D R and Machleidt R 2003 *Phys. Rev. C* **68** 041001
[12] Navràtil P 2007 *Few-Body Syst.* **41** 117
[13] Wiringa R B, Stoks V G J, and Schiavilla R 1995 *Phys. Rev. C* **51** 38
[14] Park T S*et al.* 2003 *Phys. Rev. C* **67** 055206
[15] Marcucci L E *et al.* 2012 *Phys. Rev. Lett.* **108** 052502
[16] Song Y-H, Lazauskas R, Park T-S 2009 *Phys. Rev. C* **79** 064002
[17] Pastore S *et al.* 2009 *Phys. Rev. C* **80** 034004; Pastore S *et al.* 2011 *Phys. Rev. C* **84** 024001
[18] Kölling S *et al.* 2009 *Phys. Rev. C* **80** 045502; K¨olling S*et al.* 2011 *Phys. Rev. C* **84** 054008
[19] Baroni A, Girlanda L, Pastore S, Schiavilla R, Viviani M 2016 *Phys. Rev. C* **93** 015501; Erratum in *Phys. Rev. C* **93** 049902
[20] Baroni A, Girlanda L, Kievsky A, Marcucci LE, Schiavilla R, Viviani M 2016 *Phys. Rev. C* **94** 024003
[21] Gårdestig A and Phillips D R 2006 *Phys. Rev. Lett.* **96** 232301
[22] Gazit D, Quaglioni S, Navratil P 2009 *Phys. Rev. Lett.* **103** 102502
[23] Viviani M *et al.* 2013 *Few-Body Syst.* **54** 885; Viviani M *et al.* 2013 *Phys. Rev. Lett.* **111** 172302
[24] Coraggio L *et al.* 2014 *Phys. Rev. C* **89** 044321
[25] Sammarruca F *et al.* 2015 *Phys. Rev. C* **91** 054311
[26] Schiavilla R *et al.* 1998 *Phys. Rev. C* **58** 1263
[27] Angulo C *et al.* 1999 *Nucl. Phys. A* **656** 3
[28] Cyburt R H *et al.* 2010 *Astrophys. J. Suppl. Ser.* **189** 240
[29] Pudliner B S, Pandharipande V R, Carlson J, Pieper S C, Wiringa R B 1997 *Phys. Rev. C* **56** 1720
[30] Kievsky A, Rosati S, Viviani M, Marcucci L E, Girlanda L 2008 *J. Phys. G* **35** 063101
[31] Marcucci L E, Kievsky A, Girlanda L, Rosati S, Viviani M 2009 *Phys. Rev. C* **80** 034003

IOP Conf. Series: Journal of Physics: Conf. Series **981** (2018) 012019 doi:10.1088/1742-6596/981/1/012019

[32] Marcucci L E *et al.* 2016 *J. Phys. G* **43** 023002
[33] Girlanda L *et al.* 2010 *Phys. Rev. Lett.* **105** 232502
[34] Griffiths G M, Lal M, Scarfe C D 1962 *Can. J. Phys.* **41** 724
[35] Schmid G J *et al.* 1996 *Phys. Rev. Lett.* **76** 3088
[36] Pisanti O *et al.* 2008 *Comput. Phys. Comm.* **178** 956
[37] Ade P A R *et al.* (Planck Collaboration) 2016 *A & A* **594** A13
[38] Piarulli M *et al.* 2015 *Phys. Rev. C* **91** 024003
[39] Piarulli M *et al.* 2016 *Phys. Rev. C* **94** 054007

IOP Conf. Series: Journal of Physics: Conf. Series **981** (2018) 012020 doi:10.1088/1742-6596/981/1/012020

Numerical studies of the Bethe–Salpeter equation for a two-fermion bound state

W. de Paula[a], **T. Frederico**[a], **G. Salmè**[b] **and M. Viviani**[c]

[a] Dep. de Física, Instituto Tecnológico da Aeronáutica, Centro Técnico Aeroespacial, 12.228-900 São José dos Campos, São Paulo, Brazil
[b]Istituto Nazionale di Fisica Nucleare, Sezione di Roma, P.le A. Moro 2, I-00185 Rome, Italy
[c]Istituto Nazionale di Fisica Nucleare, Sezione di Pisa, Largo Pontecorvo 3, 56100, Pisa, Italy

E-mail: `michele.viviani@pi.infn.it`

Abstract. Some recent advances on the solution of the Bethe-Salpeter equation (BSE) for a two-fermion bound system directly in Minkowski space are presented. The calculations are based on the expression of the Bethe-Salpeter amplitude in terms of the so-called Nakanishi integral representation and on the light-front projection (i.e. the integration of the light-front variable $k^- = k^0 - k^3$). The latter technique allows for the analytically exact treatment of the singularities plaguing the two-fermion BSE in Minkowski space. The good agreement observed between our results and those obtained using other existing numerical methods, based on both Minkowski and Euclidean space techniques, fully corroborate our analytical treatment.

1. Introduction

More than half a century ago, in a seminal work [1] Salpeter and Bethe presented a dynamical equation for describing bound systems within the relativistic field theory. In the subsequent years, there has been a large number of applications of their integral equation, but mainly adopting Euclidean variables or effective reduction to a 3D space. More recently, a method based on the so-called Nakanishi integral representation (NIR) of the Bethe-Salpeter (BS) amplitude (see, e.g., Ref. [2] and references therein), has allowed to make substantial steps forward in obtaining accurate numerical solutions of the actual Bethe-Salpeter equation (BSE). With massive-boson exchanges, it has been investigated: (i) two-scalar bound and zero-energy states [3, 4, 5, 6, 7] as well as two-fermion ground states [8, 9], with a ladder kernel, governing, as well-known, the tail of the momentum distributions; (ii) a two-scalar system, with a cross-ladder kernel [10].

In this contribution, we present the formally exact integration of the singularities that prevent a straightforward application of the NIR for solving the two-fermion ladder BSE in Minkowski space, as it was accomplished in the case of two-scalar systems [4, 5, 6, 7]. Then, after exactly transforming BSE in a coupled eigen-equation system, we compare our eigenvalues with both (i) the ones still obtained in Minkowski space [8], but introducing an auxiliary smoothing function, and (ii) outcomes in Euclidean space [11].

This contribution is so organized. In Sect. 2, we recall the general properties of the BS amplitudes as derived from its transformation properties under various symmetries. In Sect. 3, we present the method of the NIR, the light-front projection, and some of the obtained results.

IOP Conf. Series: Journal of Physics: Conf. Series **981** (2018) 012020 doi:10.1088/1742-6596/981/1/012020

Finally, in Sect 4, we report the conclusions and also discuss the perspectives of the present approach.

2. Properties of the Bethe-Salpeter amplitude

In this Section, we discuss about some properties of the BS amplitude for spin 1/2 fermions. To be definite, let us consider these spin 1/2 particles to be quarks. Then, the BS amplitude $\Phi_{\alpha,\beta}(k,p)$ of a bound quark-antiquark system (a meson) is defined to be

$$\Phi_{\alpha,\beta}(k,p) = \int d^4x_1\, d^4x_2 e^{i(\frac{p}{2}+k)x_1+i(\frac{p}{2}-k)x_2}\langle 0|T\{\psi^H_\alpha(x_1)\overline{\psi}^H_\beta(x_2)\}|M\rangle \tag{1}$$

where $|M\rangle$ is the state describing the bound system (the various quantum numbers defining it, as spin, parity, etc., are understood) and $\psi^H_\alpha(x)$ the quark fields in Heisenberg picture. The index α specifies one of the 4 Dirac component of the field, since it describes a spin 1/2 particle. The amplitude $\Phi_{\alpha,\beta}(k,p)$ is then a 4×4 matrix in the space of the Dirac components, and in general it can be decomposed in terms of the elements of the Clifford algebra. However, we have to take into account also the transformation properties of the Dirac fields under Lorentz, parity, and charge conjugation, as well as the quantum numbers of the meson bound state.

Let us consider first parity. Denoting with U_P the unitary operator inducing the parity transformation in the Hilbert space, it is well known that [12]

$$U_P\psi^H_\alpha(x)U_P^\dagger = \overline{\eta}_P(\gamma^0)_{\alpha,\alpha'}\psi^H_{\alpha'}(\tilde{x})\ , \tag{2}$$

where $\overline{\eta}_P$ is the quark intrinsic parity and $\tilde{x}\equiv(t,-\boldsymbol{x})$. Using this property, we find

$$\Phi_{\alpha,\beta}(k,p) = \eta_P(\gamma^0)_{\alpha,\alpha'}\Phi_{\alpha',\beta'}(\tilde{k},\tilde{p})(\gamma^0)_{\beta',\beta} \equiv \eta_P\gamma^0\Phi(\tilde{k},\tilde{p})\gamma^0\ , \tag{3}$$

where η_P is the parity of state $|M\rangle$, namely $U_P|M\rangle = \eta_P|M\rangle$.

Let us consider now the charge conjugation operator U_C. We know that [12]

$$U_C\psi^H_\alpha(x)U_C^\dagger = -i\Big(\overline{\psi}^H(x)\mathcal{C}\Big)^T_\alpha\ , \tag{4}$$

where with X^T we denote the transpose of the quantity X, and $\mathcal{C} = \gamma^0\gamma^2$. Assuming $U_C|M\rangle = \eta_C|M\rangle$, the BS amplitude has to satisfy

$$\Phi_{\alpha,\beta}(k,p) = \eta_C(\mathcal{C})_{\alpha,\alpha'}\Phi_{\alpha',\beta'}(-k,p)(\mathcal{C})_{\beta',\beta}\ . \tag{5}$$

For a $q\bar{q}$ system, charge conjugation is equivalent to the exchange of the two particles. Assuming that the $q\bar{q}$ system is in a state of relative orbital quantum number L and total spin S $(=0,1)$, then the exchange produces a factor $(-)^{L+S}$, therefore $\eta_C = (-)^{L+S}$. For example, for a pion we know that $L=S=0$, therefore $\eta_C=+1$, while for a ρ-meson $L=0$, $S=1$, and consequently $\eta_C=-1$, etc.

Finally, let us discuss the properties of the BS amplitude under an orthochronous Lorentz transformation Λ. Let us denote the operator inducing the Lorentz transformation as U_Λ. First of all,

$$U_\Lambda\psi^H_\alpha(x)U_\Lambda^{-1} = S(\Lambda)_{\alpha,\alpha'}\psi^H_{\alpha'}(\Lambda x)\ , \tag{6}$$

where $S(\Lambda)$ are well known 4×4 matrices which can be written in terms of the γ matrices [12]. The only property we need to know is the following

$$S(\Lambda)\gamma^\mu S(\Lambda)^{-1} = \Lambda^\mu{}_\nu\gamma^\nu\ . \tag{7}$$

Moreover, $U_\Lambda|M, JJ_z\rangle = \mathcal{D}^J_{J'_z,J_z}(W)|M, JJ'_z\rangle$, where W is the Wigner rotation [13] related to the Lorentz transformation Λ and now we have explicitly specified the total angular momentum quantum number J and its z-component of the meson state. The BS amplitude is found to verify

$$\Phi^{J,J_z}_{\alpha,\beta}(k,p) = \mathcal{D}^J_{J'_z,J_z}(W) S(\Lambda)_{\alpha,\alpha'} \Phi^{J,J'_z}_{\alpha',\beta'}(\Lambda k, \Lambda p) S(\Lambda)^{-1}_{\beta',\beta} \ , \tag{8}$$

For example, for a scalar particle $\mathcal{D}^J_{J'_z,J_z}(W) = 1$, and the above expression reads

$$\Phi_{\alpha,\beta}(k,p) = S(\Lambda)_{\alpha,\alpha'} \Phi_{\alpha',\beta'}(\Lambda k, \Lambda p) S(\Lambda)^{-1}_{\beta',\beta} \ , \tag{9}$$

namely $\Phi_{\alpha,\beta}(k,p)$ must be a scalar quantity. The most general expression verifying this condition is [14]

$$\Phi(k,p) = A + B\gamma^5 + C\not{p} + D\not{k} + E\not{p}\gamma^5 + F\not{k}\gamma^5 + Gi\sigma^{\mu\nu}p_\mu k_\nu + Hi\sigma^{\mu\nu}p_\mu k_\nu \gamma^5 \ , \tag{10}$$

where A, B, ... are functions of the scalar quantities p^2, $p \cdot k$, and k^2, namely $A \equiv A(k,p) \equiv A(p^2, p\cdot k, k^2)$, etc. Moreover, if we specialize to the BS amplitude of a pion, for which $\eta_P = -1$ and $\eta_C = +1$, we find that the parity condition (3) is verified only if $A = C = D = G = 0$. The charge conjugation condition (5) imposes

$$\begin{aligned} B(k,p) &= +B(-k,p) \ , \qquad E(k,p) = +E(-k,p) \ , \\ F(k,p) &= -F(-k,p) \ , \qquad H(k,p) = +H(-k,p) \ . \end{aligned} \tag{11}$$

Therefore, the BS amplitude for a scalar particle of negative parity and positive charge conjugation can be written as a sum of four terms,

$$\Phi(k,p) = B\gamma^5 + E\not{p}\gamma^5 + F\not{k}\gamma^5 + Hi\sigma^{\mu\nu}p_\mu k_\nu \gamma^5 \ , \tag{12}$$

with B, E, and H (F) even (odd) under the exchange $k \to -k$.

3. Solution and results

We discuss here the calculation of the BS amplitude for a $q\bar{q}$ system in a $J^\pi = 0^-$, positive charge-conjugation state. As in [8], the BS amplitude is written as

$$\Phi(k,p) = \phi_1(k,p)S_1 + \phi_2(k,p)S_2 + \phi_3(k,p)S_3 + \phi_4(k,p)S_4 \ , \tag{13}$$

where

$$S_1 = \gamma_5 \ , \quad S_2 = \frac{\not{p}}{M}\gamma_5 \ , \quad S_3 = \frac{k\cdot p}{M^3}/p\,\gamma_5 - \frac{1}{M}/k\gamma_5 \ , \quad S_4 = \frac{i}{M^2}\sigma^{\mu\nu}p_\mu k_\nu\,\gamma_5 \ , \tag{14}$$

and $M^2 = p^2$. The amplitudes ϕ_i are again function of p^2, $k \cdot p$, and k^2, and are combinations of the amplitudes B, E, F, and H introduced earlier, see Eq. (12). In particular

$$\phi_1 = B, \qquad \phi_2 = ME + \frac{k\cdot p}{M^2}MF \ , \qquad \phi_3 = -MF \ , \qquad \phi_4 = M^2 H \ . \tag{15}$$

The choice to express the BS amplitude as in Eq. (13) is related to the fact that the quantities S_i are "orthogonal" to each other, i.e. $Tr\left[S_i\ S_j\right] = \mathcal{N}_i(k,p)\ \delta_{ij}$.

In the following, we consider the fermions interacting via the exchange of either a scalar or a pseudoscalar boson of mass μ. The corresponding interaction Lagrangians is $\mathcal{L} = g\bar{\psi}(x)\phi(x)\psi(x)$

in case of a scalar boson and $\mathcal{L} = ig\bar{\psi}(x)\gamma^5\phi(x)\psi(x)$ for a pseudoscalar boson. The BS amplitude in both cases satisfies the following homogeneous BSE [8]

$$\Phi(k,p) = S(k+p/2)\left[\int d^4k'\ F^2(k-k')\ i\mathcal{K}(k,k')\ \Gamma_1\ \Phi(k',p)\ \bar{\Gamma}_2\right] S(k-p/2)\ , \quad (16)$$

where $S(q) = i\frac{\not{q}+m}{q^2-m^2+i\epsilon}$ is the propagator of a particle of spin 1/2 (and mass m) and $\Gamma_1 = \Gamma_2 = 1$ (γ_5) in case of the scalar (pseudoscalar) vertex. Finally, $F(k-k') = (\mu^2-\Lambda^2)/[(k-k')^2-\Lambda^2+i\epsilon]$ is a vertex form factor. The dimensionless coupling constant, g, and the momentum-dependent part of the exchanged-boson propagator are contained in $i\mathcal{K}$. In ladder approximation, in both scalar and pseudoscalar case, the kernel is $\mathcal{K} = \pm\ g^2/[(k-k')^2-\mu^2+i\epsilon]$ (plus for the scalar case and minus for the pseudoscalar case). In Eq. (16), $\bar{\Gamma}_2 = -\mathcal{C}\ \Gamma_2^T\ \mathcal{C}^{-1}$, where $\mathcal{C} = \gamma^0\gamma^2$ is the charge conjugation matrix defined above.

Inserting Eq.(13) in (16) and taking the trace, a system of four coupled integral equations are obtained:

$$\phi_i(k,p) = ig^2\sum_j \int \frac{d^4k''}{(2\pi)^4}\frac{\phi_j(k'',p)}{(k-k'')^2-\mu^2+i\epsilon}\frac{c_{ij}(k,k'',p)\ F^2(k-k'')}{\left[(\frac{p}{2}+k)^2-m^2+i\epsilon\right]\left[(\frac{p}{2}-k)^2-m^2+i\epsilon\right]}\ , \quad (17)$$

with $i,j = 1,\ 2,\ 3,\ 4$. The coefficients

$$c_{ij}(k,k'',p) = \frac{1}{\mathcal{N}_i(k,p)}Tr\left\{S_i(k)\left(\frac{\not{p}}{2}+\not{k}+m\right)\Gamma_1\ S_j(k'')\bar{\Gamma}_2\left(\frac{\not{p}}{2}-\not{k}-m\right)\right\}\ , \quad (18)$$

are explicitly given in Ref. [8], for both couplings. Notably, the numerator of each $c_{ij}(k,k'',p)$ can contain the third power of the four-momentum k, at the most.

In complete analogy with the two-scalar interacting system, where only one amplitude is present [3, 4, 5, 6, 7, 8, 10], one can introduce the NIR for each amplitudes ϕ_i, viz

$$\phi_i(k,p) = \int_{-1}^{1} dz' \int_0^\infty d\gamma' \frac{g_i(\gamma',z')}{[k^2+z'p\cdot k-\gamma'-\kappa^2+i\epsilon]^3}\ , \quad (19)$$

where $\kappa^2 = m^2 - M^2/4$ and $g_i(\gamma',z')$ are unknown real functions, called the Nakanishi weight functions, to be numerically determined through the solutions of the eigen-problem, formally generated after inserting the above NIR in the BSE. Another ingredient that greatly facilitates the numerical solution of the BSE is obtained by introducing the light-front (LF) components for the involved momenta, i.e. $k^\pm = k^0 \pm k^3$ and $\mathbf{k}_\perp$ [2, 5, 6, 7, 9], and integrating both sides of Eq. (17) over k^-.

First of all, let us consider the integration of the LHS. One gets, see Ref.[9] (in preparation) for details:

$$LHS = \int \frac{dk^-}{2\pi}\phi_i(k,p) = \frac{i}{M}\int_0^\infty d\gamma'\ \frac{g_i(\gamma',z)}{\left(\gamma+\gamma'+m^2z^2+\kappa^2(1-z^2)-i\epsilon\right)^2}\ , \quad (20)$$

where $p^\mu = (M,0,0,0)$ (we put ourselves in the CM), and

$$\gamma = (k_\perp)^2\ , \qquad z = -\frac{2}{M}k^+\ . \quad (21)$$

Regarding the RHS, first we use the generalized Feynmann trick

$$\frac{1}{A^m\ B^n\ C^\ell} = \frac{\Gamma(m+n+\ell)}{\Gamma(m)\Gamma(n)\Gamma(\ell)}\int_0^1 dv \int_0^1 dw\ \frac{v^{m-1}w^{n-1}(1-w-v)^{\ell-1}\ \theta(1-v-w)}{\left[Av+Bw+C(1-w-v)\right]^{m+n+\ell}}\ ,(22)$$

then we shift the integration over $k'' \to k'' + k(1-v) + vpz'/2$. Now it turns out that all $c_{ij}(k, k'', p)$ are either independent on k'' or linearly depending on it. Therefore,

$$c_{ij}(k, k'', p) = \bar{c}_{ij}(k, p) + \tilde{c}_{ij}(k, p)_\mu (k'')^\mu \to \bar{c}_{ij}(k, p) + \tilde{c}_{ij}(k, p)_\mu \Big(k'' + k(1-v) + \frac{vpz'}{2}\Big)^\mu \quad (23)$$

Integrating over k'', the term linear in k'' gives a vanishing contribution. Moreover it can be proved that $\tilde{c}_{ij}(k, p)_\mu p^\mu = 0$. Therefore, in practice we can assume that $c_{ij}(k, k'', p) \equiv c_{ij}(k, k(1-v), p)$. After the integration over k'', the RHS reads

$$\begin{aligned} RHS &= \sum_j \int_{-1}^{+1} dz' \int_0^\infty d\gamma' \, g_j(\gamma', z') \frac{-3g^2(\Lambda^2 - \mu^2)^2}{(4\pi)^2} \frac{1}{\Big[(\frac{p}{2} + k)^2 - m^2 + i\epsilon\Big]} \\ & \frac{1}{\Big[(\frac{p}{2} - k)^2 - m^2 + i\epsilon\Big]} \int \frac{dk^-}{2\pi} \int_0^1 dv\, v^2 \int_0^{1-v} dw\, w \frac{c_{ij}(k, k(1-v), p)}{\Big(\beta k^- + A + i\epsilon\Big)^4}, \end{aligned} \quad (24)$$

where $\beta = v(1-v)\frac{M}{2}(z' - z)$ and

$$A = -v(1-v)\gamma - v^2 \frac{M^2}{4}(z')^2 - v(1-v)\frac{M^2}{4} z z' - \mu^2(1-v) - v(\kappa^2 + \gamma') - w(\Lambda^2 - \mu^2) \,. \quad (25)$$

Moreover,

$$\Big(\frac{p}{2} + k\Big)^2 - m^2 + i\epsilon = \frac{M}{2}(1-z)(k^- - k_L^-) \,, \quad k_L^- = \frac{\gamma + m^2 - \frac{M^2}{4}(1-z) - i\epsilon}{\frac{M}{2}(1-z)} \,, \quad (26)$$

$$\Big(\frac{p}{2} - k\Big)^2 - m^2 + i\epsilon = -\frac{M}{2}(1+z)(k^- - k_U^-) \,, \quad k_U^- = \frac{-\gamma - m^2 + \frac{M^2}{4}(1+z) + i\epsilon}{\frac{M}{2}(1+z)} \,, \quad (27)$$

with $\Im(k_U^-) > 0$ and $\Im(k_L^-) < 0$, and we can write in general

$$c_{ij}(k, k(1-v), p) = \sum_{n=0}^{3} c_{ij}^{(n)}(v, \gamma, z)(k^-)^n \,. \quad (28)$$

At this point, we perform the integration over k^-, which has to be carefully done due to the extra powers of k^- in the numerator. In Ref. [8], a numerical treatment has been employed, i.e. by multiplying both sides of the BSE equation by a non-negative smoothing function of k^- before performing the integration. Here, following Ref. [9], we present an alternative method, where the singular terms arising from the direct integration over k^- are correctly taken into account. This can be achieved by using the following results (here $n = 0, 1, 2, 3$),

$$\begin{aligned} & \int \frac{dk^-}{2\pi} \frac{1}{k^- - k_U^-} \frac{1}{k^- - k_L^-} \frac{(k^-)^n}{(\beta k^- + A + i\epsilon)^4} = \\ &= \theta(\beta) \frac{i}{k_U^- - k_L^-} \frac{(k_U^-)^n}{(\beta k_U^- + A + i\epsilon)^4} + \theta(-\beta) \frac{i}{k_U^- - k_L^-} \frac{(k_L^-)^n}{(\beta k_L^- + A + i\epsilon)^4} \\ & \quad - \frac{i}{3}\delta(\beta) \frac{(k_U^-)^{n-1} - (k_L^-)^{n-1}}{k_U^- - k_L^-} \frac{1}{(A + i\epsilon)^3} \delta_{n\geq 2} \\ & \quad - \frac{i}{6} \frac{\delta(\beta)}{\beta} \frac{(k_U^-)^{n-2} - (k_L^-)^{n-2}}{k_U^- - k_L^-} \frac{1}{(A + i\epsilon)^2} \delta_{n\geq 3} \,, \end{aligned} \quad (29)$$

where $\delta_{n\geq m} = 1$ for $n \geq m$ and 0 otherwise, obtaining

$$\begin{aligned}
RHS &= \sum_j \int_{-1}^{+1} dz' \int_0^\infty d\gamma'\, g_j(\gamma', z') \frac{-3ig^2(\Lambda^2-\mu^2)^2}{(4\pi)^2} \int_0^1 dv\, v^2(1-v)^2 \\
&\Bigg\{ \theta(z'-z)(1+z)^4 \frac{\sum_{n=0}^3 c_{ij}^{(n)}(v,\gamma,z)(k_U^-)^n}{MD_0} \frac{3D_- + (1-v)(1+z)(\Lambda^2-\mu^2)}{6D_-^2\Big(D_- + (1-v)(1+z)(\Lambda^2-\mu^2)\Big)^3} \\
&+\theta(z-z')(1-z)^4 \frac{\sum_{n=0}^3 c_{ij}^{(n)}(v,\gamma,z)(k_L^-)^n}{MD_0} \frac{3D_+ + (1-v)(1-z)(\Lambda^2-\mu^2)}{6D_+^2\Big(D_+ + (1-v)(1-z)(\Lambda^2-\mu^2)\Big)^3} \\
&+\frac{\delta(z'-z)}{\frac{M}{2}v(1-v)} \frac{c_{ij}^{(2)}(v,\gamma,z) + c_{ij}^{(3)}(v,\gamma,z)(k_U^- + k_L^-)}{6\widetilde{D}\Big(\widetilde{D} + (1-v)(\Lambda^2-\mu^2)\Big)^2} \frac{1}{-\frac{M^2}{4}(1-z^2)} \\
&+\frac{\frac{d\delta(z'-z)}{dz'}}{\Big[\frac{M}{2}v(1-v)\Big]^2} \frac{c_{ij}^{(3)}(v,\gamma,z)}{6(\Lambda^2-\mu^2)^2} \frac{1}{-\frac{M^2}{4}(1-z^2)} \frac{1}{(1-v)^2} \left[\ln\left(1+\frac{\widetilde{B}}{\widetilde{A}}\right) - \frac{\widetilde{B}}{\widetilde{A}+\widetilde{B}}\right]\Bigg\}, \quad (30)
\end{aligned}$$

where we have used the relation $\delta(\beta)/\beta = -d\delta(\beta)/d\beta$. Moreover, $\widetilde{B} = (1-v)(\Lambda^2-\mu^2)$,

$$\begin{aligned}
D_0 &= \gamma + m^2z^2 + \kappa^2(1-z^2)\,, & (31)\\
D_+ &= v(1-v)(1-z')\gamma + vm^2\left[(1-v)(1-z')z^2 + vz'^2(1-z)\right] \\
&+ v\kappa^2(1-z)(1-z')\left[1+z-v(z-z')\right] + (1-z)[(1-v)\mu^2 + v\gamma']\,, & (32)\\
\widetilde{A} &= v(1-v)\gamma + v^2\frac{M^2}{4}(z')^2 + v(1-v)\frac{M^2}{4}zz' + \mu^2(1-v) + v(\kappa^2+\gamma')\,, & (33)\\
\widetilde{D} &= v(1-v)\gamma + v\gamma' + m^2vz^2 + \kappa^2v(1-z^2) + \mu^2(1-v)\,, & (34)
\end{aligned}$$

and $D_- = D_+(\gamma, -z, \gamma', -z', v)$. The final expression for the coupled equations is:

$$\begin{aligned}
&D_0 \int_0^\infty d\gamma' \frac{g_i(\gamma', z)}{\Big(\gamma + \gamma' + m^2z^2 + \kappa^2(1-z^2) - i\epsilon\Big)^2} \\
&= -\frac{g^2}{4\pi}\frac{m^2(\Lambda^2-\mu^2)^2}{2\pi} \sum_j \int_{-1}^{+1} dz' \int_0^\infty d\gamma'\, g_j(\gamma', z') \int_0^1 dv\, v^2(1-v)^2 \\
&\Bigg\{\theta(\beta)(1+z)^4 \frac{\sum_{n=0}^3 c_{ij}^{(n)}(v,\gamma,z)(k_U^-)^n}{4m^2} \frac{3D_- + (1-v)(1+z)(\Lambda^2-\mu^2)}{D_-^2\Big(D_- + (1-v)(1+z)(\Lambda^2-\mu^2)\Big)^3} \\
&+\theta(-\beta)(1-z)^4 \frac{\sum_{n=0}^3 c_{ij}^{(n)}(v,\gamma,z)(k_L^-)^n}{4m^2} \frac{3D_+ + (1-v)(1-z)(\Lambda^2-\mu^2)}{D_+^2\Big(D_+ + (1-v)(1-z)(\Lambda^2-\mu^2)\Big)^3} \\
&- 2\frac{\delta(z'-z)}{v(1-v)} \frac{D_0}{m^2M^2(1-z^2)} \frac{c_{ij}^{(2)}(v,\gamma,z) + c_{ij}^{(3)}(v,\gamma,z)(k_U^- + k_L^-)}{\widetilde{D}\Big(\widetilde{D} + (1-v)(\Lambda^2-\mu^2)\Big)^2} \\
&-4\frac{\frac{d\delta(z'-z)}{dz'}}{v^2(1-v)^4} \frac{D_0}{m^2M^3(1-z^2)} \frac{c_{ij}^{(3)}(v,\gamma,z)}{(\Lambda^2-\mu^2)^2} \left[\ln\left(1+\frac{\widetilde{B}}{\widetilde{A}}\right) - \frac{\widetilde{B}}{\widetilde{A}+\widetilde{B}}\right]\Bigg\}\,. \quad (35)
\end{aligned}$$

The last term can be integrated by parts taking into account that $g_j(\gamma', z' = \pm 1) = 0$, obtaining a term depending on dg_j/dz'. It is easy to prove that $v \to 1$ the integrands are not singular. The

Table 1. (After Ref. [9]) The squared scalar coupling constant vs the binding energy for two masses of the exchanged particle $\mu/m = 0.15$ and $\mu/m = 0.50$. First column: binding energy. Second column: coupling constant g^2 for $\mu/m = 0.15$, obtained in Ref. [9] using the method discussed in this contribution. Third column: results for $\mu/m = 0.15$, from Ref. [8] with a numerical treatment of the singularities. Fourth column: the same as the second one, but for $\mu/m = 0.50$. Fifth column: the same as the third one, but for $\mu/m = 0.50$. Sixth column: results in Euclidean space from Ref. [11]. In the vertex form factor it is taken $\Lambda = 2$, as in [8] and [11].

	$\mu/m = 0.15$		$\mu/m = 0.50$		
B_e/m	g^2_{dPFSV}	g^2_{CK}	g^2_{dPFSV}	g^2_{CK}	g^2_E
0.01	7.844	7.813	25.327	25.23	-
0.02	10.040	10.05	29.487	29.49	-
0.04	13.675	13.69	36.183	36.19	36.19
0.05	15.336	15.35	39.178	39.19	39.18
0.10	23.122	23.12	52.817	52.82	-
0.20	38.324	38.32	78.259	78.25	-
0.40	71.060	71.07	130.177	130.7	130.3
0.50	88.964	86.95	157.419	157.4	157.5
1.00	187.855	-	295.61	-	-
1.40	254.483	-	379.48	-	-
1.80	288.31	-	421.05	-	-

coupled equations reported above correctly take into account of the singular behavior related to the k^- integration.

The solution of the coupled integral equations (35) is obtained in our case by expanding the Nakanishi weight functions $g_i(\gamma', z')$ on a suitable basis. As in Ref. [5] for two-scalar bound states, we have chosen a basis composed by Laguerre and Gegenbauer polynomials (with the needed weights). The derivative dg_j/dz' is easily obtained in terms of derivatives of the Gegenbauer polynomials. In our basis, we have up to 44 Laguerre polynomials (with the same parameters as in Ref. [5]) and 44 Gegenbauer ones. In the studies of BSE, it is customary to assign a value to the binding energy $B_e = 2m - M$, and, in correspondence, look for an eigenvalue g^2. Then, an eigen-problem of the type $\mathcal{B}\ G = g^2\ \mathcal{A}\ G$, (with $\mathcal{B}$ and $\mathcal{A}$ suitable matrices and G a vector of coefficients to be determined) is obtained.

If the eigenvalue exists then the whole procedure is validated. Tables 1 (scalar coupling) and 2 (pseudoscalar coupling) show the comparison between the values of g^2 obtained within our approach, where the singularities have been singled out and analytically evaluated, and both (i) the calculations by Ref. [8], where a non trivial numerical treatment of the singular behaviors was introduced (without recognizing the possibility of a systematic analysis of the singularities as in [15]) and (ii) the available numerical results in Euclidean space [11], with a suitable number of digits. Notably, we were also able to extend our calculation up to $B_e/m \sim 2$, namely when the expected critical behavior of a ϕ^3 theory manifests itself [16].

4. Conclusion and perspectives

The achieved final agreement, within the adopted numerical accuracy, strongly supports the validity of our analytical method for treating the singularities that plague the ladder BSE, when an interacting two-fermion system is considered. The most severe singularity is met when the third power of k^- appears in the numerator of the kernel in Eq. (24). The powers of k^- are generated only by the external propagators and the structure of the BS amplitude, present in

Table 2. (After Ref. [9]) The same as in Table 1, but for a pseudoscalar coupling.

	$\mu/m = 0.15$		$\mu/m = 0.50$	
B_e/m	g^2_{dPFSV}	g^2_{CK}	g^2_{dPFSV}	g^2_{CK}
0.01	225.7	224.8	422.6	422.3
0.02	233.2	232.9	430.5	430.1
0.04	243.1	243.1	440.9	440.4
0.05	247.1	247.0	444.9	444.3
0.10	262.1	262.1	460.4	459.9
0.20	282.9	282.9	482.1	480.7
0.40	311.7	311.8	513.3	515.2
0.50	322.9	323.1	525.8	525.9
1.00	362.3	-	570.9	-
1.40	380.1	-	591.8	-
1.80	388.7	-	602.1	-

the RHS of (16). For instance, in the case of a two-vector system, this simple counting rule leads to expect derivatives of the Dirac delta-function not too high (≥ 2, depending only upon the complexity of the BS amplitude, like in Eq. (13)), and therefore still manageable within our approach.

The robustness of the technique based on NIR for solving the BSE with spin degrees of freedom encourages to extend this novel tool to many areas, since old limitations constraining the calculations to an unphysical space can be removed. The approach can deal with further dynamical effects, since the analytical structure of BS kernels, truncated at any power of the coupling constant, is made explicit as in the ladder case (see, e.g., [2] for the half-off-shell T-matrix), allowing the LF projection.

Acknowledgments

WdP and TF thank the partial support from Conselho Nacional de Desenvolvimento Científico e Tecnológico (CNPq) and Coordenação de Aperfeiçoamento de Pessoal de Nível Superior (CAPES) of Brazil.

References

[1] Salpeter EE and Bethe HA 1951 *Phys. Rev.* **84** 1232
[2] Frederico T, Salmè G, and Viviani M 2012 *Phys. Rev.* D **85** 036009
[3] Kusaka K, Simpson K, and Williams AG 1997 *Phys. Rev.* D **56** 5071
[4] Karmanov VA and Carbonell J 2006 *Eur. Phys. Jou.* A **27** 1
[5] Frederico T, Salmè G, and Viviani M 2014 *Phys. Rev.* D **89** 016010
[6] Gutierrez C, Gigante V, Frederico T, Salmè G, Viviani M, and Tomio L 2016 *Phys. Lett.* B **759**, 131
[7] Frederico T, Salmè G, and Viviani M 2015 *Eur. Phys. J.* C **75** 398
[8] Carbonell J and Karmanov VA 2010 *Eur. Phys. J.* A **46** 387
[9] de Paula W, Frederico T, Salmè G, and Viviani M 2016 *Phys. Rev.* D **94** 071901(R)
[10] Carbonell J and Karmanov VA 2006, *Eur. Phys. J.* A **27** 11
[11] Dorkin SM, Beyer M, Semikh SS, and Kaptari LP 2008 *Few-Body Syst.* **42** 1
[12] Peskin ME and Schroeder DV 1995 *An Introduction To Quantum Field Theory* (Addison-Wesley Publishing Company)
[13] Weinberg S 1995 *The Quantum Theory of Fields I* (Cambridge University Press)
[14] Llewellyn Smith CH 1969 *Ann. Phys.* **53** 521
[15] Yan TM 1973 *Phys. Rev.* D **7** 1780
[16] Baym G 1960 *Phys. Rev.* **117** 886
[17] Ji X, Ma JP, and Yuan F 2003 *Phys. Rev. Lett.* **90** 241601

IOP Conf. Series: Journal of Physics: Conf. Series **981** (2018) 012021 doi:10.1088/1742-6596/981/1/012021

1/2 spin-isospin fermions close to the unitary limit

A. Kievsky[1] and M. Gattobigio[2]

[1]Istituto Nazionale di Fisica Nucleare, Largo Pontecorvo 3, 56127 Pisa, Italy
[2] Université de Nice-Sophia Antipolis, Institut Non-Linéaire de Nice, CNRS, 1361 route des Lucioles, 06560 Valbonne, France

E-mail: alejandro.kievsky@pi.infn.it

E-mail: mario.gattobigio@inln.cnrs.fr

Abstract. The equal mass three-fermion system having 1/2 spin-isospin symmetry is study around the unitary limit. The two body system has two different scattering lengths, a_0 and a_1, corresponding to the spin singlet state $S = 0$ and the spin triplet state $S = 1$, respectively. The unitary limit is defined when the two quantities $a_0, a_1 \to \infty$. The three-nucleon system is located very close to this limit, the singlet and triplet $n - p$ scattering lengths are large with respect to the range of the nuclear interaction. The ratio of the two is about $a_0/a_1 \approx -4.31$. This value defines a plane in which a_0 and a_1 can be varied. Using a nucleon-nucleon spin dependent potential with variable strength it is possible to study the behavior of the three-nucleon binding energy along that plane. This analysis can be considered an extension of the Efimov plot for three bosons to the case of three 1/2-spin-isospin fermions.

1. Introduction

Weakly bound states have always attracted some attention because the particles have a large probability to be outside the interaction range and the system has universal characteristics. The limiting case is the unitary limit in which the two-body scattering length diverges, $1/a = 0$. In this case the two-body system is resonant and can be described reasonable well in the zero-range approximation. In a series of papers [1, 2] V. Efimov has shown that a system of three identical bosons interacting through a two-body short-range potential has a geometrical series of energy levels at the unitary limit accumulating at zero energy. The ratio between the energies of two consecutive states is constant and, remarkably it results independent of the particular form of the interaction. This behavior has been denoted Efimov effect and its universal characteristic has given to this effect a particular relevance. In fact in the last two decades an enormous amount of work has been dedicated to study this effect in different fields as molecular, atomic, nuclear and particle physics.

The spectrum of the three-boson system close to the unitary limit, in the zero-range limit, is described by the Efimov radial law

$$\begin{aligned} E_3^n/(\hbar^2/ma^2) &= \tan^2\xi \\ \kappa_* a &= \mathrm{e}^{(n-n^*)\pi/s_0}\,\frac{\mathrm{e}^{-\Delta(\xi)/2s_0}}{\cos\xi}\,, \end{aligned} \tag{1}$$

where $\Delta(\xi)$ is an universal function. A parametrization of this function in the range $-\pi < \xi < -\pi/4$ can be found in Ref. [3]. The quantity $s_0 \approx 1.00624$ is a universal number and κ_*

defines the energy $\hbar^2\kappa_*^2/m$ of the level $n = n^*$ at the unitary limit. Knowing the value of κ_* the complete spectrum is determined as a function of a. The above equations have been derived in the zero-range limit in which the two-body energy is $E_2 = \hbar^2/ma^2$. The three-body sector shows a discrete scale invariance (DSI), the two-body low energy observables can be written in terms of a, making this quantity a control parameter (see for instance Ref. [3] and references therein). When the two-body potential has a finite range, the Efimov radial law can be modified as it was discussed in Refs. [4, 5]. In the following the case in which the particles are 1/2-spin-isospin fermions [6] is discussed.

2. Effective description of 1/2-spin-isospin fermions close to the unitary limit

The zero-range energy spectrum given by Eq.(1) is unbounded from below as has been shown by Thomas [7]. Finite-range potentials cure this pathology limiting the energy spectrum of the three-body system from below being the lowest state the ground state of the system. For values of a much bigger than the range of the interaction the three-body energy spectrum has many excited state with this number diverging as $a \to \infty$. Eq.(1) is still valid for the highest levels, however the ground and first excited levels show significant finite-range effects. In order to take into account these corrections the Efimov radial law is modified as follow [4, 5],

$$E_3^n/E_2 = \tan^2\xi$$
$$\kappa_n^3 a_B + \Gamma_n^3 = \frac{\mathrm{e}^{-\Delta(\xi)/2s_0}}{\cos\xi}\,. \tag{2}$$

Two modifications have been introduced, the first one takes into account the fact that in the case of finite range interactions the two-body scattering length a and the energy length, a_B, defined by $E_2 = \hbar^2/ma_B^2$, with E_2 the two-body binding energy if $a > 0$, or the two-body virtual-state energy in the opposite case, $a < 0$, are slightly different. The second modification consists in the introduction of a finite-range parameter, the shift Γ_n^3, depending on the energy level. Recently it has been shown that the value of the shift is almost the same for very different potentials as the atomic helium-helium LM2M2 interaction of Aziz [8] and a two-parameter gaussian potential in which the strength and range have been fixed to reproduce the values of E_2 and a given by that interaction. Other examples are a combination of yukawians, as the MT-III interaction [9], describing the s-wave triplet $n-p$ state or a nonlocal gaussian potential. Interestingly the value of the shift for the ground state in the three cases results to be almost the same, $\Gamma_0^3 \approx 0.8$ [10, 11].

Following those findings we would like to discuss the case of fermions having 1/2 isospin symmetry in which the interaction is different in the singlet and triple state as is the case of the two-nucleon system. The experimental data for the zero energy $p-n$ system are $a_0 = -23.740 \pm 0.020$ fm and $a_1 = 5.424 \pm 0.003$ fm and the respective effective range values are $r_{eff}^0 = 2.77 \pm 0.05$ fm and $r_{eff}^1 = 1.753 \pm 0.008$ fm (as reported in Ref. [12]). To describe these data it is possible to use a gaussian potential

$$V(r) = V_S\mathrm{e}^{-r^2/r_S^2}\,, \tag{3}$$

with different strength in the singlet ($S = 0$) and triplet ($S = 1$) states. To simplify the description, the same range (in the following we consider $r_0 = r_1 = 1.65$ fm) is considered. The values $V_0 = -37.90$ MeV and $V_1 = -60.575$ MeV approximate well the experimental data and, for $S = 1$, the deuteron binding energy is well described too. Varying the strengths V_S it is possible to explore the three-nucleon binding energy as the values of a_S move toward the unitary limit. This study is done fixing the ratio $a_0/a_1 = -4.31$ to its experimental value, accordingly both strengths are related. It should be noticed that there are other possibilities to make this study. For example, the ratio $a_0/a_1 = 1$ corresponds to explore the three-boson system. In Fig.1 selected planes (the three-body energy defines the third axis not shown in the figure) are

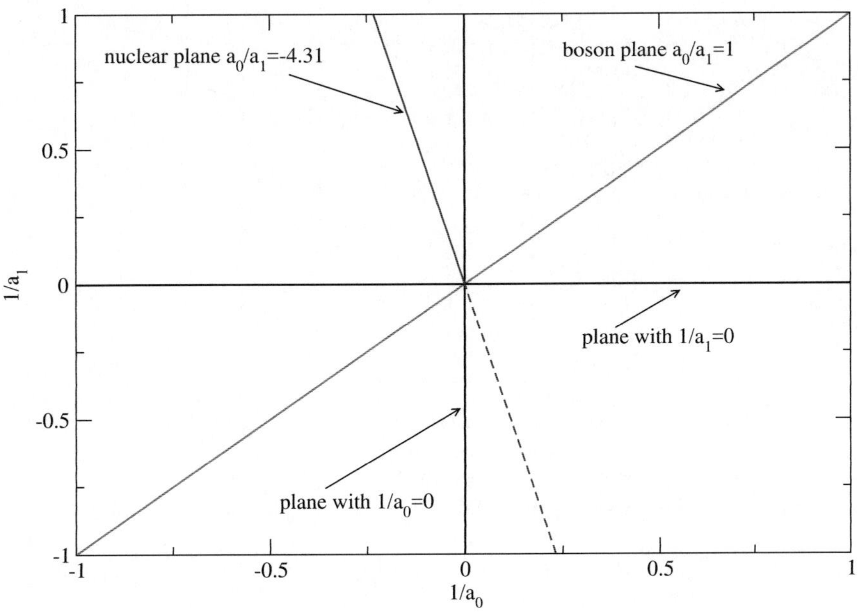

Figure 1. Selected planes to study the three-body spectrum as function of the inverse of the a_0 and a_1 scattering lengths.

shown determined by the values of a_0 and a_1. As mentioned, when the two scattering lengths are equal, $a_0/a_1 = 1$, the corresponding plane is the boson plane for three equal particles very well studied (see for instance Ref. [4]). In the figure this is indicated by the (red) solid line. The (blue) solid line corresponds to a plane in which the nuclear physical point is included, this plane is called the nuclear plane. The continuation to the fourth quadrant given by the (blue) dashed line does not give new information since in this case the singlet and triplet potentials are exchanged and the system has the same binding energy. The two axes correspond to planes in which the singlet or triplet scattering lengths are on the unitary limit. The energy values E_3^n could be different in the different planes, however when the two scattering lengths are at the unitary limit the energy value is unique and it is that of the boson spectrum.

3. Three-body spectrum of 1/2-spin-isospin fermions close to the unitary limit

The spin-dependent potential defined in Eq.(3) can be used to explore the spectrum of the three-nucleon system close to the unitary limit. With the strengths and range defined in the previous section the low energy two-body data are reasonable well described. Extending the analysis to the three-body case a binding energy of $E_3 = -10.2$ MeV is obtained to be compared to the experimental value of the triton $E(^3\text{H}) = -8.48$ MeV. The observed over binding is common characteristic when attractive gaussian potentials are used. In order to have a quantitative agreement with the experimental value a three-body force has to be included (see for instance Ref. [13]). Here we are interested to study the variation of the three-body binding energy as a function of a_1 with the ratio a_0/a_1 fixed to the nuclear value, thus we do not include such a three-body force.

The results of the present study are given in terms of the binding momentum, defined from the relation $(\hbar^2/m)(K_3^n)^2 = E_3^n$, with $n = 0, 1$ for the ground and first excited state respectively. In Fig.2 the binding momentum, K_3^n, is shown as a function of a_1 (in units of r_*, the value of the effective range at the unitary limit) stressing that we have fixed the ratio a_0/a_1 around -4.31. The (blue) solid line describes the ground state whereas the (blue) dot-dashed line describes the

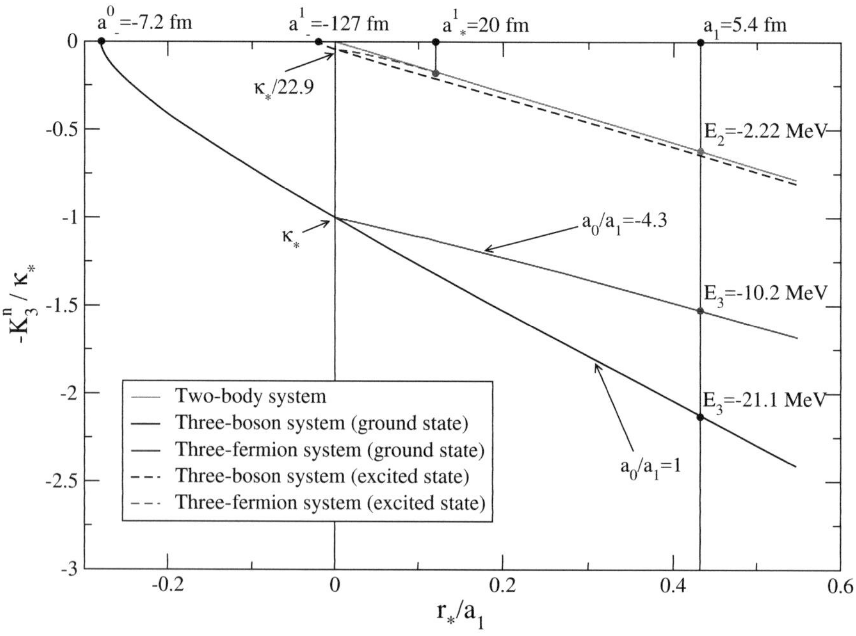

Figure 2. The binding momentum K_3^n for the ground state ($n = 0$) and first excited state ($n = 1$) as a function of the inverse of the triplet scattering length a_1.

first excited state. For comparison the binding momentum values in the case of three bosons are given too by the (black) solid and dashed lines. The (red) solid line is the two-body biding momentum. The binding momentum K_3^n is shown in units of κ_*, the binding momentum of the ground state at the unitary limit and both, the fermion and boson values, are the same and correspond to a binding energy of $(\hbar^2/m)\kappa_*^2 = 3.6$ MeV. The excited state in this limit has a binding momentum equal to $K_*/22.9$ showing a slight finite-range correction (the zero-range theory predicts the value $\kappa_*/22.7$). In the figure the (negative) values of the triplet scattering lengths, a_-^0 and a_-^1, at which the three-boson ground and excited states disappear into the three-nucleon continuum are shown. Their ratio of about 17.6 is well below the prediction of 22.7 given by the zero-range theory. In the case of the three-fermion system it is interesting to notice that the first excited state disappears into the $n - d$ continuum at the value $a_*^1 \approx 20$ fm, well before matching the nuclear physical point corresponding to $a_1 = 5.4$ fm). Conversely, in the case of the three-boson system, the excited state remains bound and always below E_2. This fact has been studied before in the case of a system of three helium atoms. The present analysis shows that in the case of three nucleons the difference in the single and triple potential strengths is such that the excited state disappear into the $n - d$ continuum before reaching the physical point. Evidence of the presence of this state embedded in the $n - d$ continuum has been given in the literature [14, 15] and the curvature of the $n - d$ effective range function close to the threshold is one of this [16, 17]. Finally, at the physical value, the two- and three-body energies are explicitly displayed. As it is well known, in order to reproduce the triton binding energy a three-body force has to be included.

The spectrum of the three-fermion system can be described by Eq.(2) with a value of the shift depending on the ratio a_0/a_1. For the ground state level the numerical results can be described with $\Gamma_3^0 \approx -0.2$, a negative value. If we compare this value to the value of about 0.8 obtained in the case of three bosons we can conclude that there is a particular value of the ratio at which the shift $\Gamma_3^0 \approx 0$. In this case Eq.(2) will be equivalent to the zero-range equation. However this would valid for the ground state level, the shift depends on the level and, as we have shown, the ratio with the excited state at the unitary limit is 22.9 (not 22.7) for a gaussian potential independently of the shift value.

4. Conclusions and Outlook

The present study analyzes the particularities of the Efimov plot in the case of 1/2-spin-isospin fermions. In order to make this study a particular way of constructing the Efimov plot has been selected. The ratio of the singlet and triplet scattering lengths has been kept fixed, accordingly the binding momentum K_3^n and a_1 define a plane similar to the boson case, in particular the values at the unitary limit coincide. The present work has to be consider a first step in the study of the Efimov plot for three nucleons. A deeper analysis including some results in the four-nucleon case and considering a three-body force are given in Ref. [13]. Extensions of the present study consist in the investigation of universal characteristic of the virtual state of the triton and the breakup reaction close to the deuteron threshold [18, 19].

5. References

[1] Efimov V 1970 *Phys. Lett. B* **33** *563*
[2] Efimov V 1971 *Sov. J. Nucl. Phys.* **12** *589*
[3] Braaten E and Hammer H W 2006 *Phys. Rep.* **428** *259*
[4] Gattobigio M and Kievsky A 2014 *Phys. Rev.* **A90** *012502*
[5] Kievsky A, Timofeyuk N K and Gattobigio M 2014 *Phys. Rev.* **A90** *032504*
[6] Efimov V 1981 *Nuc. Phys. A* **362** *45*
[7] Thomas L H 1935 *Phys. Rev.* **47** *903*
[8] Aziz R A and Slaman M J 1991 *J. Chem. Phys.* **94** *8047*
[9] Malfliet R A and Tjon J A 1969 *Nuc. Phys.* **A 127** *161*
[10] Kievsky A and Gattobigio M 2015 *Phys. Rev.* **A92** *062715*
[11] Álvarez-Rodríguez R, Deltuva A, Gattobigio M and Kievsky A 2016 *Phys. Rev.* **A93** *062701*
[12] Piarulli M, Girlanda L, Schiavilla R, Navarro Pérez R, Amaro J E and Ruiz Arriola E 2015 *Phys. Rev.* **C91** *024003*
[13] Kievsky A and Gattobigio M 2016 *Few-Body Syst.* **57** *217*
[14] Adhikari S K, Fonseca A C and Tomio L 1982 *Phys. Rev.* **C 26** *77*
[15] Frederico T, Goldman I D and Delfino A 1988 *Phys. Rev.* **C 37** *497*
[16] Chen C R, Payne G L, Friar J L and Gibson B F *Phys. Rev. C* **39** *1261*
[17] Kievsky A and Gattobiggio M 2014 *Phys. Rev.* **A87** *052719*
[18] Kievsky A, Rosati S and Viviani M 1999 *Phys. Rev. Lett.* **82** *3759*
[19] Garrido E, Gattobigio M and Kievsky A 2013 *Phys. Rev.* **A88** *032701*